Recent Trends in Designing Novel Foods: Nutritional Profiles and Consumer Perceptions

Recent Trends in Designing Novel Foods: Nutritional Profiles and Consumer Perceptions

Guest Editors

Raquel P. F. Guiné
Ofélia Anjos
Maria João Barroca

MDPI

Basel • Beijing • Wuhan • Barcelona • Belgrade • Novi Sad • Cluj • Manchester

Guest Editors
Raquel P. F. Guiné
CERNAS Research Centre
Polytechnic University of Viseu
Viseu
Portugal

Ofélia Anjos
CERNAS-IPCB Research Centre
Polytechnic University of Castelo Branco
Castelo Branco
Portugal

Maria João Barroca
Agriculture School
Polytechnic University of Coimbra
Coimbra
Portugal

Editorial Office
MDPI AG
Grosspeteranlage 5
4052 Basel, Switzerland

This is a reprint of the Special Issue, published open access by the journal *Foods* (ISSN 2304-8158), freely accessible at: https://www.mdpi.com/journal/foods/special_issues/08L1E865PO.

For citation purposes, cite each article independently as indicated on the article page online and as indicated below:

Lastname, A.A.; Lastname, B.B. Article Title. *Journal Name* **Year**, *Volume Number*, Page Range.

ISBN 978-3-7258-7787-4 (Hbk)
ISBN 978-3-7258-7788-1 (PDF)
https://doi.org/10.3390/books978-3-7258-7788-1

Cover image courtesy of Raquel P. F. Guiné

Contents

About the Editors

Raquel P. F. Guiné

Raquel P. F. Guiné is a Full Professor in the Food Industry Department at Polytechnic University of Viseu. She has a PhD in Chemical Engineering, a Specialization in Food Engineering, and a Habilitation in Food Science. Has been president of the scientific board, president of the board of representatives, director of the department, and director of bachelor's and master courses. Has been coordinator of the CERNAS-IPV research unit and co-coordinator of CI&DETS research centre. The scientific areas of interest include food science, food engineering, food processing, food chemistry, nutrition, diets and food consumption. Global project coordinator of international projects like the EATMOT (involving 18 countries, EISuFood (involving 18 countries), and Food-WISE (involving 17 countries). Participation in the organization of 32 events/conferences and member of the Scientific Committee of 66 congresses/Meetings. Authorship of 27 books, 119 chapters, 413 scientific articles and 398 conference proceedings/abstracts. Communications at scientific and knowledge transfer meetings: 41 lectures/plenaries/Keynotes by invitation, 259 oral presentations and 215 posters.

Ofélia Anjos

Ofélia Anjos is a Full Professor in Food Science and Technology at the Agrarian School of the Polytechnic University of Castelo Branco. She was also a teacher collaborator at UBI University and Aberta University. Ofélia Anjos has a PhD in Materials Science and a Habilitation in Ciências Agrárias e Florestais, with a specification in food science. She has been involved in many organisational activities within her institution and is involved in the development of ISO standards in the food sector, having served as convenor for 8 years and currently is a member of working groups on bee products. She is the leader of the pollen group of the International Honey Commission. She is currently the coordinator of the CERNAS-IPCB research unit. Her scientific interests include food science and technology, and forest science and technology, with a focus on vibrational spectroscopy and chromatography. Authorship of books and books chapters, 144 scientific articles and more than 300 conference proceedings/abstracts.

Maria João Barroca

Maria João Barroca is an Adjunct Professor at the Polytechnic University of Coimbra in the field of Food Technology and Biotechnology, and a researcher at the Molecular Physical Chemistry Unit of the Department of Chemistry of the University of Coimbra (UC). She holds a degree in Chemical Engineering and a PhD in Chemical Engineering, with a specialization in Chemical Processes, both from UC. Her research focuses on the study of bioactive compounds in food matrices and the development of novel functional foods through the incorporation of maritime plants and plant extracts. She has published several papers in indexed scientific journals and book chapters, and has participated in various research projects, as well as in the organization of scientific events and conferences, and as a member of the Scientific Committee of congresses and meetings.

foods

MDPI

Editorial

Recent Trends in Designing Novel Foods: Nutritional Profiles and Consumer Perceptions

Raquel P. F. Guiné [1,*], Maria João Barroca [2] and Ofélia Anjos [3]

[1] CERNAS-IPV, Polytechnic University of Viseu, Campus Politécnico, Repeses, 3404-510 Viseu, Portugal
[2] ESAC—Agriculture School of Coimbra, IPC—Polytechnic University of Coimbra, Bencanta, 3045-601 Coimbra, Portugal; mjbarroca@esac.pt
[3] CERNAS—IPCB, Polytechnic University of Castelo Branco, Quinta da Senhora de Mércules, 6001-909 Castelo Branco, Portugal; ofelia@ipcb.pt
* Correspondence: raquelguine@esav.ipv.pt

1. Introduction

Designing novel foods is a pivotal issue for companies to remain active and competitive in today's demanding food market. This design of novel food products entails significant challenges from diverse perspectives: technology, sustainability, food security, composition, nutritional profile, health effects, and modern market trends, to name a few.

The world population continues to rise while people's standard of living is also improving. This results in an unceasing growing demand for food, exacerbating the pressure on food resources. Therefore, there is an urgent need to develop new food resources through a holistic, multifaceted approach [1].

Sustainability is a current underpinning of many new food development trends across several fundamental areas. In the area of sustainable packaging, the focus is on bio-based packaging [2], edible packaging [3], active [4] and intelligent packaging [5], or 3D printing packaging [6]. Also, there is a widespread emphasis on sustainable ingredients and valorisation of resources like food waste [7–9], sources of alternative and unconventional proteins [10], including animal proteins like edible insects [11] or plant-based proteins from fruits and vegetable by-products [12], algae [13], or mushrooms [14], for example.

The design of healthier foods, with improved composition and nutritional profiles that confer health benefits, has also been in the spotlight due to the increasing consumer demand for foods that not only satiate but also have positive health effects when consumed [15]. The challenges of improving food composition to achieve balanced nutritional profiles and greater richness in bioactive compounds have been addressed from many different perspectives. Fermented products [16], food design based on gut microbiota [17], functional foods [18] and nutraceuticals enhanced by nanotechnology [19] are some examples.

The unquestionable role of the consumer must not be seen only from the point of view of a future potential buyer, but also as an active element in the design of products throughout the development stages [20]. Consumer-based food design is, at present, a more reassuring approach for anticipating the success of the marketing phase of new foods. Some chefs and food designers have been developing collaborative activities together with academic professionals and sensory analysis experts [21]. Also, consumers have been integrating teams to develop alternative and innovative packaging solutions [22,23].

Research on novel food technology acceptance has seen significant advances in recent decades. The Unified Theory of Acceptance and Use of Technology (UTAUT) is undoubtedly one of the most comprehensive models to date. This model has advanced the

 https://doi.org/10.3390/foods15050795

Technology Acceptance Model (TAM) by assimilating insights from different theoretical approaches, enabling a better understanding of consumers' acceptance and use of technology. The most relevant contributions comprise the incorporation of social influencing factors and facilitating circumstances [24].

2. Overview of Published Articles

This Special Issue gathered studies focusing on the development of new food products across all the domains this complex process involves, from conceptualisation to commercialisation and consumer acceptance.

The study by Nina G. Kon'kova and collaborators (Contribution 1) explores the potential of chufa tuber flour to enhance the functional properties of bread. They conducted diverse evaluations (biochemical, farinographic, and baking properties) on chufa tubers and bread samples with 30% replacement of wheat flour with chufa tuber flour. Their results revealed that most assessments improved with chufa tuber flour and bread containing chufa flour, compared with the control samples (whole-grain wheat flour and wheat bread). The improved biochemical indicators were oil content, fibre content, total phenolic compounds, and antioxidant activity, while no improvement was observed in protein and starch content. Regarding the technological properties, viscoamylography, farinographic sedimentation, and baking quality measurements confirmed that the bread dough incorporating chufa flour was stronger, more resistant to kneading, and showed a lower degree of liquefaction than the control. Also, a sensorial evaluation was conducted to compare the improved breads with the control samples. The authors concluded that chufa tuber powder enhances the rheological and sedimentation properties of bread dough, making it a possible ingredient to improve baking properties and the manufacture of functional food products.

The study by Hassan Barakat and collaborators (Contribution 2) evaluated the potential of sukkari date paste and mixed nuts as ingredients to improve the nutritional and biochemical profiles of high-energy bars. They compared five formulations with variable date-to-nut ratios ranging from 40:60 to 80:20. The results showed that higher percentages of date increased moisture and carbohydrates, whereas higher percentages of nut enhanced protein, fat, and energy. Regarding dietary minerals, they observed increases in calcium, phosphorus, magnesium, and trace elements as the percentage of date decreased. However, the phenolic compounds and antioxidant activity were highest in the bars with a higher percentage of nuts, and so were the ratio of total to essential amino acids and protein quality. Their findings evidence the potential of optimised date-nut bars as nutrient-dense functional snacks.

The study by Daniela Magalhães and collaborators (Contribution 3) is in line with the previous work and also explored the enrichment of a food, in this case, it was mortadella (a bologna-type sausage), through the addition of lemon powder rich in dietary fibre. The authors demonstrated that the functional lemon dietary fibre possessed antioxidant and antibacterial activities and reduced sodium levels and residual nitrite levels. Although the reduction in nitrites led to a slight increase in lipid oxidation, it remained below the rancidity threshold ($\geq$1.0). The addition of lemon powder resulted in a paler colour and increased hardness, but the sensory evaluation received positive feedback.

The study by Rocio Lopez-Navarro and collaborators (Contribution 4) examined the consumer perceptions towards cashew nuts based on a neuroscientific and sensory approach, as a way to analyse both conscious and unconscious responses that allow understanding consumer preferences. Their results showed consumer preference for one of the analysed samples, which was confirmed by electroencephalogram metrics and frontal alpha asymmetry for flavour (showing greater activity in the left frontal lobe, associated with positive emotions, for the preferred sample). Task engagement measurements revealed higher engagement with the same sample during flavour evaluation. The authors believe that their findings have implications for product optimisation, market segmentation, and the development of marketing strategies to improve product commercialisation.

The study by Natália L. Seixas and collaborators (Contribution 5) examined the addition of fermented elderberries as a bread ingredient to enhance bread quality and biological value. They evaluated how incorporating fermented elderberries into a traditional bread affected its bioactivity and shelf life. They confirmed that fermentation increased the bioactive compounds and antioxidant activity of the elderberries, resulting in an extended shelf life for the studied bread.

The study by Pedro Coelho and collaborators (Contribution 6) investigated how a mixed system composed of white *Chlorella vulgaris* and a disruptive emulsifier (yeast protein extract), could be used to transform traditional coriander mayonnaise into an analogue product. The mixed system was found to confer stability to the mayonnaise, and the introduction of microalgae enhanced bioactivity, particularly total phenolic compounds and antioxidant activity. Sensorial evaluation confirmed the similarity of the products, except for the parameter colour.

The study by Alessandro Bianchiand and collaborators (Contribution 7) focused on enriching salty biscuits with bee pollen, in either fresh or dried form, to improve their functional characteristics. The authors studied the composition and sensory characteristics of the biscuits. Their results showed that biscuits with 5% fresh pollen had the best combination of chemical and sensory properties. They concluded that biscuits fortified with bee pollen can serve as a vehicle for the ingestion of phenolic compounds and carotenoids, while maintaining an acceptable sensory profile.

The study by Marcelina Maciejewska and collaborators (Contribution 8) analysed edible insect hydrolysates as sustainable sources of protein and investigated consumers' attitudes towards insect-based foods in two European countries (Poland and Spain). Their approach was based on alignment between technical advancements and consumer preferences, to intertwine laboratory innovation with market feasibility, thereby contributing to the development of sustainable functional foods. They optimised the enzyme hydrolysis process to increase the antioxidant activity of the hydrolysates. On the other hand, consumer surveys confirmed that demographic and cultural characteristics significantly influence consumer acceptance of insect-based foods.

The last contribution (Contribution 9) is a review article by Hassan Muzaffar and collaborators that discusses the therapeutic potential of white kidney beans as a functional food. In their review, they examine the mechanistic roles of white kidney bean in weight regulation, though suppression of starch digestion, attenuation of postprandial glycaemia, modulation of appetite and satiety, and hypolipidemic effects. They concluded that white kidney bean has potential as a nutraceutical for metabolic health, supported by clinical evidence demonstrating its impact on reducing body fat mass, glycaemic excursions, and overall weight.

3. Final Remarks

The continuous interest in developing successful new food products that meet consumer expectations and needs has been the motor of change over the past decades, leading to innovative food products, new and improved ingredients, more sustainable and efficient production and transformation technologies, and new integrated food design approaches.

The domain of novel foods development also continuously opens new doors to future work, leads to unforeseen approaches, and constitutes a field where technological advancement allied to multifaceted conceptualisation will undoubtedly have a positive impact in the future.

https://doi.org/10.3390/foods15050795

Author Contributions: Conceptualisation, R.P.F.G.; methodology, R.P.F.G.; resources, R.P.F.G.; writing—original draft preparation, R.P.F.G.; writing—review and editing, M.J.B. and O.A.; supervision, R.P.F.G. All authors have read and agreed to the published version of the manuscript.

Funding: This research was funded by FCT—Foundation for Science and Technology through CERNAS Research Centre (Ref. UID/00681/2025).

Data Availability Statement: Not applicable.

Acknowledgments: This work was supported by the FCT—Foundation for Science and Technology through CERNAS Research Centre (doi: 10.54499/UID/00681/2025) and the Polytechnic University of Viseu.

Conflicts of Interest: The authors declare no conflicts of interest.

List of Contributions:

1. Kon'kova, N.G.; Khoreva, V.I.; Popov, V.S.; Yakusheva, T.V.; Kibkalo, I.A.; Malyshev, L.L.; Solovyeva, A.E.; Shelenga, T.V. Technological Assessment of Bread with the Addition of Cyperus esculentus L. Accessions Flour Grown in the Kuban–Azov Plain. *Foods* **2025**, *14*, 3680. https://doi.org/10.3390/foods14213680.
2. Barakat, H.; Alfheeaid, H.A.; Aljutaily, T.; Alayouni, R.; Alharbi, H.F.; Alsanei, W.A. Elucidating the Nutritional Profile and Biochemical Characterization of High-Energy Nutritional Bar Formulated with Sukkari Date Paste and Mixed Nuts. *Foods* **2025**, *14*, 3661. https://doi.org/10.3390/foods14213661.
3. Magalhães, D.; Rodrigues, C.V.; Botella-Martinez, C.; Muñoz-Tebar, N.; Pérez-Álvarez, J.A.; Viuda-Martos, M.; Teixeira, P.; Pintado, M. Lemon Dietary Fibre-Based Powder as a Promising Ingredient for the Food Industry: Enhancing Mortadella Nutritional Quality. *Foods* **2025**, *14*, 1693. https://doi.org/10.3390/foods14101693.
4. Lopez-Navarro, R.; Montero-Vicente, L.; Escriba-Perez, C.; Buitrago-Vera, J.M. Implicit and Explicit Consumer Perceptions of Cashews: A Neuroscientific and Sensory Analysis Approach. *Foods* **2025**, *14*, 1213. https://doi.org/10.3390/foods14071213.
5. Seixas, N.L.; Paula, V.B.; Dias, T.; Dias, L.G.; Estevinho, L.M. The Effect of Incorporating Fermented Elderberries (*Sambucus nigra*) into Bread: Quality, Shelf Life, and Biological Enhancement. *Foods* **2025**, *14*, 724. https://doi.org/10.3390/foods14050724.
6. Coelho, P.; Serrano, C.; Komora, N.; Raymundo, A. From a Coriander Mayonnaise to a Vegan Analogue: Assessing pH and Salt Influence in a *Saccharomyces cerevisiae* Yeast Protein Extract and *Chlorella vulgaris* Mixed System. *Foods* **2025**, *14*, 587. https://doi.org/10.3390/foods14040587.
7. Bianchi, A.; Capparelli, S.; Taglieri, I.; Sanmartin, C.; Pistelli, L.; Venturi, F. Salty Biscuits Enriched with Fresh and Dried Bee Pollen: Chemical, Technological, and Sensory Characterisation. *Foods* **2025**, *14*, 527. https://doi.org/10.3390/foods14030527.
8. Maciejewska, M.; Dąbrowska, A.; Cano-Lamadrid, M. Sustainable Protein Sources: Functional Analysis of *Tenebrio molitor* Hydrolysates and Attitudes of Consumers in Poland and Spain Toward Insect-Based Foods. *Foods* **2025**, *14*, 333. https://doi.org/10.3390/foods14020333.
9. Muzaffar, H.; Jehangir, M.; Hu, J.; Yu, Y.; Yu, M.; Hu, Y. Therapeutic Potential of White Kidney Beans (*Phaseolus vulgaris*) for Weight Management. *Foods* **2025**, *14*, 3940. https://doi.org/10.3390/foods14223940.

References

1. Liu, G.; Liu, M.; Zhao, J.; Zhang, L.; Yan, W. Development of Novel Aquatic Food Resources: Current Status, Future Potential, and Safety Considerations. *J. Future Foods* **2025**, *in press*. [CrossRef]
2. Khalid, M.Y.; Arif, Z.U. Novel Biopolymer-Based Sustainable Composites for Food Packaging Applications: A Narrative Review. *Food Packag. Shelf Life* **2022**, *33*, 100892. [CrossRef]
3. Trajkovska Petkoska, A.; Daniloski, D.; D'Cunha, N.M.; Naumovski, N.; Broach, A.T. Edible Packaging: Sustainable Solutions and Novel Trends in Food Packaging. *Food Res. Int.* **2021**, *140*, 109981. [CrossRef] [PubMed]
4. Zhang, Y.; Liu, H.; Wang, Q.; Zhang, J.; Liu, C.; Wang, C.; Lu, L.; Qiao, Y. Advances in pH-Responsive Controlled-Release Systems for Antimicrobial Active Packaging in Food Preservation. *Trends Food Sci. Technol.* **2026**, 105554. [CrossRef]

5. Rahman, M.; Sobhan, A.; Sadak, O. Emerging Trends in Intelligent Packaging for Tackling Food Waste in the Modern Food Supply Chain. *Trends Food Sci. Technol.* **2026**, *167*, 105436. [CrossRef]
6. Li, C.; Li, L.; Zheng, X.; Yan, F. 3D Printing Technology Applied in Smart Packaging for Food Preservation. *Trends Food Sci. Technol.* **2025**, *165*, 105329. [CrossRef]
7. Pant, M.; Bisen, D.; Kewlani, P.; Srivastav, A.L.; Bhatt, I.D.; Chakma, S. Review of Food Waste Valorization Technologies: A Sustainable Approach to Resource Recovery and Utilisation. *Biomass Futures* **2026**, *1*, 100001. [CrossRef]
8. Lima de Souza, R.; Pérez-Gago, M.B.; Palou, L. From Waste to Health: Valorization of Agri-Food by-Products for the Control of Fresh Fruit Fungal Postharvest Decay. *Postharvest Biol. Technol.* **2026**, *233*, 114054. [CrossRef]
9. Soon, W.L.; Foo, J.L.; Chang, M.W. Microbial Food Waste Valorization: Advances, Challenges, and Perspectives. *Curr. Opin. Biotechnol.* **2025**, *94*, 103323. [CrossRef] [PubMed]
10. Fang, B.; Peng, Z.; Chen, B.; Rao, J. Unconventional Sources of Vegetable Proteins: Technological Properties. *Curr. Opin. Food Sci.* **2024**, *57*, 101150. [CrossRef]
11. Guiné, R.P.F.; Correia, P.; Coelho, C.; Costa, C.A. The Role of Edible Insects to Mitigate Challenges for Sustainability. *Open Agric.* **2021**, *6*, 24–36. [CrossRef]
12. Khalid, L.; Jabeen, I.; Ahmad, T.; Inam-ur-Raheem, M.; Shahid, A.; Cirillo, T.; Esposito, F. Valorization of Fruit and Vegetable By-Products for Protein Extraction and Their Functional Applications in Food and Non-Food Sectors. *Food Bioprod. Process.* **2025**, *154*, 198–215. [CrossRef]
13. Rather, J.A.; Akhter, N.; Punoo, H.A.; Haddad, M.; Ghnamat, S.A.; Manzoor, N.; Goksen, G.; Dar, B.N. Sustainable Algal Proteins, Novel Extraction Techniques and Applications in the Bakery, Dairy and Pharmaceutical Industries: A Comprehensive Review. *Food Chem.* **2025**, *465*, 141828. [CrossRef]
14. Hanafi, M.A.; Mohamad Asri, N.; Auwal, S.M.; Brishti, F.H.; Saari, N. Edible Mushroom Proteins for Future Foods: Integrative Insights into Nutrition, Technofunctionality, Extraction, and Safety. *Food Res. Int.* **2026**, *227*, 118158. [CrossRef]
15. Yang, Y.; Sun, W.; Li, Y.; Huang, J. Taste the Joy: How Multisensory Simulation Enhances Young Adults' Hedonic Expectations for Healthy Foods. *Food Qual. Prefer.* **2026**, *139*, 105858. [CrossRef]
16. Mu, H.; Yao, Y.; Gong, Y.; Yang, T. The Role of Fermented Foods in Healthy Longevity: A Review of Potential Anti-Aging Mechanisms. *Curr. Res. Food Sci.* **2026**, *12*, 101300. [CrossRef]
17. Yan, F.; Xiang, S.; Ying, J.; Chen, J.; Shi, L.; Zhang, Y.; Han, J.; Zhu, X. Healthy Food Design Based on Gut Microbiota Metabolic Flux. *Trends Food Sci. Technol.* **2024**, *154*, 104790. [CrossRef]
18. Marcìa-Fuentes, J.A.; Aleman, R.S.; Areche, F.O.; Flores, D.C.; Roman, A.V.; Martín-Vertedor, D.; Montero-Fernández, I. Functional Foods: A Review of Foods Ingredient and Their Health Benefits. *Food Humanit.* **2026**, *6*, 100953. [CrossRef]
19. Balusamy, S.R.; Vijayalakshmi, S.; Sundaravadivelu, S.; Huq, M.A.; Nag, S.; Mohanto, S.; Sukweenadhi, J.; Oh, D.H.; Perumalsamy, H. Nutraceuticals Enhanced by Nanotechnology: A New Frontier for Obesity Treatment. *Process Biochem.* **2025**, *156*, 217–235. [CrossRef]
20. Guiné, R.P.F.; Florença, S.G.; Barroca, M.J.; Anjos, O. The Link between the Consumer and the Innovations in Food Product Development. *Foods* **2020**, *9*, 1317. [CrossRef]
21. Schifferstein, H.N. Supporting Food Design with Consumer Research: From Inspiration and Validation to Participation and Integration. *Curr. Opin. Food Sci.* **2023**, *51*, 101020. [CrossRef]
22. Yokokawa, N.; Amasawa, E.; Hirao, M. Design Assessment Framework for Food Packaging Integrating Consumer Preferences and Environmental Impact. *Sustain. Prod. Consum.* **2021**, *27*, 1514–1525. [CrossRef]
23. Hamlin, R.F. The Consumer Testing of Food Package Graphic Design. *Br. Food J.* **2016**, *118*, 379–395. [CrossRef]
24. Jung, I.; Sharma, A. Role of Trust and Hedonism in Driving Novel Food Technology Acceptance. *Future Foods* **2025**, *12*, 100824. [CrossRef]

https://doi.org/10.3390/foods15050795

foods

MDPI

Article

Technological Assessment of Bread with the Addition of *Cyperus esculentus* L. Accessions Flour Grown in the Kuban–Azov Plain

Nina G. Kon'kova *, Valentina I. Khoreva, Vitaliy S. Popov, Tamara V. Yakusheva, Ilya A. Kibkalo, Leonid L. Malyshev, Alla E. Solovyeva and Tatiana V. Shelenga

Nikolai Ivanovich Vavilov All-Russian Institute of Plant Genetic Resources, Bolshaya Morskaya Street 42-44, 190000 Saint-Petersburg, Russia; horeva43@mail.ru (V.I.K.); v.popov@vir.nw.ru (V.S.P.); kos-vir@yandex.ru (T.V.Y.); i.kibkalo@vir.nw.ru (I.A.K.); l.malyshev@vir.nw.ru (L.L.M.); alsol64@yandex.ru (A.E.S.); t.shelenga@vir.nw.ru (T.V.S.)
* Correspondence: ninakonkova.1@mail.ru

Abstract

Due to increased interest in new functional food products, 20 accessions of chufa tubers from the collection of the N.I. Vavilov Institute of Plant Industry, grown in the Kuban–Azov Plain in 2022, as well as bread samples made from mixed flour (70% whole-grain wheat flour, 30% chufa tuber flour) were studied. Biochemical, farinographic, and baking evaluations were carried out. Differences between the properties of dough with the addition of flour from various accessions of chufa tubers were recorded. According to the results of comparative, dispersion, and principal component analysis, all biochemical indicators (oil, fiber, sum of phenolic substances, antioxidant activity) of chufa tuber flour and bread with added chufa flour surpassed control samples (whole-grain wheat flour and wheat bread), with the exception of protein and starch content. Viscoamylographic, farinographic sedimentation, and baking quality evaluations indicated that the dough made from mixed flour was stronger than the control (dough from whole-grain wheat flour), more resistant to kneading, and had a lower degree of liquefaction. In terms of organoleptic properties, differences were also identified, and the accessions that enhance the taste of mixed bread were selected. Therefore, a preliminary conclusion can be drawn that chufa tubers grown in the conditions of the Kuban–Azov Plain with high rheological properties and high sedimentation values of the mixed dough can be recommended for improving the baking properties not only of wheat but also of other bread cereals. Chufa is also a promising crop in the manufacture of functional food products in the Krasnodar region and for the food industry in general as a potential thickener.

Keywords: *Cyperus esculentus*; oil content; protein; starch; fiber; phenolic substances; antioxidant capacity; baking qualities; functional food products; composite flour

1. Introduction

Cyperus esculentus L. is a valuable agricultural crop that is widespread in Africa (Nigeria, Mali, Senegal, Ghana), Italy (Sicily), and Spain. The largest commercial plantings in Europe are located in Spain, in the province of Valencia [1]. The yield of chufa is quite high; depending on the region of cultivation and the growing method, the tuber yield of chufa varies widely from 3 to 8.5 tons per hectare. In Russia, at the moment, there are no significant commercial plantings; the cultivation of chufa is carried out by individual farmers on small areas.

Chufa is a heat- and moisture-loving plant; for its cultivation, light soils with good drainage are suitable. Chufa possesses remarkable adaptability, which allows obtaining high yields when cultivated on less fertile sandy soils [2]. The vegetation period is 110–120 days. However, our previous work has shown that chufa can also be grown in more northern regions of the world, such as the southwestern and Central-Chernozem regions of the Russian Federation [3]. For food purposes, the tubers that formed on the plant's roots are used. The tubers contain 20–25% oil, 25–30% starch, 20% carbohydrates, 5–7% protein, and 7–11% fiber [4,5]. The oil also possesses valuable food properties. The most important fatty acid in chufa oil is oleic acid, the content of which reaches more than 70% [6,7]. Chufa oil is characterized by a high content of vitamin E [8]. The high biomass yield and the content of nutrients in chufa tubers make them a promising source of food, oil, and feed to meet the growing needs of the food industry.

Currently, functional foods receive great attention due to their nutritional value. These products contain all the ingredients necessary for a healthy diet and have a beneficial effect on human health [9,10]. Therefore, increasing the use of various food products based on chufa is very promising [11]. Chufa is used in various branches of the food industry; for example, it is widely used in the confectionery industry for the production of chocolate, sweets, cookies, and cupcakes [12–16]. The traditional Spanish drink "Horchata de Chufa" made from chufa tubers is very popular [17–19]. Based on "Horchata de Chufa", many drinks are produced with the addition of soy milk, fruit juices, and extracts from the leaves of various plants [20,21]. Lactic acid bacteria are also used to prepare milk from chufa tubers [22,23].

Due to its high biologically active properties and its content of phenolic compounds, fatty acids, and fiber, chufa flour is used for the production of combined bakery products. Bread prepared from a mixture of wheat flour with chufa tuber flour has good nutritional properties with high dietary fiber content and reduced starch content [24]. It has been shown that chufa is a promising ingredient for the preparation of gluten-free bread [25] and special products for diabetics. The number of people suffering from these diseases is increasing over time [26]. Due to the trends towards climate change and the demand for an expanding the range of functional foods, there is a necessity to investigate the crops that were previously not considered economically significant or typical for production in certain regions of the world [3]. Bread is a staple food in the daily human diet. It is possible to increase the nutritional value and taste of bread using functional components from edible chufa tuber flour. The aim of this research is to investigate 20 accessions of chufa tubers of various origins from the VIR collection for biochemical composition and technological parameters of bread prepared from a mixture of whole-grain wheat flour (70%) and chufa tuber flour (30%). This study will help to confirm that the chufa accessions available in the VIR collection have good potential for creating new functional food products of regional importance (Krasnodar region of the Russian Federation).

2. Materials and Methods

2.1. Material

2.1.1. Plant Material

Twenty accessions of chufa from the VIR collection (Table 1) were grown at the Kuban Experimental Station (KES VIR 45°13′ N, 40°47′ E, Kuban–Azov plain) in 2022. The Kuban Experimental Station of the All-Russian Institute of Plant Industry is situated in the steppe zone of Krasnodar region (65 m elevation). It features thick-layered Priazov chernozem soils with 130–140 cm humus horizons. The region has a temperate continental climate and variable precipitation, with average annual temperatures of 10.4 °C. The whole-grain flour of the spring soft wheat variety Pamyati Suslyakova from the VIR collection was used for the composite bread preparation and as a control. This variety was previously used as a control for composite bread evaluation [27].

Table 1. List of *C. esculentus* accessions from the VIR collection used as a research material.

Catalog Number	Origin
1	Russia
2	Russia
7	Poland
8	Bulgaria
9	Bulgaria
10	Bulgaria
11	Mali
12	Benin
13	Germany
14	Ivory Coast
15	Ivory Coast
16	Ivory Coast
17	Ivory Coast
19	Russia
20	Ukraine
21	Belarus
23	France
24	Ukraine
26	Russia
27	Russia

2.1.2. Bread Material

Twenty model samples of bread were made from mixed flour (30% flour from chufa tuber accessions and 70% whole-grain wheat flour). The bread made from 100% whole-grain wheat flour was used as a control.

2.2. Biochemical Analysis

The following parameters were investigated: protein content, oil content, starch content, fiber content, total phenolic compounds, and antioxidant activity. The analysis was performed in two biological replicates, and for each replicate, a separate grain aliquot was taken. All parameters were calculated on a dry weight basis.

2.2.1. Preparation of Plant Material for Study

Chufa tubers and bread samples were preliminarily dried at room temperature, then ground using a laboratory disc mill CM 290 Cemotec (FOSS, Höganäs, Sweden) to particles of about 10 μm in size. To prevent changes in the properties of plant material when exposed to light and oxygen, the flour of chufa tubers was stored in metal containers with airtight lids. Before analysis, the flour was thoroughly homogenized, and two portions were selected for each analysis. An experimental procedure was carried out in the Department of Biochemistry and Molecular Biology using methods for screening collection's accessions adopted at VIR [3].

2.2.2. Protein Content Analysis

Protein/nitrogen content was determined by the Kjeldahl method (conversion factor for chufa—5.5) using an automatic Kjeltec 2200 protein analyzer (FOSS, Höganäs, Sweden); protein content was expressed as % of the dry weight of the sample.

2.2.3. Oil Content Analysis

Oil content—by the mass of dry fat-free residue using a Soxhlet apparatus with petroleum ether (40–70 °C) as solvent; oil content was expressed as % of the dry weight of the sample.

2.2.4. Starch Content Analysis

Starch content—by the Ewers polarimetric method (conversion factor for chufa—203) using an automatic SAC-i polarimeter/saccharimeter (ATAGO, Kobe, Japan); starch content was expressed as % of the dry weight of the sample.

2.2.5. Fiber Content Analysis

Fiber content—by the gravimetric Weende method on a Dosi-Fiber FIWE-6 extraction unit (VELP, Usmate Velate, Italy); fiber content was expressed as % of the dry weight of the sample.

2.2.6. Dry Matter Content (Moisture) Analysis

Dry matter content (moisture) was determined by weighing a portion of the ground material (chufa, wheat flour, and ground material from composite bread and wheat bread) before and after drying at 100–102 °C to constant weight using a Memmert drying oven (Memmert GmbH , Schwabach, Germany); dry matter was expressed as % of the weight of the sample before drying.

2.2.7. Total Phenolic Compounds Analysis

Total phenolic compounds (PC) were determined spectrophotometrically using Folin–Ciocalteu reagent. The results were expressed as mg gallic acid equivalents/100 g sample [26].

2.2.8. Antioxidant Activity Analysis

Antioxidant activity (AOA) was measured spectrophotometrically using DPPH reagent as a free radical donor according to the method [28]. The results were expressed as mg ascorbic acid equivalents/100 g sample.

2.3. *Technological Assessment*

2.3.1. Rheological Assessment

The rheological properties of composite dough were studied using a Brabender Farinograph (Brabender GmbH & Co, Duisburg, Germany).

Farinographic Evaluation

Sample Preparation

An amount of 10 g of the flour sample was placed in a mixer of the same volume, which is an integral part of the Brabender Farinograph (Germany). Then, distilled water was added to the flour to achieve the desired water absorption capacity. The dough was then kneaded until it reached a standard consistency (500 units). The registration of the rheological properties of the flour sample continued for 12 min from the beginning of the dough liquefaction or from the beginning of dough consistency drop. The farinographic properties of the dough samples were automatically recorded using a Brabender Farinograph (Germany).

Farinographic Indicators Evaluation

The following indicators were studied: water absorption capacity (WAC) of the dough, dough formation time or dough development time (DDT) and dough stability (DS), dough softening or dilution of the dough (DD), and valorimeter number (VN). WAC characterizes the technological properties of the raw materials—the amount of water necessary to obtain a dough of standard consistency. In the mixed bread samples, WAC substantially decreased. DDT and DS characterize the strength of the flour. It is believed that higher indicators correspond to better quality. In the mixed bread samples, the dough substantially strengthened. DD shows how strongly the weakening of the dough occurs during its prolonged mixing. The higher indicator corresponds to the weaker dough. For the mixed samples, softening significantly decreased. VN—an indicator generalizing the assessment of rheological properties on this device. The higher this value is, the better the quality of the bread sample. Units of farinographic indicators measurement: WAC—%, DDT, DS—min, and DD—device units.

Viscoamylographic Assessment

Sample Preparation

A total of 100 mL of distilled water was added to 15 g of flour sample to prepare a homogeneous suspension. The prepared suspension was placed in the measuring module of the micro visco-amylograph device (Brabender GmbH & Co, Duisburg, Germany). The measurements were carried out using a ready-made protocol for potato starch in automatic mode. The measurement time was 20 min. The measurement results were recorded as visco-amylograms.

Viscoamylographic Indicators Evaluation

The study of the starch thermodynamic properties was conducted on a micro visco-amylograph from Brabender (Germany). The essence of the method consists of changing the temperature regime of the suspension of the ground material during constant stirring and recording the load on the paddle. Thus, the viscosity of the paste is measured, and the time and temperature of the onset of a particular phase of gelatinization were recorded. For the tests, a protocol for potato starch was used. The temperature regime consisted of gradually heating the suspension to 92 °C, a holding period at this temperature, gradually cooling the suspension to 50 °C, and a short holding period at this temperature at the end of the test. Thus, two or three thickening peaks were recorded. The protocol was intended for pure starch, where a sample weight of 5 g was assumed. Since our material is not pure starch, we had to increase the sample weight to 15 g. The following viscoamylographic parameters were detected: pasting time (PT), time to peak viscosity (TPV); pasting temperature start (PTS), peak viscosity temperature (PVT); initial pasting viscosity (IPV), peak viscosity in the hot state (PVHS), minimum hot paste viscosity (MHPV), cold paste peak viscosity (CPPV), viscosity at the end of measurement (VEM), viscosity drop at peak temperature (VDPT), viscosity increase upon cooling (VIUC), viscosity increase from hot to cold (VIHC). PT and TPV were measured in min, PTS and PVT in °C, and IPV, PVHS, MHPV, CPPV, VEM, VDPT, VIUC, and VIHC in device units.

2.3.2. Sedimentation Analysis (Swelling Capacity of Flour)

Sample Preparation

An amount of 1 g of the flour sample was suspended in graduated test tubes with cork stoppers along with 6 mL of a 2% acetic acid solution in distilled water. After the suspension was allowed to stand for 4–5 min, the volume of the precipitate was measured (first phase). Then, a 1.9% aqueous solution of sodium dodecyl sulfate (SDS) was added to the same sample. The mixture was resuspended, was allowed to stand for 2 min, and the volume of the precipitate was measured (second phase).

Sedimentation Indicators Evaluation

The two phases of this methodology—flour swelling capacity phase 1 (PSC1) and flour swelling capacity phase 2 (PSC2)—show the ability of protein structures to resist increasing physicochemical loads or, thanks to them, to realize their swelling potential. Accordingly, the ratio of the indicators of the two phases—sedimentation index or phase 2/phase 1 (P2-1)—characterizes the stability of the protein structures. If it is close to 1, then the sample is stable and has realized its potential. If it is less than 1, the sample is less stable to loads; if it is more than 1, the sample carries greater swelling potential. PSC1 and PSC2 parameters were measured in mL.

2.3.3. Flour Color Evaluation

The color indicators were assessed using a special flour module of the Precision Colorimeter NR20XE, 3nh (Shenzhen, Guangdong China). The color indicators were expressed as parameters: *L* (flour color brightness), *a* (red–green range of the spectrum), *b* (yellow–blue range of the spectrum), *C* (chromaticity index), and *h* (shade angle). The values of *L* are measured in the range from 0 (black) to 100 (white); *a* represents the red color at positive values or green at negative values; *b* represents yellow at positive values or blue at negative values. For a more complete assessment of flour color, the calculation of the hue angle ($h°$) and the chroma index *C*, analogous to saturation or color intensity, are used. They are obtained by calculation from the values *a* and *b*. *C* is calculated as ½ (a*2 + b*2) and represents the hypotenuse of a right triangle created by connecting the points (a, b) and (a, 0). The hue angle *h* is calculated as the arctangent of b/a, the angle between the hypotenuse and the 0° axis (blue–green/red–purple). The values of all color indicators are calculated automatically in accordance with the software of the Precision Colorimeter NR20XE Sw version: 3nh_NR20XE_V00.01 .

2.3.4. Baking Evaluation

Bread Sample Preparation

Bread samples were baked from dough made from 15 g of mixed and whole-grain wheat flour. Dough preparation for baking or proofing in a CRT-8-01 proofing cabinet (Abat, Cheboksary, Russia) was carried out at a temperature of 35 °C and a relative humidity of 75–85%. Hearth bread (hereinafter referred to as bread) was baked in a laboratory convection oven KPP-4M (Abat, Cheboksary, Russia) at a temperature of 270 °C for 10–15 min.

Baking Indicators Evaluation

One of the main parameters of bread quality is its shape stability (SS) indicator h/d: the ratio of its height to diameter. The higher the indicator, the higher the technological quality of the sample. Another indicator of bread quality, porosity, was not evaluated, as it did not differ significantly between samples. The physical properties of the crumb were evaluated as follows: density (CD) (mass per unit volume), and hardness (CH) using the structural analyzer ST-2 (Quality Laboratory, Moscow , Russia), and the staling of the crumb over 24 h (S24), defined as the increase in sample hardness over 24 h. CD was determined as the ratio of the mass of a cylindrical crumb sample to the volume occupied by this cylinder in g/mL; CH and S24 were measured in g load. Taste quality assessment or taste quality score (TQS) was conducted on a five-point system (1—very bad, 2—bad, 3—fair, 4—good, 5—very good).

2.4. Statistical Processing

Statistical processing was carried out using the Statistica 12.0 application software package (Dell Software Company, Palo Alto, CA, USA,StatSoft, Inc., Tulsa, OK, USA (2019),

STATISTICA (data analysis software system), version 12. www.statsoft.com (accessed on 1 October 2020) and included the following:

Calculation of basic descriptive statistics for the studied samples (mean, error of mean, coefficient of variation); calculated averages were used for further analysis; analysis of variance to assess the significance of the influence of the "sample" and "experimental treatment" factors; factor analysis of the correlation system of characters.

3. Results and Discussion

3.1. Biochemical Assessment

Despite the increased interest in new functional food products, there is not much research on products based on chufa, which makes our experiment more relevant. As a control for the chufa flour samples, whole-grain wheat flour was used, and for the samples of bread from mixed flour, bread made only from wheat flour was used. In Figure 1, data on the protein content in the chufa samples and bread are presented. Thanks to the high protein content in wheat flour (21.09%), the samples of wheat bread with chufa also have a higher protein content compared to the samples of chufa tubers. The average protein content in the chufa samples is 6.15%, and in the bread samples—16.21% (bread control—21.68%). The highest protein content was found in the chufa sample k-23 (7.31%) and bread sample k-10 (17.94%), and the lowest content was noted in the chufa sample k-9 (5.41%) and bread k-27 (13.55%). The maximum protein values established for chufa in the current study are slightly higher than we previously established (5.4–6.5%) [3] and were consistent with the data [29]. The protein level in bread made from mixed flour with the addition of 30% chufa, studied by Shaban et al., has lower protein values (8.73%), which is apparently related to the use of lower protein wheat flour for the bread preparation (11.80%) [29]. Our results correspond to the data obtained by Özcan for mixed bread with the addition of 20% chufa flour (15.61%) [24].

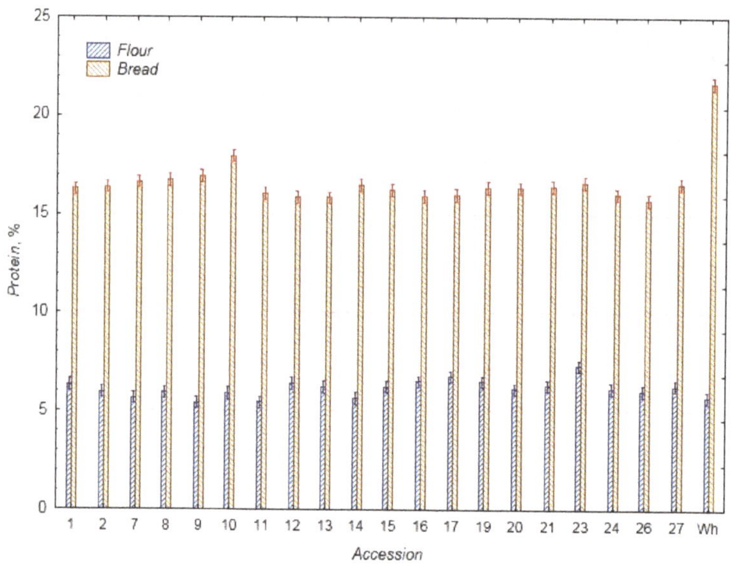

Figure 1. Protein content (%) in flour samples of chufa tubers, whole-grain wheat (Wh), composite bread, and whole-grain wheat bread (Wh).

The results for starch content in the chufa and bread samples are presented in Figure 2. In this case, a similar situation is observed, and thanks to the high starch content in wheat flour (54.42%), the samples of bread from mixed flour also have a higher starch content compared to the chufa flour samples. The average starch content in the chufa flour samples is 24.53%, and in the bread samples—42.71% (control—54.42%, 52.69% for flour and bread, respectively). The highest starch content was found in the flour of chufa sample k-21 (26.89%) and the bread samples k-7, 9, 14, and 21 (45.69%, 45.68%, 45.33%, and 45.29%, respectively), and the lowest content was noted in the chufa sample k-9 (21.52%) and bread k-13 and 17 (37.19% and 26.08%), which is consistent with our previous data [3].

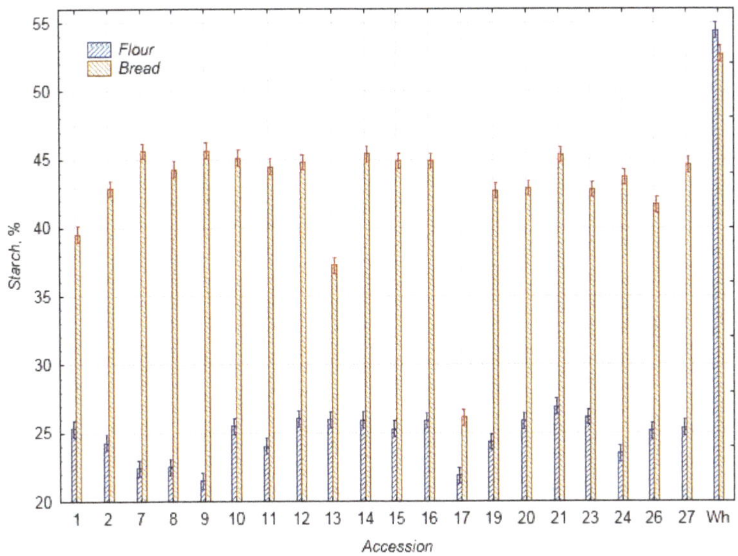

Figure 2. Starch content (%) in flour samples of chufa tubers, whole-grain wheat (Wh), composite bread, and whole-grain wheat bread (Wh).

In Figure 3, data on the oil content in the chufa and bread samples are presented. Thanks to the high oil content in chufa tubers, which significantly exceeds that of wheat flour (2.28%), the bread with chufa addition also has approximately 2.5 times higher oil content than the control bread (2.29%). The average oil content in the chufa flour samples is 22.03%, and in the bread samples, 8.10%. The highest oil content was found in the chufa flour sample k-12 (26.64%) and bread samples k-10, 12, and 27 (9.42, 9.24, and 9.02%, respectively). Conversely, the lowest content was observed in chufa flour k-17 (17.28%) and bread k-9 (6.41%). The results obtained for the 2022 chufa reproductions do not differ significantly from those for 2020–2021 [3]. Shaban et al. note a higher oil content in chufa tubers (28.50%). This is probably due to the conditions of the reproduction region [29]. Other researchers found a higher oil content (16.81%) in bread samples with 30% chufa addition [29], and a lower content (4.90%) in bread with 40% chufa addition [24]. The difference in biochemical parameters is obviously due to the characteristics of the chufa accessions chosen for making bread and the reproduction conditions.

In Figure 4, data on the fiber content in the chufa flour samples and bread samples are presented. Since the chufa flour samples contain approximately two to three times more fiber than the wheat flour control (2.61%), the samples of bread with chufa addition also have an increased fiber content. The fiber content of bread from whole grain wheat flour was almost the same as that of wheat flour. The average fiber content in the chufa flour samples was 7.88%, and in the bread samples, 4.19%. The highest fiber content was found in the chufa flour k-9 (11.11%) and the bread samples k-9, 16, and 27 (5.16, 4.54, and

4.40% respectively), and the lowest content was noted in the chufa flour k-19 (6.95%) and bread k-11 (3.93%) and k-19 (3.91%). Our data obtained for chufa tuber flour is consistent with Shaban et al.'s data (8.50%). However, the amount of fiber in bread with 30% chufa addition was significantly lower (2.43%) [29].

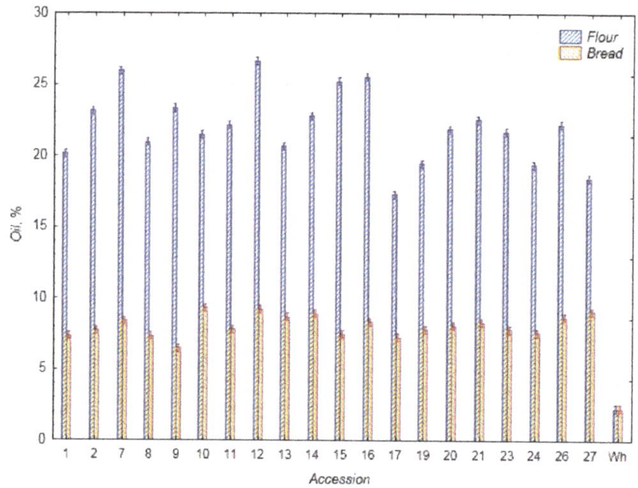

Figure 3. Oil content in flour samples of chufa tubers, whole-grain wheat (Wh), composite bread, and whole-grain wheat bread (Wh) (%).

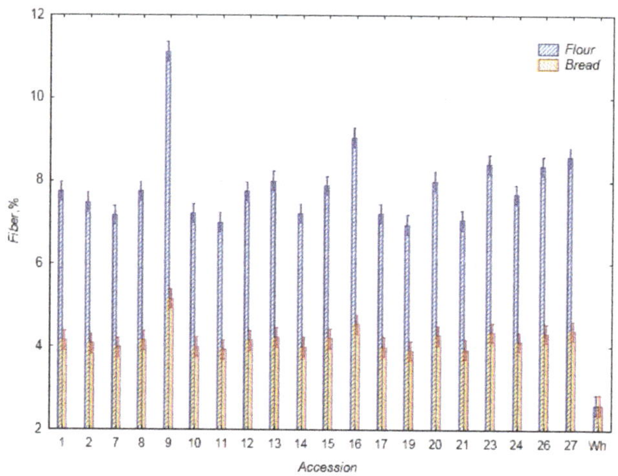

Figure 4. Fiber content in flour samples of chufa tubers, whole-grain wheat (Wh), composite bread, and whole-grain wheat bread (Wh) (%).

The results on the content of phenolic substances and AOA in the chufa samples and bread are presented in Figures 5 and 6. The average content of phenolic substances and AOA in the chufa flour samples was 125.6 mg% and 19.6 mg%, respectively, and in the bread samples—191.0 mg% and 22.2 mg%, respectively. The highest content of phenolic substances and AOA was found in the chufa flour k-9 (298.1 mg% and 34.7 mg%), and the lowest content of phenolic substances—in flour k-12 (93.4 mg%) and AOA—in k-16 (12.1 mg%). In the bread samples, the highest content of phenolic substances was found in samples k-9, 17, and 24 (213.8, 208.1, and 212.1 mg%, respectively), and AOA—in samples k-2, 9, 19, and 26 (25.6, 24.2, 24.4, and 24.4 mg%, respectively). The lowest content of phenolic substances was detected in samples k-10 and 14 (174.3 and 175.5 mg%) and AOA—in samples k-7 and 13 (20.2 and 19.5 mg%). The amount of phenolic compounds in the chufa tubers selected for baking mixed bread, as reported by Shaban et al. (210.5 mg%), falls within the range of variability in the current study [29]. Ozkan's PC results for bread with 40% chufa were significantly lower than our results (92.42 mg%), while the AOA results were higher (3.15 mmol/kg or 55.4 mg%) [24].

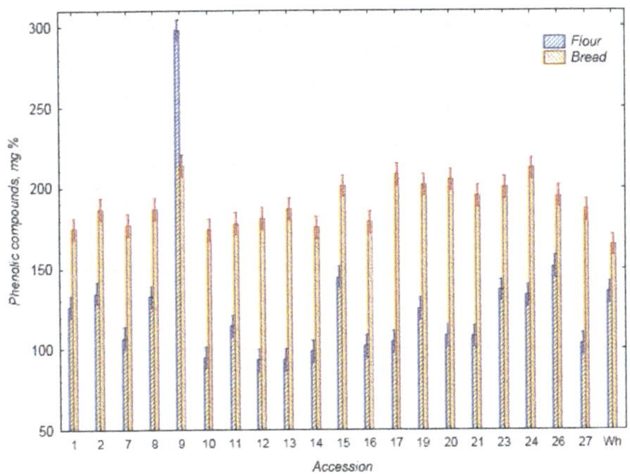

Figure 5. Content of total phenolic compounds (mg%) in flour samples of chufa tubers, whole-grain wheat (Wh), composite bread, and whole-grain wheat bread (Wh).

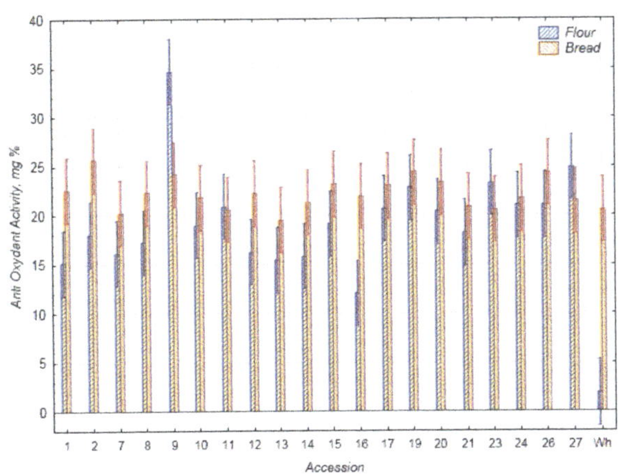

Figure 6. Antioxidant activity (AOA) in flour samples of chufa tubers and whole-grain wheat (Wh), composite bread, and wheat bread (Wh) (mg%).

In terms of all biochemical indicators, except protein and starch, all samples of chufa tuber flour and bread with chufa flour addition surpass control samples: whole-grain wheat flour and wheat bread. Among all the bread samples with chufa tuber flour addition, k-9 stood out with a high fiber content (5.16%), sum of phenolic substances (213.8 mg%), and AOA (24.2 mg%). K-10—high protein and oil content (17.94 and 9.42%, respectively); k-27—oil and fiber content (9.42 and 4.40%, respectively); k-16—high fiber content (4.54%); k-24—high sum of phenolic substances (212.1 mg%); k-19, 26—high AOA

(24.2 and 24.4 mg%, respectively). Thus, we can assume that these chufa accessions (k-9, 10, 16, 19, 24, 27) can be used as raw materials for the production of functional food.

3.2. Statistical Evaluation of Biochemical Parameters

3.2.1. Comparative Analysis of the Biochemical Parameters

Comparative analysis of the biochemical composition of flour and bread showed that the oil and fiber content of chufa is higher in flour than in bread, while in wheat, these levels are consistent and lower than those of chufa (Table 2, Figure 7A,D). Conversely, the protein and phenolic substance content increases in the bread (Table 2, Figure 7B,E). The starch content in wheat remains at the same level, whereas for chufa, it increases sharply in the bread; overall, starch levels are higher in wheat (Table 2, Figure 7C). Regarding AOA, wheat exhibits low AOA in flour, which increases sharply in bread, while chufa shows stable AOA that is slightly higher in bread compared to wheat (Table 2, Figure 7F).

Table 2. Biochemical parameters of chufa tuber flour, whole-grain wheat flour samples, and bread from mixed flour and whole-grain wheat flour.

Level	Flour				Bread			
	Chufa		Wheat		Chufa 30%/ Wheat 70%		Wheat 100%	
	X	Sx	X	Sx	X	Sx	X	Sx
Oil, %	22.0	0.39	2.3	0.02	8.1	0.12	2.3	0.00
Protein, %	6.2	0.07	5.7	0.00	16.4	0.08	21.7	0.08
Starch, %	24.7	0.25	54.4	0.57	42.7	0.70	52.7	0.00
Fiber, %	7.9	0.15	2.6	0.00	4.2	0.05	2.6	0.00
Phenolic substances, mg%	125.6	6.96	135.2	0.35	191.0	2.07	164.8	2.95
Antioxidant activity, mg%	19.6	0.78	1.9	0.09	22.3	0.39	20.6	0.35

X—Average value of biochemical parameters. Sx—standard deviation

3.2.2. Correlation Analysis

Correlation analysis of the biochemical composition of chufa tuber flour revealed highly significant positive correlations between oil and protein content (r = 0.728), starch and phenolic compounds (r = 0.767), oil and fiber (r = 0.728), and fiber and AOA (r = 0.775), as well as highly significant negative correlations between oil and starch content (r = −0.826), starch and fiber (r = −0.784) and starch and AOA (r = −0.698). Correlation analysis of the biochemical composition of composite bread showed highly significant positive correlations between phenolic compounds and oil content (r = 0.708), and AOA and oil content (r = 0.952). A highly significant positive correlation was also observed between the fiber content in composite bread and oil content (r = 0.728) and AOA (r = 0.774) in chufa flour, as well as between phenolic compounds in composite bread and AOA in chufa flour (r = 0.722). Additionally, a highly significant negative correlation was found between starch content in chufa flour and fiber content in composite bread (r = −0.783) (Table 3).

3.2.3. Analysis of Variance

The results of the analysis of variance showed that all traits differed significantly among the variants (flour, bread) and (except for phenolic compound content) by the culture factor (pure wheat, chufa and wheat mixture), as well as by the interaction of factors (Variant*Culture) (Table 4).

In terms of protein content, the main share of influence is attributed to the "Variant" factor (Figure 8A); in terms of oil, starch, and fiber content—to the "Culture" factor

(Figure 8B–D); in terms of AOA—it is approximately equal for both factors (Figure 8F); and in terms of phenolic compounds—it is mainly associated with unaccounted factors (Figure 8E).

For protein content, the main share of influence is attributed to the "Variant" factor (Figure 8B); for oil, starch, and fiber content—to the "Culture" factor (Figure 8A,C,D); for AOA—it is approximately equal for both factors (Figure 8F); and for PC—it is primarily associated with unaccounted factors (Figure 8E).

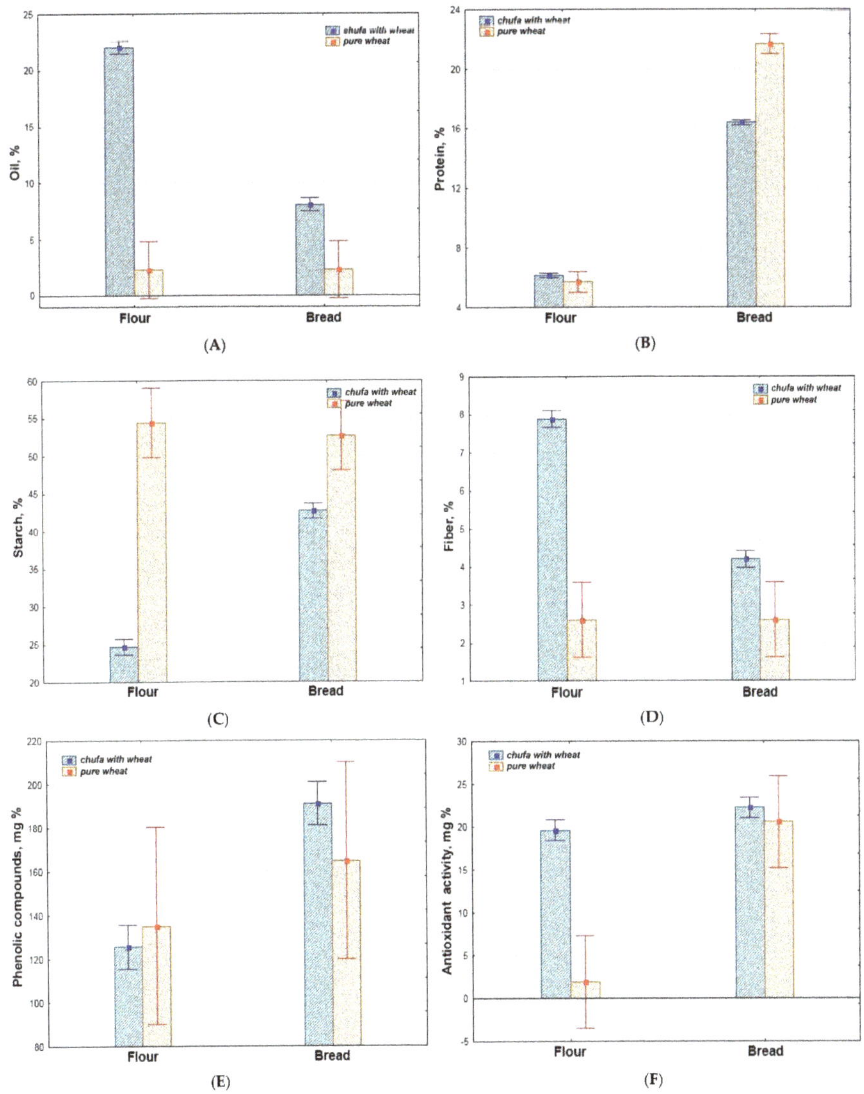

Figure 7. The share of influence of variation sources on the biochemical traits of chufa tuber flour, whole-grain wheat flour; and bread made from mixed flour and whole-grain wheat flour. (**A**)—oil, (**B**)—protein, (**C**)—starch, (**D**)—fiber, (**E**)—phenolic compounds, (**F**)—antioxidant activity.

3.2.4. Factorial Analysis

Analysis of the system of correlations showed that the factor structure of the variation of traits of the biochemical composition of flour (chufa, wheat) and composite bread (30% chufa/70% wheat) is practically identical (similar) regarding the content of oil, fiber, phenolic substances, and AOA. The following factors were identified. Factor 1—oil content, starch, and fiber vary (chufa flour, wheat flour); oil, protein, and fiber vary (composite bread); Factor 2—phenolic substances and AOA vary (chufa flour, wheat flour) and phenolic substances and AOA vary (composite bread); Factor 3—protein varies (chufa flour, wheat flour) and starch varies (composite bread). AOA—antioxidant activity; B/Oil—oil content of composite bread; B/Pr—protein content of composite bread; B/St—starch content of composite bread; B/F—fiber content of composite bread; B/PC—phenolic compounds content of composite bread; B/AOA—antioxidant activity of composite bread. Regarding the protein content, a positive structure of variation was noted in chufa flour and wheat flour, and a negative one in the composite bread, which indicates a strong influence of thermal processing on the protein content during the baking of bread (Table 5, Figure 9).

Table 3. Correlation map of the biochemical traits of chufa tuber flour, composite bread (30% chufa/70% wheat), and their significance levels.

	Oil	Protein	Starch	Fiber	PC	AOA	B/Oil	B/Pr	B/St	B/F	B/PC	B/AOA
Oil	1.000	0.068	−0.826	0.728	−0.001	0.453	0.800	−0.811	−0.136	0.728	0.213	0.189
Protein	0.771	1.000	−0.134	0.124	−0.313	0.036	0.201	−0.295	−0.459	0.125	0.312	−0.004
Starch	0.000	0.563	1.000	−0.784	−0.069	−0.698	−0.766	0.887	0.490	−0.783	−0.469	−0.285
Fiber	0.000	0.593	0.000	1.000	0.443	0.775	0.568	−0.708	−0.252	1.000	0.524	0.315
PC	0.995	0.167	0.767	0.044	1.000	0.564	−0.379	0.112	0.172	0.439	0.468	0.429
AOA	0.039	0.878	0.000	0.000	0.008	1.000	0.384	−0.509	−0.249	0.774	0.722	0.389
B/Oil	0.000	0.383	0.000	0.007	0.090	0.086	1.000	−0.784	−0.277	0.568	0.087	0.014
B/Pr	0.000	0.194	0.000	0.000	0.628	0.019	0.000	1.000	0.534	−0.708	−0.433	−0.224
B/St	0.557	0.037	0.024	0.269	0.456	0.277	0.225	0.013	1.000	−0.253	−0.404	−0.177
B/F	0.000	0.589	0.000	0.000	0.047	0.000	0.007	0.000	0.269	1.000	0.527	0.318
B/PC	0.353	0.169	0.032	0.015	0.032	0.000	0.708	0.050	0.069	0.014	1.000	0.435
B/AOA	0.412	0.987	0.210	0.165	0.052	0.081	0.952	0.329	0.442	0.160	0.049	1.000

PC—phenolic compounds, AOA—antioxidant activity, B/Oil—oil content of composite bread, B/Pr—protein content of composite bread, B/St—starch content of composite bread, B/F—fiber content of composite bread, B/PC—phenolic compounds content of composite bread, B/AOA—antioxidant activity of composite bread. Correlation color map: r >= scarlet: −1.00–−0.80; red: −0.80–−0.20; pink: −0.20–−0.00; grey: 0.00–−0.20; light blue: 0.20–0.40; blue: 0.40–1.00.. p-value color map for correlations: p <= scarlet: 0.000–0.001; red: 0.001–0.010; light red: 0.010–0.050; pink: 0.050–0.150; light blue: 0.150–0.500; blue: 0.500–1.000.

Table 4. Reliability and proportion of the source influence of the biochemical traits of chufa tubers and whole-grain wheat flour samples, composite bread (30% chufa/70% wheat) and whole-grain wheat bread samples, and their significance levels.

Sign	Reliability (p)			(η^2) The Proportion of The Source's Influence			
	Variant	Culture	Variant*Culture	Variant	Culture	Variant*Culture	y
Oil, %	0.000	0.000	0.000	14.8	49.6	14.8	20.8
Protein, %	0.000	0.000	0.000	90.0	3.1	4.4	2.6
Starch, %	0.000	0.000	0.000	8.5	50.4	12.5	28.7
Fiber, %	0.000	0.000	0.000	11.8	40.9	11.8	35.5
Phenolic substances, mg%	0.005	0.613	0.279	9.3	0.3	1.3	89.1
Antioxidant activity, mg%	0.000	0.000	0.000	19.5	16.2	11.0	53.4

Variant—flour, bread. Culture—chufa, pure wheat. Variant*Culture—different behavior of the trait under different combinations of variant and culture. U—unaccounted factors.

In the system of factors identified during the analysis of the composition of flour and bread, pure wheat and samples of composite bread baked from accessions of chufa k-9 and k-17 stand out (Figure 10).

3.3. Technological Assessment Indicators

One of the main objectives of the current investigation was to identify differences between various chufa accessions in their baking application. At the preliminary stage of research on the rheology of the mixed dough, these differences were recorded. It was also noted that these differences are most evident when the ratio of wholegrain wheat flour to chufa tuber flour is 70:30%.

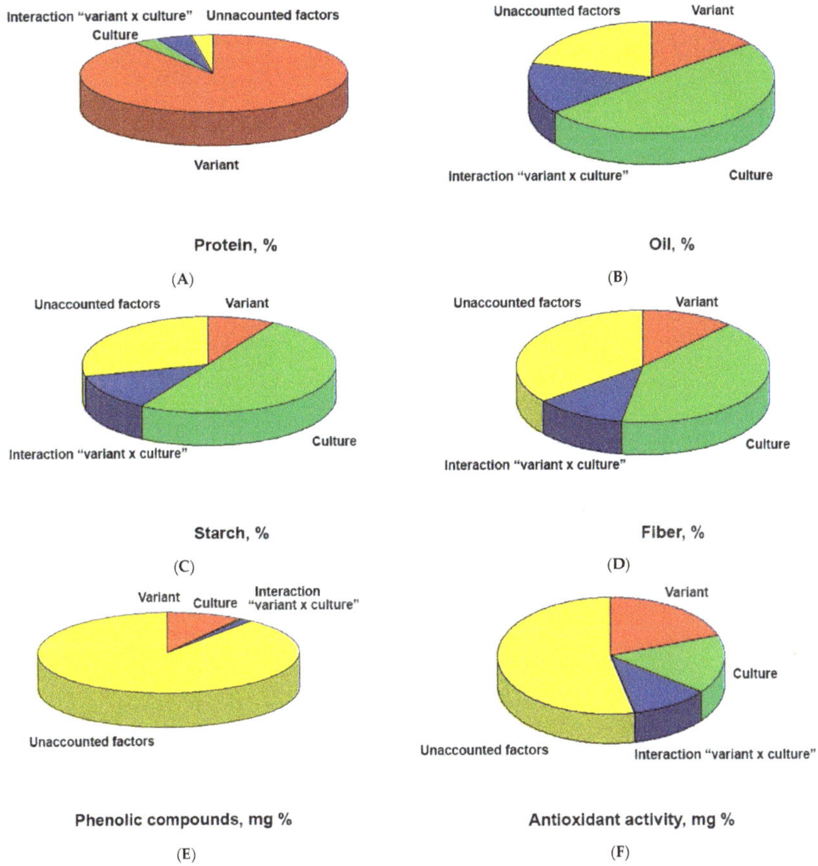

Figure 8. Sectoral diagrams showing the share of influence of factors: Variant—flour—bread; Culture—chufa, pure wheat; Variant*Culture—different behavior of the trait under different combinations of variant and culture, on biochemical indicators. **(A)**—protein content, **(B)**—oil content, **(C)**—starch content, **(D)**—fiber content, **(E)**—phenolic compounds content, **(F)**—OAO of chufa tuber flour and composite bread.

3.3.1. Sedimentation Analysis

Sedimentation analysis shows the ability of the experimental material to sustain stable swelling. All major methods of processing cereal crops are based on this ability. In wheat, this indicator is directly linked to the quality of gluten. Thus, its expressiveness is mainly influenced by the properties of the protein—its structural quality, and in some cases to its quantity as well. For the sedimentation analysis, a two-phase methodology recommended for non-wheat crops was used [30,31]. This methodology shows the ability of protein structures to resist increasing physicochemical loads or, thanks to them, to realize their swelling potential. Accordingly, the ratio of the indicators of the two phases characterizes the stability of the protein structures. If it is close to 1, then the sample is stable and has realized its potential. The further the sedimentation value is from 1, the less resistant the sample is to loads; the higher it is compared to 1, the greater swelling potential it carries. Almost all sedimentation values for the flour samples were equal to (k-9, 14, 21, 24), tended towards (k-12, 13, 15, 19, 20, 23), or slightly exceeded 1. The largest part of the samples formed the last group (k-1, 2, 7, 8, 10, 11, 16, 17, 26, 27). The maximum sedimentation index was demonstrated by the flour of the chufa samples k-11 (1.11).

Table 5. Factor structure of the variation of biochemical traits of chufa tuber flour and whole-grain wheat flour samples, composite bread (30% chufa/70% wheat), and whole-grain wheat bread samples.

	Factor 1	Factor 2	Factor 3
Oil	0.937	0.068	−0.034
Protein	0.051	−0.044	0.828
Starch	−0.878	−0.286	−0.237
Fiber	0.768	0.555	0.022
PC	−0.098	0.885	−0.375
AOA	0.510	0.746	0.061
B/Oil	0.913	−0.206	0.176
B/Pr	−0.856	−0.131	−0.387
B/St	−0.215	−0.140	−0.815
B/F	0.768	0.555	0.024
B/PC	0.169	0.774	0.432
B/AOA	0.070	0.640	0.101
The variance proportion by factor	39.6	26.1	16.1

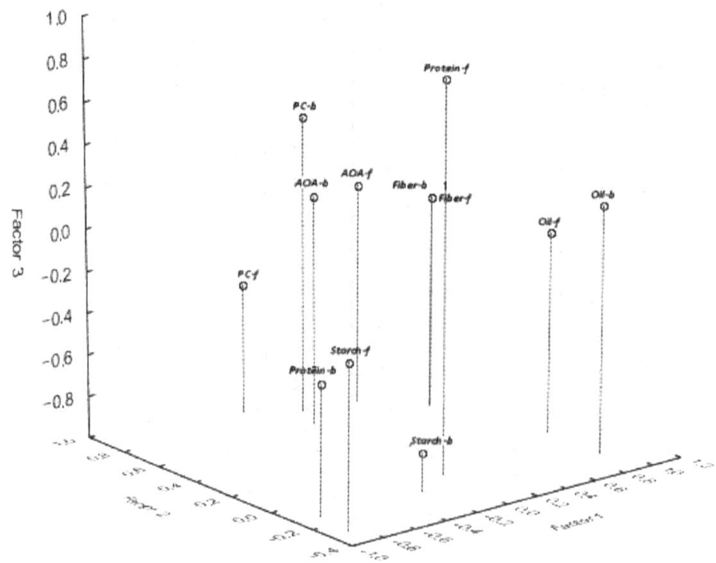

Figure 9. Factor structure of the variation of biochemical traits of chufa and whole-grain wheat flour (f) and bread samples (b).

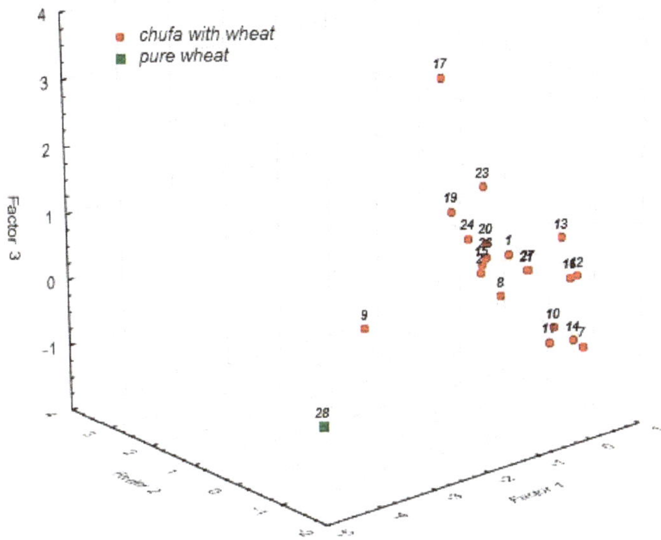

Figure 10. Differentiation of chufa and whole-grain wheat samples in the factor space.

3.3.2. Rheological Assessment Indicators

Viscoamylographic Visco-Mylographic

Viscoamylographic indicators allow us to characterize the thermodynamic properties of the carbohydrate complex of various flour samples. During the analysis of chufa tuber flour samples 2 (Figure 11J–M,S) or 3 (Figure 11A–I,N–R,T), thickening peaks were recorded. PT and PTS for all analyzed flour samples were approximately the same and averaged 6.26 min and 77.47 °C. Based on IPV, the flour samples could be divided.

Into groups with a viscosity of less than 80 (k-12, 14, 17, 19, 20, 27, Figure 11H,J,M–O,T) and more than 80 (k-1, 2, 7, 8, 9, 10, 11, 13, 15, 16, 21, 23, 24, 26, Figure 11A–G,I,K,L,P–S) units. The second group turned out to be more numerous. The time of maximum viscosity, or TPV, occurred on average at the ninth minute from the start of the experiment; the difference between the maximum and minimum values for the flour samples was 2.07 min. The temperature at which maximum viscosity or PVT was reached was either close to 90 °C or slightly exceeded this value, averaging 91.86 °C. The difference between the minimum or MHPV (297 units for k-1, Figure 11A) and maximum or PVHS (718 units for k-26, Figure 11S) values of hot maximum viscosity among the chufa flour samples was significant—421 units. The difference between the indicators of minimum viscosity in the hot state for the studied chufa flour samples decreased to 239 units; the minimum values (265 units) were noted for k-1 (Figure 11A), and the maximum (504 units) for k-14 (Figure 11J). The highest value of maximum viscosity in the cold state was noted for sample k-14 (Figure 11J), and the lowest (462 units) for k-1 (Figure 11A); the difference between the values was 734 units. At the end of the measurement, the minimum and maximum viscosity values or VEM were noted for samples k-14 (1108 units, Figure 11J) and k-1 (446 units, Figure 11A), with a slightly less significant difference of 662 units. The drop in viscosity at the maximum temperature, or VDPT, shows the resistance of the paste viscosity to physical load (mixing) at the maximum test temperature. Accordingly, the greater the drop, the lower the resistance [32]. This indicator in the samples ranged from 32 (k-1, Figure 11A) to 147 (k-13, Figure 11I), showing quite large variability. The rise in viscosity during cooling, or VIUC, and CPPV are associated with a change in the rheological properties of the water–flour mixture as the temperature decreases. The range of VIUC variability was significant. It ranged from 197 units (k-1, Figure 11A) to 692 units (k-14, Figure 11J), and the difference was 495 units. According to this indicator, the flour samples can be conditionally divided into a group with the lowest VIUC values (from 197 to 311 units), which included ten composite flour samples (k-1, 2, 7, 8, 9, 10, 11, 12, 24, 27, Figure 11A–H,R,T), with average values (from 364 units to 396 units) of four samples (k-16, 19, 20, 21, Figure 11L,N–P), and with high (from 434 units to 692 units) VIUC values of six composite flour samples (k-13, 14, 15, 17, 23, 26, Figure 11I–K,M,Q,S). The indicator of the rise in viscosity from the maximum values in the hot state to the maximum values in the cold state, or VIHC, shows the difference between the rise in viscosity during cooling and the drop in viscosity at the maximum temperature. The grouping of composite flour samples based on all three aforementioned indicators was identical. As can be seen from the above results, the range of variability among the samples for a number of visco-amylography indicators was quite high. Obviously, the result may be influenced by the amount of carbohydrates in the tubers of various chufa accessions and their quality. The difference between the minimum (26.08% for k-17) and maximum (45.69% for k-7) starch content was 19.61%. On average, for most samples, the starch content was 44.37%.

Farinographic Studies

WAC characterizes the technological properties of the raw material and determines how much water is needed to obtain a dough of standard consistency. Depending on the quality and grade of flour, the WAC can range from 40 to 75%. For the control sample, the WAC was 68%. In the composite samples (70% wheat flour and 30% chufa tuber flour), the WAC significantly decreased and averaged 53% (from 52% for k-14 to 55% for k-27). This is higher than the values obtained for bread with a 30% chufa content (49%) and corresponds to the WAC values for bread with a 20% chufa content (52%) [29]. DDT and DS characterize the strength of the flour. It is believed that the higher they are, the better the quality. In the composite flour samples, this indicator significantly increased compared to the control (3.68 and 1.3 min, respectively). The dough development time for the composite flour varied in the range from 6.17 min (k-26) to 18.48 min for flour with the addition of sample k-1. Dough stability ranged from 3.8 min (k-21) to 16.85 min (k-23). As a result, the dough in the mixed bread samples we studied became much stronger, which was significantly different from the values for bread with the same ratio of wheat flour and chufa tuber flour obtained by Shaban et al. (DDT-3.0 min and DS-9.5 min, respectively) [29]. DD shows how much the dough weakens during its prolonged mixing. The higher this indicator, the weaker the dough [33–37]. The control sample had a value of 80 units. In the composite samples, softening significantly decreased. The indicator value ranged from eight units (k-12) to thirty-four units (k-13). The VN indicator summarizes the assessment of the rheological properties of various flour samples obtained using the device. It is believed that the higher it is, the better the quality of the flour sample. The VN of the composite flour samples exceeded the control value of 51. For samples of wheat flour blended with flour from tubers of different chufa accessions, the VN ranged from 68 (k-26) to 94 (k-1), indicating that the quality of the composite flour in terms of rheological properties is much higher than that of the control wheat flour. Significant differences were noted for all studied indicators among the composite dough samples. As a result of the study, specific samples were highlighted: k-24 for WAC (57%), k-23 (15.03 min) and k-8 (16.9 min) for DDT; k-19 (16.8 min) and k-7 (16.03 min) for DS; and k-7 (91), k-2 (93), k-8 (93), and k-23 (91) for VN.

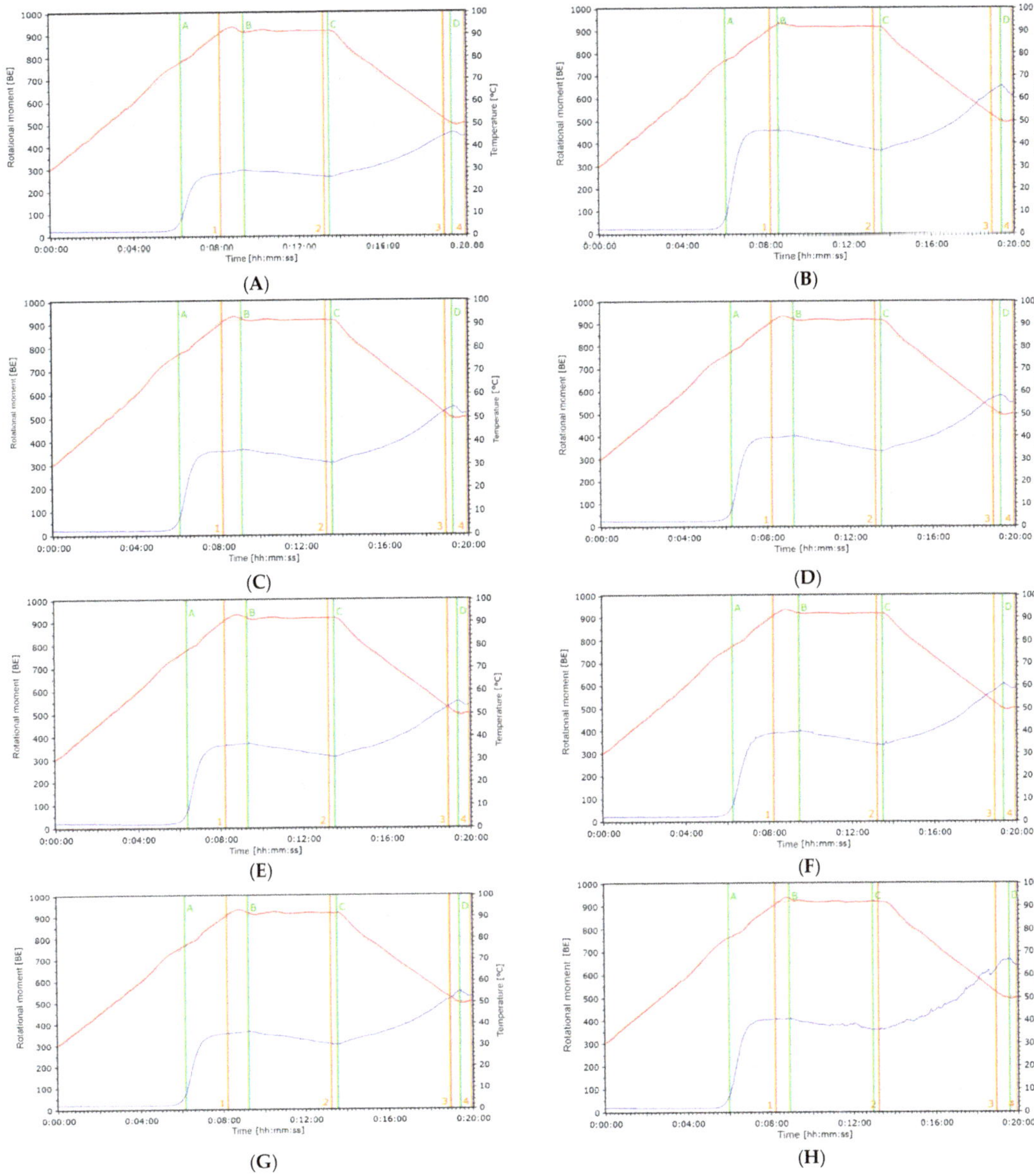

(A) (B) (C) (D) (E) (F) (G) (H)

Figure 11. *Cont.*

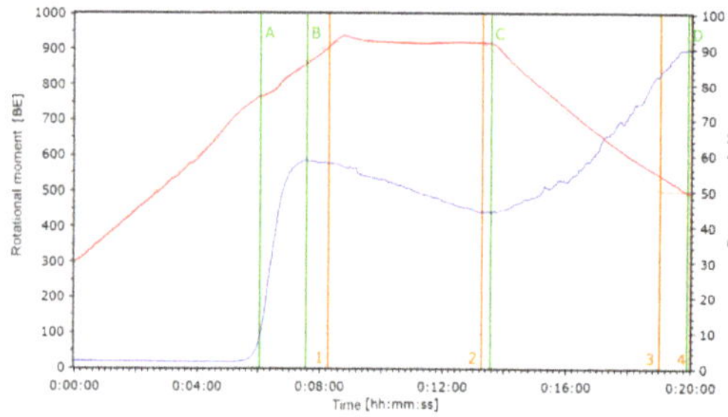

(I)

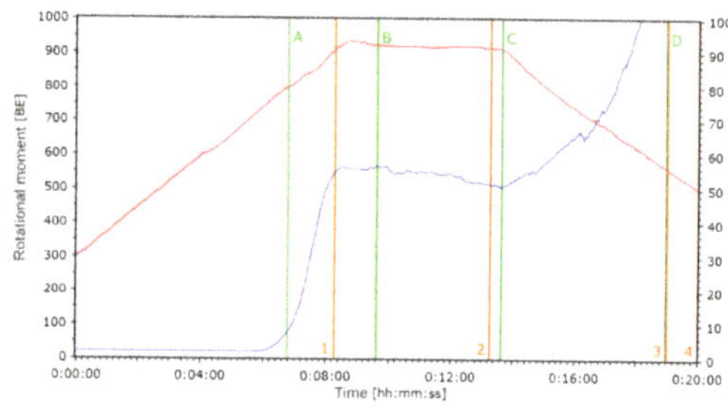

(J)

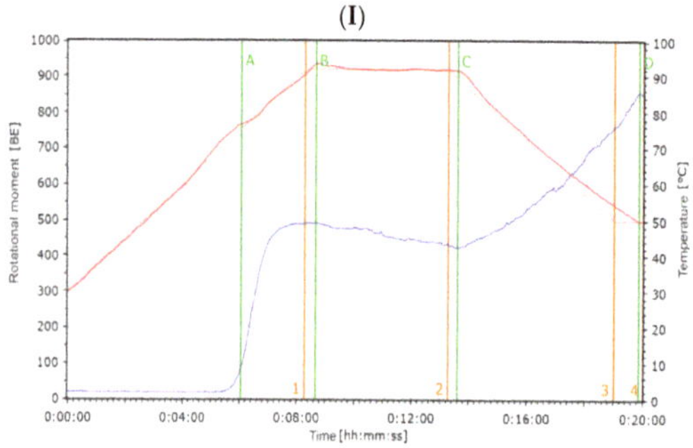

(K)

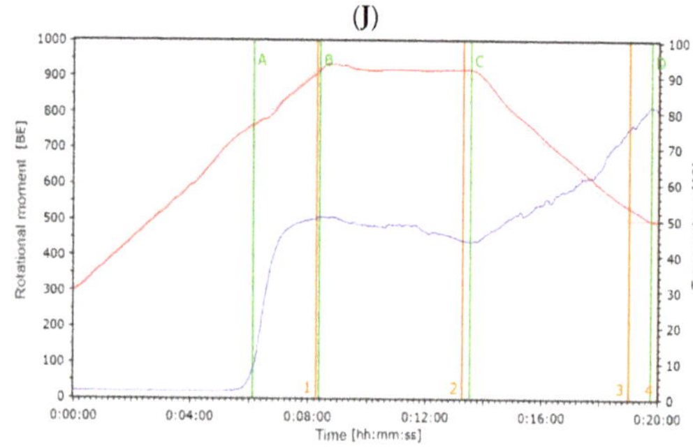

(L)

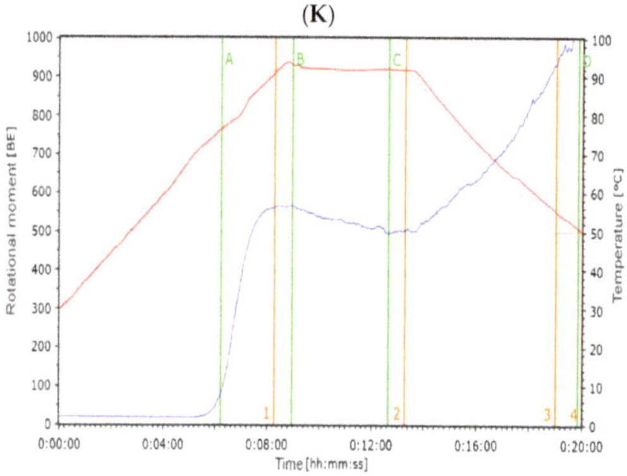

(M)

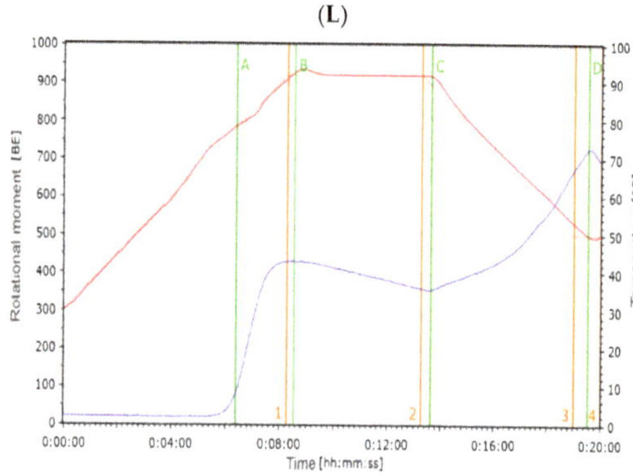

(N)

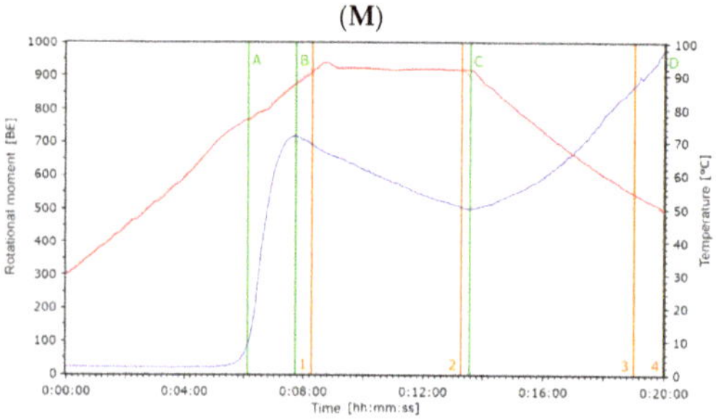

(O)

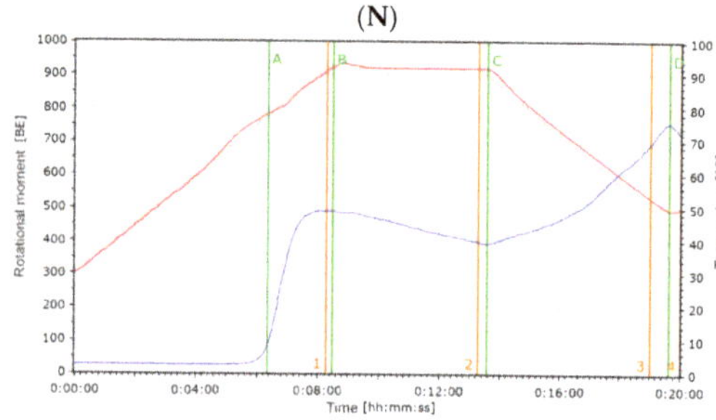

(P)

Figure 11. *Cont.*

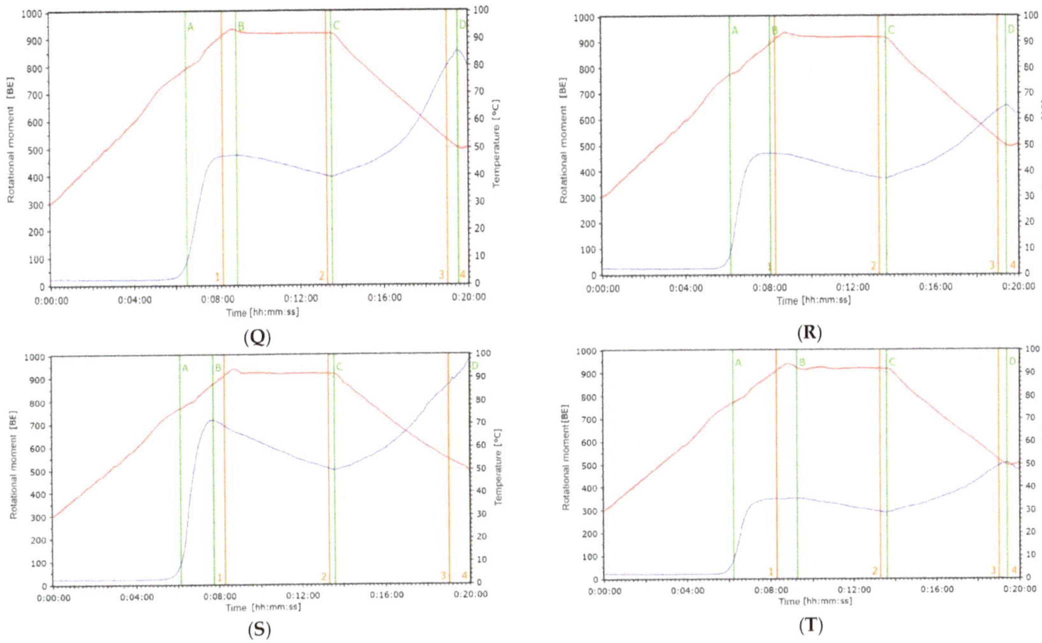

Figure 11. Viscoamylograms of flour samples with chufa addition (blue line indicates rotational moment; orange line—number of revolutions; red line—temperature; numbers 1, 2, 3, 4—phases of thermodynamic tests; letters **A, B, C, D**—indicators of thermodynamic tests). The green line indicates the main phases of the sample's rheological transformations: A—onset of gelatinization; B—maximum viscosity in the hot state; C—minimum viscosity in the hot state; D—maximum viscosity upon cooling.

3.3.3. Evaluation of Flour Color Indicators

The color of composite flour is determined by the presence of coloration from the endosperm and seed coats in the ground material of chufa tubers. The study of the color characteristics of flour from chufa tubers was carried out in the RGB color space, a color system that defines the final color as the result of adding three basic colors: red, green, and blue, within the device's color model [38,39]. The least variable indicator for the flour samples was the brightness indicator *L* (*CV* less than 10%), from the minimum (43.92 for k-21) to the maximum (65.28 for k-19). For the main part of the samples, *L* was more than 50 units, on average, 59.5; i.e., the flour had different shades of beige. The indicators of the blue–yellow range *b*, the chroma index C, and the hue angle ($h°$) showed medium variability (*CV* from 10 to 20%). The values of the indicator b were positive with a maximum of 17.25 for k-11 and a minimum of 4.73 for k-21, with an average value of 15.3; i.e., all samples had a yellow color of varying intensity. The minimum chroma index C was 13.61 for k-1, the maximum was 28.49 for k-21, and the average was 16.6. The hue angle ($h°$) changed from 9.3° for k-21 to 88.07° for k-10, with an average of 81.8°. The main part of the flour samples had shades of yellow, while k-21 exhibited a red–purple hue. The most variable color indicator (*CV* more than 100%) for the studied flour samples was *a* (red–green range), from a minimum of 0.475 for k-10 to a maximum of 28.09 for k-21; the average was 2.6. The values of *a* for all samples were positive, i.e., they had shades of red of varying intensity. The color assessment of chufa tuber flour obtained by us is partially consistent with the data of Shaban et al. for bread with 30% chufa addition, where the *L* value was 58.95, *a* ws 2.15, but *b* was significantly higher (22.82). This is probably because Shaban et al. studied the color of baked bread from mixed flour, whereas in our case we provide values for chufa tuber flour, and also due to the difference in the material chosen for making bread [29].

3.3.4. Baking Quality

Baking quality determines the commercial appeal of bakery products. SS corresponds to the ratio of the height (H) of the baked product to its diameter (D). SS less than 0.35 is considered weak; bread with such indicators has a flattened shape. Medium—from 0.35 to 0.45 and strong—more than 0.45. Bread with strong shape stability is characterized by an optimal ratio of the diameter and height of the baked product and is considered the most attractive to the consumer [39,40]. The first group consisted of three samples (k-15, 20, 26, Figure 12L,P,T), the second consisted of the majority of bread samples (k-2, 7, 10, 11, 12, 13, 14, 16, 17, 21, 23, 24, Figure 12C,D,G–K,M,N,Q–S) and the third consisted of k-8, 9, 19, 27 (Figure 12E,F,O–U). It should be noted that almost all values obtained for bread made from composite flour were higher than the shape stability of the control (Figure 12A), except samples k-15, 20 (Figure 12L,P). CD, g/mL, was determined as the ratio of the mass of a cylindrical crumb sample to the volume occupied by this cylinder. For almost all bread samples made from composite flour, except k-13 and k-26, the CD was higher than the control (0.53 g/mL) and ranged from 0.57 to 0.67 g/mL. However, the obtained results were lower than the CD values of bread made from wholegrain wheat flour (1.21 g/mL) [41,42]. CH (g load) for the control was 4901 g. For bread samples made from composite flour k-2, 10, 12, 13, 15, 16, 20, CH was lower (from 411 g for k-2 to 4686 g for k-15), while for the remaining bread samples, it was higher. S24 (g load) for the control was 2770 g. For bread samples with the addition of chufa k-1, 11, 12, 15, 16, 17, 21, 23, 24, the indicator was lower than the control (from 649 g for k-24 to 2663 g for k-12), while for the remaining samples it was higher. When comparing the values of the CH and S24 indicators, we can identify bread samples with the addition of chufa (k-12, 15, 16) in which both indicators were lower than the control, meaning that the staling rate was lower. TQS, bread with added chufa flour from various accessions was divided into three groups. The first group included products (k-17, 23, Figure 12N,R) with a score above four points, the second (k-1, 2, 7, 9, 10, 11, 14, 19, 20, 21, 24, 26, 27, Figure 12B–D,F–H,K,O–Q,T,U) was three points and above, and the third (k-8, 12, 13, 15, 16, Figure 12E,I,J,L,M) was below three points, which corresponded to good, satisfactory, and unsatisfactory taste.

As a result of the viscoamylographic analysis of the mixed dough, it was found that the maximum viscosity upon cooling was recorded for the sample k-14 (1108 units), which makes it possible to use it as a thickener in food production. Farinography showed that optimal parameters for high-quality mixed dough (DDT, VN, DS, DD) were recorded for dough samples with the addition of chufa k-1, k-23, k-12, and k-1, exhibiting values of 18.48 min, 94 units, 16.85 min, and 8 units, respectively. Good baking quality indicators such as shape stability were noted for the samples of mixed bread k-8, k- 9, k-19, k-27; CD—for k-13, k-26; CH and S24—for k-12, k-15, k-16; good TQS—for k-17, k-23. Thus, the flour of chufa accessions improves the mixed dough and bread quality. Hence, chufa tubers can be recommended for improving the baking properties of not only wheat but also other bready cereals, and for functional food production.

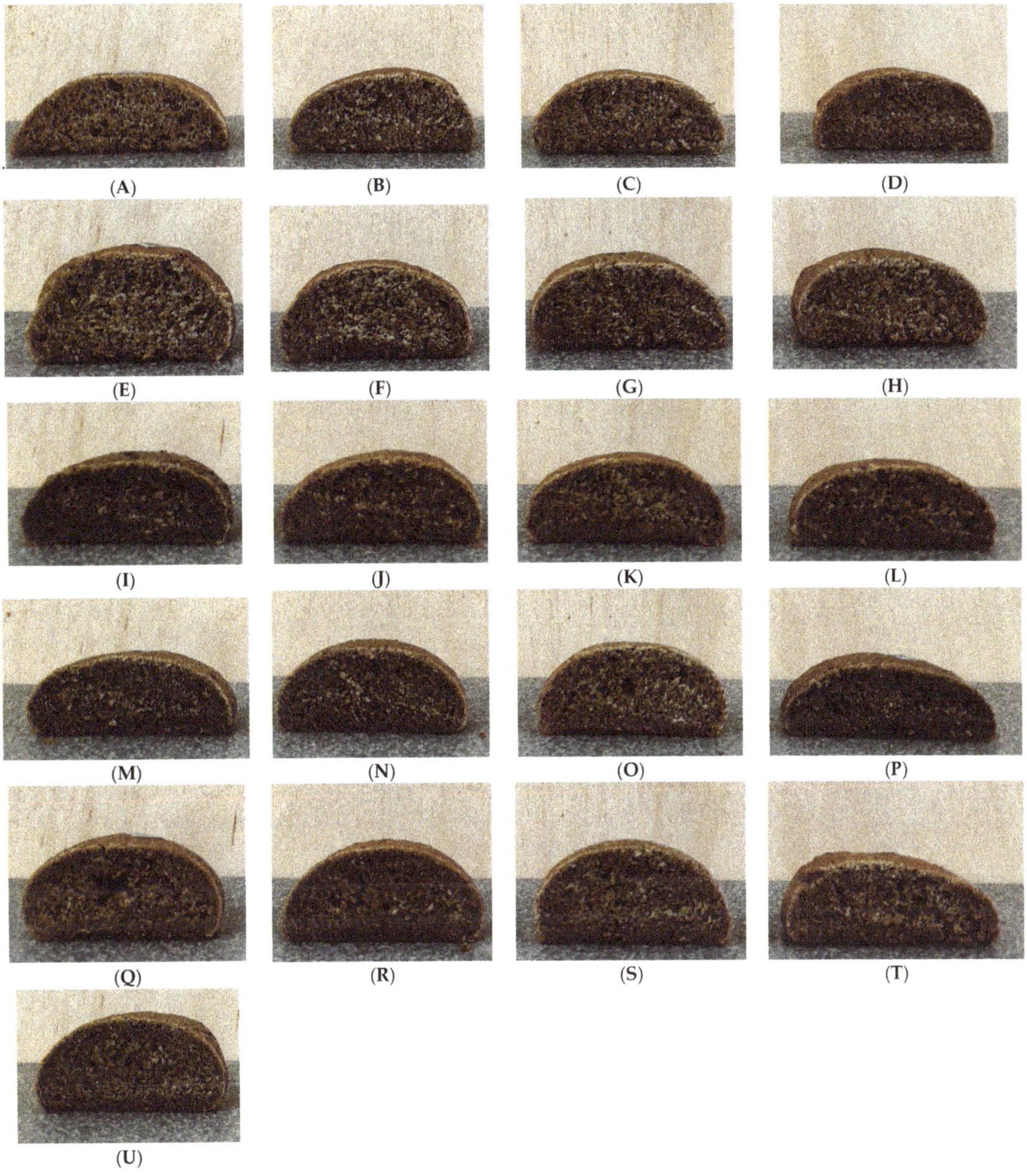

Figure 12. External view of bread from 100% whole-grain wheat flour as control, (**A**) bread from composite of 70% whole-grain wheat flour and 30% from chufa tubers accessions. (**B**)—k-1, (**C**)—k-2, (**D**)—k-7, (**E**)—k-8, (**F**)—k-9, (**G**)—k-11, (**H**)—k-10, (**I**)—k-12, (**J**)—k-13, (**K**)—k-14, (**L**)—k-15, (**M**)—k-16, (**N**)—k-17, (**O**)—k-19, (**P**)—k-20, (**Q**)—k-21, (**R**)—k-23, (**S**)—k-24, (**T**)—k-26, (**U**)—k-27.

3.4. Results of Statistical Analysis of Technological Assessment

3.4.1. Correlation Analysis of Technological Assessment Parameters

The matrix of correlation coefficients and their significance are presented in Tables S1 and S2, respectively. A direct mutual influence (ρ = 0.000) was established between the baking evaluation indicators SS and CD (0.681); between the farinographic evaluation indicators (DDT) and the sedimentation analysis (PSC2) (0.191); and between the sedimentation analysis indicators PSC1 and PSC2 (0.798). Positive correlations were observed between the viscoamylography indicators PT and PTS; PVHS and MHPV; MHPV and CPPV; CPPV and VEM; TPV and PVT (0.952; 0.953; 0.959; 0.995; 0.742, respectively), while negative correlations were observed between TPV and PVHS; TPV and VDPT; PVT and VDPT (−0.699; −0.715; −0.809). An inverse relationship was established between the color indicators: brightness (*L1*, *L2*), yellow hue (*b1*, *b2*), hue angle (*h1*, *h2*) of the composite flour and the fiber content (−0.830; −0.824; −0.910; −0.910; −0.762; −0.765, respectively); a direct relationship was established between the indicators of red hue of the flour (*a1*, *a2*) and the fiber content (0.766; 0.765, respectively). With a reliability of $0.01 < \rho < 0.045$, a direct influence was observed of oil content on the farinographic evaluation of dough DD (0.593); of AOA on baking qualities CD (0.589); of farinography indicators (DDT, DS, VN) on the amylography indicator TPV (0.500; 0.493; 0.529, respectively); of DD on VIUC (0.445); and of farinography indicators (DDT, DS, VN) on baking qualities (SS and CD) (r was approximately 0.5 or greater). The oil content negatively affected baking qualities SS (−0.473) and CD (−0.594); the fiber content affected sedimentation index PSC2 (−0.464), viscoamylographic PTS (−0.454), baking evaluation TQS (−0.492). With a reliability of $0.05 < \rho < 0.097$, a direct relationship was established between the protein content and the farinography indicators DDT (0.382) and VN (0.391) between the PS and AOA content (0.390). Based on the results of the correlation analysis, the influence of biochemical composition characteristics (protein, oil, PS, AOA) on the technological evaluation data is evident. The taste qualities (TQS) reliably depended on the fiber content, sedimentation indicators (PSC1, PSC2), farinographic analysis (WAC), visco-amylography (MHPV), other baking qualities (S24, SS), and flour color indicators.

3.4.2. Factor Analysis

Factor analysis of the system of correlations of traits of biochemistry and technological assessment allowed the isolation of eight factors, which collectively cover 90.7% of the total variability of the complex (Table 6). Factor 1 varies the content of oil, fiber, and the color traits of the chufa flour. Factor 2 varies mostly the traits of viscoamylograph assessment as follows: "maximum viscosity of starch in the hot state" (PVHS), "minimum viscosity in the hot state" (MHPV), "maximum viscosity in the cold state" (CPPV), "viscosity at the end of measurement" (VEM), "rise in viscosity during cooling" (VIUC), "rise in viscosity from maximum hot to maximum cold" (VIHC). Factor 3 varies fiber content in the flour samples, starch gelatinization onset temperature (PTS), chufa flour color traits, shape stability (SS). Factor 4 varies the starch content in the flour and bread staling after 24 h (S24). Factor 5 also varies the traits of viscoamylograph assessment as follows: starch gelatinization onset viscosity Time (IPV), maximum viscosity in the hot state (PVHS), maximum viscosity temperature (PVT), drop in viscosity at maximum temperature (VDPT). Factor 6 varies rheological properties of composite dough as follows: dough development time (DDT), dough stability (DS), and valorimeter number (VN). Factor 7 varies sedimentation analysis traits as follows: flour swelling capacity phase 1 (PSC1) and flour swelling capacity phase 2 (PSC2). Factor 8 varies antioxidant activity of the flour samples and the ratio phase 2/phase 1 (Phase 2/Phase 1).

Table 6. The factor structure of variation in technological assessment traits of chufa tubers and whole-grain wheat dough samples, composite bread (30% chufa/70% wheat) and whole-grain wheat bread samples.

	Factor 1	Factor 2	Factor 3	Factor 4	Factor 5	Factor 6	Factor 7	Factor 8
Oil	0.648	−0.121	−0.263	−0.407	−0.357	−0.130	−0.252	−0.007
Protein	−0.403	0.261	0.523	−0.349	−0.290	0.259	−0.098	0.125
Starch	−0.101	−0.088	−0.199	−0.621	−0.071	0.212	−0.048	0.028
Fiber	−0.628	0.153	−0.652	0.054	0.124	0.013	−0.110	−0.081
PC	−0.454	0.215	−0.140	0.509	−0.028	−0.003	0.549	−0.177
AOA	−0.275	0.215	0.113	0.257	0.080	0.258	0.025	−0.718
WAC	0.191	0.569	−0.359	−0.216	−0.150	−0.135	0.593	−0.132
DDT	0.017	0.307	0.210	−0.051	0.169	0.889	0.005	0.009
DS	−0.169	0.319	−0.013	−0.070	0.258	0.809	0.079	−0.120
DD	0.361	−0.362	−0.010	−0.363	−0.413	0.049	0.073	0.217
VN	−0.105	0.375	0.116	−0.037	0.213	0.871	−0.000	−0.020
PSC1	0.263	−0.054	0.233	0.038	0.033	0.042	0.896	−0.116
PSC2	0.221	0.140	0.482	0.190	0.097	0.063	0.684	0.331
P2-1	0.041	0.323	0.496	0.255	0.124	0.054	−0.059	0.696
PT	0.110	−0.507	0.549	0.043	0.570	0.112	0.193	−0.051
IPV	0.363	0.174	−0.033	0.257	−0.628	−0.030	0.205	−0.340
PTS	0.010	−0.464	0.684	0.051	0.503	0.075	0.150	−0.107
TPV	−0.055	0.223	0.155	−0.200	0.799	0.288	0.141	0.197
PVHS	0.050	−0.764	0.011	0.107	−0.532	−0.208	−0.157	0.101
PVT	−0.098	0.208	−0.178	0.131	0.886	0.135	0.027	0.031
MHPV	0.034	−0.882	−0.071	0.093	−0.307	−0.196	−0.211	0.096
CPPV	0.056	−0.959	−0.068	−0.001	−0.176	−0.188	−0.057	0.009
VEM	0.080	−0.934	−0.114	0.001	−0.236	−0.194	−0.095	0.004
VDPT	0.073	−0.328	0.084	0.083	−0.833	−0.101	0.010	0.169
VIUC	0.066	−0.967	−0.063	−0.055	−0.093	−0.176	0.034	−0.042
VIHC	0.050	−0.940	−0.125	−0.095	0.176	−0.130	0.044	−0.076
L1	0.421	0.191	0.849	0.069	−0.094	0.077	0.097	0.079
L2	0.410	0.166	0.857	0.063	−0.083	0.103	0.046	0.092
a1	−0.959	0.036	−0.241	−0.009	0.052	0.036	−0.090	−0.037
a2	−0.958	0.034	−0.240	−0.014	0.054	0.041	−0.096	−0.033
b1	0.716	−0.007	0.678	0.029	−0.057	0.027	0.104	0.034
b2	0.720	−0.011	0.670	0.040	−0.041	0.008	0.121	0.046
C1	−0.970	0.059	0.106	0.001	0.056	0.079	−0.077	−0.056
C2	−0.975	0.047	0.081	0.018	0.066	0.074	−0.055	−0.050
h1	0.959	−0.036	0.240	0.012	−0.049	−0.039	0.091	0.027
h2	0.959	−0.033	0.241	0.011	−0.048	−0.037	0.093	0.031
SS	−0.003	0.407	0.713	0.121	0.257	0.232	0.274	−0.169
CD	−0.157	0.244	0.462	0.361	0.563	0.323	−0.106	−0.255
CH	−0.002	−0.294	0.130	0.523	−0.229	0.546	0.308	−0.216
S24	0.053	0.266	0.104	−0.703	0.406	−0.133	−0.058	−0.095
TQS	0.428	0.284	0.219	0.190	0.127	0.242	0.552	0.064
share of total	21.8	17.9	14.3	5.9	11.8	8.1	6.8	4.1

The results of the analysis of variance showed that a significant influence of the origin of the chufa tubers accessions on the trait value was noted for the trait "gelatinization onset viscosity" (p = 0.005) and close to significant for three traits: "dough development time" (p = 0.107), "dough stability" (p = 0.101), and "valorimeter number" (VN) (p = 0.098) (Table 7) .

Table 7. Reliability of the influence of chufa tuber accession origin on the technological assessment traits of bread (30% chufa/70% wheat) samples.

Trait	Abbreviation	*p*
Oil	Oil	0.975
Protein	Protein	0.098
Starch	Starch	0.817
Fiber	Fiber	0.510
Phenolic compounds	PC	0.576
Antioxidant activity	AOA	0.162
Water absorption capacity	WAC	0.250
Dough development time	DDT	0.107
Dough stability	DS	0.101
Dilution of the dough	DD	0.612
Valorimeter number	VN	0.098
Flour swelling capacity phase 1	PSC1	0.148
Flour swelling capacity phase 2	PSC2	0.541
Phase 2/Phase 1	P2-1	0.778
Pasting temperature	PT	0.350
Initial pasting viscosity	IPV	0.005
Pasting temperature start	PTS	0.188
Time to peak viscosity	TPV	0.934
Peak viscosity in the hot state	PVHS	0.791
Peak viscosity temperature	PVT	0.380
Minimum hot paste viscosity	MHPV	0.554
Cold paste peak viscosity	CPPV	0.433
Viscosity at the end of measurement	VEM	0.500
Viscosity drop at peak temperature	VDPT	0.528
Viscosity increase upon cooling	VIUC	0.388
Viscosity increase from hot to cold	VIHC	0.269
L1 Flour color brightness	*L1*	0.832
L2 Flour color brightness	*L2*	0.925
a1 The red–green range of the spectrum	*a1*	0.300
a2 The red–green range of the spectrum	*a2*	0.282
b1 Yellow–blue range of the spectrum	*b1*	0.393
b2 Yellow–blue range of the spectrum	*b2*	0.349
C1 Chromaticity index	*c1*	0.416
C2 Chromaticity index	*c2*	0.437
h1 Shade angle	*h1*	0.272
h2 Shade angle	*h2*	0.280
Shape stability	SS	0.824
Crumb density	CD	0.807
Crumb hardness	CH	0.186
Staling in 24 h	S24	0.421
Taste quality score	TQS	0.114

3.4.3. Discriminant Analysis by Origin

As a result of the discriminant analysis, classification functions were constructed (Table 8). Based on these functions, the chufa tuber accessions were differentiated by origin into three groups: Eastern Europe, Southern Europe, and Central Europe. These functions provide 100% correct classification (Table 9).

Table 8. Classification functions for groups of chufa accessions of different origin from the VIR collection based on biochemical and technological assessment indicators.

	Eastern Europe	South and Central Europe	Africa
Initial pasting viscosity (IPV)	−166.4	−172.1	−174.2
Dough stability (DS)	−445.3	−452.3	−458.2
Peak viscosity temperature (PVT)	260.6	259.5	263.9
Pasting temperature Start (PTS)	2718.7	2740.6	2775.1
Pasting temperature (PT)	−11,199.4	−11,304.5	−11,443.1
Valorimeter number (VN)	450.0	457.9	463.1
Oil	1022.1	1034.7	1047.6
Dough development time (DDT)	−750.5	−762.2	−771.1
Constant number	−90,778.1	−91,787.8	−94,197.1

Table 9. The results of the classification analysis based on different geographical origins of chufa tuber accessions.

	Correct Decisions	Eastern Europe	South and Central Europe	Africa
Eastern Europe	100.0	8	0	0
South and Central Europe	100.0	0	6	0
Africa	100.0	0	0	6
Total	100.0	8	6	6

For two accessions, k-2 (Russia) $p = 0.994$ and k-16 (Africa) $p = 0.985$, the reliability of assignment to the observed group is below 1.00 (Table 10).

Table 10. Posterior probabilities of chufa tuber accessions belonging to groups of different geographical origins.

Catalog	The Observed Classification	Russia	Europe	Africa
k-1	Russia	1.000	0.000	0.000
k-2	Russia	0.994	0.006	0.000
k-7	Europe	0.000	1.000	0.000
k-8	Europe	0.000	1.000	0.000
k-9	Europe	0.000	1.000	0.000
k-10	Europe	0.000	1.000	0.000
k-11	Africa	0.000	0.000	1.000
k-12	Africa	0.000	0.000	1.000
k-13	Europe	0.000	1.000	0.000
k-14	Africa	0.000	0.000	1.000
k-15	Africa	0.000	0.000	1.000
k-16	Africa	0.000	0.015	0.985
k-17	Africa	0.000	0.000	1.000
k-19	Russia	1.000	0.000	0.000
k-20	Russia	1.000	0.000	0.000
k-21	Russia	1.000	0.000	0.000
k-23	Europe	0.000	1.000	0.000
k-24	Russia	1.000	0.000	0.000
k-26	Russia	1.000	0.000	0.000
k-27	Russia	1.000	0.000	0.000

3.4.4. Canonical Analysis

In the canonical analysis, two axes were identified—Root 1 (85.8% of variance), differentiating European and African accessions, and Root 2 (14.2%), differentiating accessions from Eastern, Central, and Southern Europe (Table 11).

Table 11. Standardized coefficients for canonical variables based on the results of biochemical and technological assessment.

Sign	Root 1	Root 2
Initial pasting viscosity (IPV)	−2.115	1.221
Dough stability (DS)	−5.110	0.494
Peak viscosity temperature (PVT)	0.635	1.600
Pasting temperature start (PTS)	5.423	1.814
Pasting temperature (PT)	−4.474	−0.961
Valorimeter number (VN)	10.659	−2.759
Oil	2.146	0.103
Dough development time (DDT)	−7.375	1.127
Variance Explained by the Factor	85.8	14.2

3.4.5. Discriminant Analysis

Discriminant analysis of chufa accessions of different origins based on baking qualities showed that k-2 stands out in the group from Eastern Europe, and k-16 in the group from Africa (Figure 13). K-2 differs from other accessions of Eastern European origin by its high VN (93) and IPV of more than 80 units. K-16 differs from other African accessions by its DS (9.18 min) and PTS (76.9 °C), and VN (79).

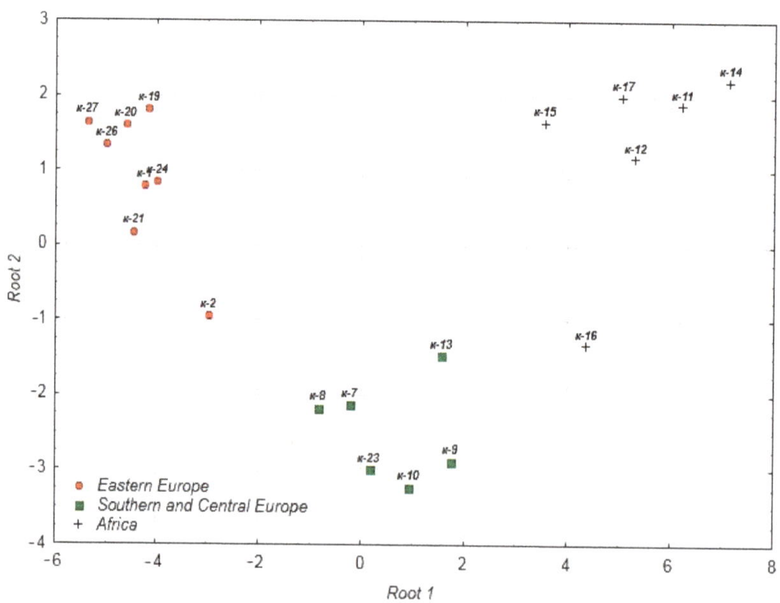

Figure 13. Discrimination of chufa tuber accessions of different origins by technological qualities.

4. Conclusions

According to biochemical data, all samples of chufa tuber flour and bread with chufa flour addition surpassed the control, except for protein and starch content. The k-9 bread sample stood out with a high fiber content (5.16%), sum of phenolic substances (213.8 mg%), and AOA (24.2 mg%); k-10 with a high protein and oil content (17.94% and 9.42%, respectively); k-27 with oil and fiber content (9.42% and 4.40%, respectively); k-16 with a high fiber content (4.54%); k-24 with high sum of phenolic substances (212.1 mg%); k-19 and k-26 with high AOA (24.2 mg% and 24.4 mg%, respectively). As a result of the viscoamylographic analysis of the mixed dough, it was found that the maximum viscosity upon cooling was recorded for the sample k-14 (1108 units), which makes it possible to use it as a thickener in food production. Farinography estimation showed that the optimal parameters for high-quality mixed dough (DDT, VN, DS, DD) corresponded to the samples with the addition of chufa accessions k-1, k-23, k-12, and k-1 (18.48 min, 94 units, and 16.85 min, 8 units, respectively). Good baking quality indicators included SS for samples k-8, k-9, k-19, k-27; CD for k-13, k-26; CH and S24 for k-12, k-15, k-16; good TQS—for k-17, k-23. Thus, these chufa accessions improve the quality of mixed dough and bread and, therefore, can be used for the production of functional foods. Based on the results of the correlation analysis, the influence of biochemical composition characteristics (protein, oil, PS, AOA) on the technological evaluation data is evident. The taste qualities (TQS) reliably depended on the fiber content, sedimentation indicators (PSC1, PSC2), farinographic analysis (WAC), viscoamylography (MHPV), other baking qualities (S24, SS), and flour color indicators. According to the results of the farinographic study, it was found that the dough made from mixed flour was stronger than the control: it was more resistant to kneading and had a lower degree of liquefaction. Therefore, we can conclude that chufa tuber flour can improve the baking properties of whole-grain wheat flour. In terms of organoleptic properties, differences were also identified, and the samples that improved the taste of mixed bread were selected. Chufa samples with high rheological properties and high sedimentation values of the mixed dough can be recommended for improving the baking properties of wheat and other bread cereals. The presented results characterize the properties and potential of chufa tuber accessions grown in the conditions of the Kuban–Azov Plain as a material for regional functional food manufacturing and, in general, for the food industry.

In general, the data from the current study are preliminary. We have confirmed that chufa is a promising crop for bakery production in the region, given the unusual reproduction conditions for chufa. The next step will be the quality evaluation of mixed bread with different ratio of chufa flour addition, including chufa accessions grown in other regions of the Russian Federation (the Central Chernozem and Northwestern regions) as a part of the research program.

Supplementary Materials: The following supporting information can be downloaded at: https://www.mdpi.com/article/10.3390/foods14213680/s1, Table S1: Parameters of biochemical and technological evaluation of samples of chufa tuber and whole grain wheat flour, samples of mixed bread and whole-grain wheat bread; Table S2: The magnitude of the correlation coefficients between features of biochemical and technological evaluation indicators.

Author Contributions: Conceptualization and methodology, N.G.K., V.I.K., V.S.P., A.E.S. and T.V.S.; resources, T.V.Y.; writing—original draft preparation, N.G.K. and L.L.M.; writing—review and editing, N.G.K., V.I.K., V.S.P., T.V.Y., L.L.M., A.E.S., T.V.S. and I.A.K.; software and visualization, L.L.M. All authors have read and agreed to the published version of the manuscript.

Funding: The study was performed within the framework of the Development Program of the National Center for Plant Genetic Resources under agreement with the Ministry of Education and Science of Russia dated 26 February 2025 No. 075-02-2025-1584. The plant material for the experiment was obtained within the framework of the state task according to the theme plan of VIR. Project No. FGEM-2022-0005.

Institutional Review Board Statement: Not applicable.

Informed Consent Statement: Not applicable.

Data Availability Statement: The original contributions presented in this study are included in the article/Supplementary Material. Further inquiries can be directed to the corresponding author.

Conflicts of Interest: The authors declare no conflicts of interest.

References

1. Pascual-Seva, N.; Pascual, B. Determination of crop coefficient for chufa crop (*Cyperus esculentus* L. var. sativus Boeck.) for sustainable irrigation scheduling. *Sci. Total Environ.* **2021**, *768*, 144975. [CrossRef]
2. Zheng, X.; Liu, J.; Cheng, Z.; Sun, Y.; Li, L.; Wang, J. Improving Tuber Yield of Tiger Nut (*Cyperus esculentus* L.) through Nitrogen Fertilization in Sandy Farmland. *Plants* **2024**, *13*, 1063. [CrossRef]
3. Yu, Y.; Lu, X.; Zhang, T.; Zhao, C.; Guan, S.; Pu, Y.; Gao, F. Tiger nut (*Cyperus esculentus* L.): Nutrition, processing, function and applications. *Foods* **2022**, *11*, 601. [CrossRef]
4. Kon'kova, N.G.; Safina, G.F. Valuable agronomic traits of chufa (*Cyperus esculentus* L.) accessions from the VIR collection: Methods of preparing nodules for long-term storage. *Proc. Appl. Bot. Genet. Breed.* **2021**, *182*, 34–44. [CrossRef]
5. Kon'kova, N.G.; Khoreva, V.I.; Popov, V.S.; Yakusheva, T.V.; Malyshev, L.L.; Solovyeva, A.E.; Shelenga, T.V. Variability of the Main Economically Valuable Characteristics of *Cyperus esculentus* L. in Various Ecological and Geographical Conditions. *Plants* **2024**, *13*, 308. [CrossRef]
6. Zommara, M.A.; Swelam, S.; Raya-Álvarez, E.; Imaizumi, K.; Elmahdy, A.; Alkhudhayri, D.A.; Aljehani, A.A.; Agil, A.; Elmahallawy, E.K. Nutritional and potential health benefits of chufa oil, olive oil, and anhydrous milk fat against gallstone disease in a C57BL/6N mouse model. *Front. Nutr.* **2024**, *11*, 1445484. [CrossRef]
7. Zhang, Y.; Sun, S. Tiger nut (*Cyperus esculentus* L.) oil: A review of bioactive compounds, extraction technologies, potential hazards and applications. *Food Chem.* **2023**, *19*, 100868. [CrossRef] [PubMed]
8. Suleiman, M.S.; Olajide, J.E.; Omale, J.A.; Abbah, O.C.; Ejembi, D.O. Proximate composition, mineral and some vitamin contents of tiger nut (*Cyperus esculentus*). *Clin. Investig.* **2018**, *8*, 161–165. [CrossRef]
9. Essa, M.M.; Bishir, M.; Bhat, A.; Chidambaram, S.B.; Al-Balushi, B.; Hamdan, H.; Govindarajan, N.; Freidland, R.P.; Qoronfleh, M.W. Functional foods and their impact on health. *J. Food Sci. Technol.* **2023**, *60*, 820–834. [CrossRef] [PubMed]
10. Hernández-Olivas, E.; Asensio-Grau, A.; Calvo-Lerma, J.; García-Hern'andez, J.; Heredia, A.; Andrés, A. Content and bioaccessibility of bioactive compounds with potential benefits for macular health in tiger nut products. *Food Biosci.* **2022**, *49*, 101879. [CrossRef]
11. Sanchez-Zapata, E.; Fernandez-Lopez, J.; Sendra, E.; Perez-Alvarez, J.A. Tigernut (*Cyperus esculentus*) commercialization: Health aspects, composition, properties, and food applications. *Compr. Rev. Food Sci. Food Saf.* **2012**, *11*, 366–377. [CrossRef]
12. Ani, I.F.; Adeoya, B.K.; Ngori, E.O.; Kehinde, Z.A. Nutritional Composition and Quality Acceptability of Soft Candy (Toffee) Made from Tiger nut. *Int. J. Res. Innov. Appl. Sci.* **2019**, *4*, 1–4.
13. Onwuzuruike, U.A.; Ukegbu, P.; Obasi, N.E.; Ogah, M.; Okereke, I.; Uche, P.C.; Echendu, C. Physical Properties, Dietary Fibre Profile and Peroxide Value of Biscuits Produced from Wheat-Tigernut Flours with Avocado Paste as A Fat Substitute. *Indones. Food Sci. Technol. J.* **2023**, *7*, 67–74. [CrossRef]
14. Kwaghsende, G.S.; Ikala, G.U.; Ochelle, P.O. Quality evaluation of biscuit from wheat-tigernut composite flour. *Int. J. Agric. Nutr.* **2019**, *1*, 24–29. [CrossRef]
15. Akajiaku, L.O.; Kabuo, N.O.; Alagbaoso, S.O.; Orji, I.G.; Nwogu, A.S. Proximate, mineral and sensory properties of cookies made from Tiger-Nut Flour. *J. Nutr. Diet. Pract.* **2018**, *2*, 1–5.
16. Chinma, C.E.; Avu, J.O.; Abubakar, Y.A. Effect of tigernut (*Cyperus esculentus*) flour addition on the quality of wheat-based cake. *Int. J. Food Sci. Technol.* **2010**, *45*, 1746–1752. [CrossRef]
17. Martín-Esparza, E.; González-Martínez, C. Horchata de chufa: A traditional Spanish beverage with exceptional organoleptic, nutritive, and functional attributes. *Funct. Prop. Tradit. Foods* **2016**, *12*, 371–375. [CrossRef]
18. Roselló-Soto, E.; Barba, F.J.; Putnik, P.; Bursać Kovačević, D.; Lorenzo, J.M.; Cantavella-Ferrero, Y. Enhancing Bioactive Antioxidants' Extraction from "Horchata de Chufa" By-Products. *Foods* **2018**, *7*, 161. [CrossRef]
19. Rios, M.; Tinitana, F.; Jarrín, P.; Donoso, N.; Romero-Benavides, J.C. "Horchata" drink in Southern Ecuador: Medicinal plants people's well-being. *J. Ethnobiol. Ethnomed.* **2017**, *13*, 18. [CrossRef]
20. Orhevba, B.A.; Bankole, O.S. Effect of process treatments on the proximate composition of tigernut-soy milk blends. *Afr. J. Food Sci.* **2019**, *13*, 261–280. [CrossRef]
21. Adebayo-Oyetoro, A.O.; Ogundipe, O.O.; Adeyeye, S.A.O.; Akande, E.A.; Akinyele, A.B. Production and evaluation of tiger nut (*Cyperus esculentus*) milk flavoured with moringa oleifera leaf extract. *Curr. Res. Nutr. Food Sci.* **2019**, *7*, 265–271. [CrossRef]
22. Sobhy, H.; El-Nawawy, M.; Hegazi, N.; El-Dieb, S.; Abd El-Razik, F.; Elkashef, H. Tiger nut tuber milk: Using dairy byproducts and probiotic bacteria. *Acta Sci. Pol. Technol. Aliment.* **2022**, *21*, 271–280. [CrossRef]

23. Madsen, M.; Thulesen, E.; Mohammadifar, M.; Bang-Berthelsen, C. Chufa drink: Potential in developing a new plant-based fermented dessert. *Foods* **2021**, *10*, 3010. [CrossRef]
24. Özcan, M.M. Quality Evaluation of Bread Prepared from Wheat–Chufa Tuber Composite Flour. *Foods* **2023**, *12*, 444. [CrossRef] [PubMed]
25. Aguilar, N.; Albanell, E.; Minarro, B.; Guamis, B.; Marta Capellas, M. Effect of tiger nut-derived products in gluten-free batter and bread. *Food Sci. Technol. Int.* **2014**, *1*, 323–331. [CrossRef] [PubMed]
26. King, H.; Aubert, R.E.; Herman, W.H. Global burden of diabetes, 1995–2025: Prevalence, nu-merical estimates, and projections. *Diabetes Care* **1998**, *21*, 1414–1431. [CrossRef] [PubMed]
27. Radchenko, E.E.; Anisimova, I.N.; Ryazanova, M.K.; Kibkalo, I.A.; Alpatieva, N.V. Newly Developed Restorer Lines of Sorghum [*Sorghum bicolor* (L.) Moench] Resistant to Green-bug. *Plants* **2024**, *13*, 425. [CrossRef]
28. Pozdnyakov, V.V.; Vasilenko, A.A. Use of test systems for evaluation of total antioxidant activity of seeds. *Plant Breed. Seed Prod.* **2017**, *112*, 153–163. [CrossRef]
29. Shaban, H.H.; El-Hadidy, G.S.; Hamouda, A.M. Evaluation of Breadsticks Prepared from Chufa Tubers as Partial Substitute of Wheat Flour. *Asian J. Food Res. Nutr.* **2023**, *2*, 73–84. [CrossRef]
30. Kibkalo, I. Effectiveness of and perspectives for the sedimentation analysis method in grain quality evaluation in various cere-al crops for breeding purposes. *Plants* **2022**, *11*, 1640. [CrossRef]
31. Kibkalo, I.A.; Loskutov, I.G.; Voitsutskaya, N.P.; Solovyova, M.V.; Obukhova, N.S.; Blinova, E.V. Development of methodo-logical approaches to assessing the technological properties of oat grain. *Plant Biotechnol. Breed.* **2024**, *7*, 6–15. (In Russian) [CrossRef]
32. Chernykh, V.Y.; Kostyuchenko, M.N.; Baluyan, K.A.; Smetanin, D.O.; Kandrokov, R.K. Reducing the staling rate of bakery products made from wheat flour. *Polzunovsky Bull.* **2024**, *2*, 96–106. (In Russian) [CrossRef]
33. Anisimova, L.V.; Serebrennikova, E.S.; Bondarenko, V.E.; Basov, V.Y. Rheological properties of dough from a mixture of wheat and lupin flour. *Polzunovsky Bull.* **2018**, *4*, 40–44. (In Russian) [CrossRef]
34. Chikpah, S.K.; Korese, J.K.; Hensel, O.; Sturm, B.; Pawelzik, E. Rheological properties of dough and bread quality characteristics as influenced by the proportion of wheat flour substitution with orange-fleshed sweet potato flour and baking conditions. *LWT* **2021**, *147*, 111515. [CrossRef]
35. Anisimova, L.V.; Soltan, O. Rheological properties of dough from a mixture of wheat and whole oat flour. *Polzunovsky Bull.* **2017**, *3*, 9–13. (In Russian)
36. Diosi, G.; More, M.; Sipos, P. Role of the farinograph test in the wheat our quality determination. *Acta Univ. Sapientiae Aliment.* **2015**, *8*, 104–110. [CrossRef]
37. Gómez, A.; Ferrero, C.; Calvelo, A.; Añón, M.C.; Puppo, M.C. Effect of Mixing Time on Structural and Rheological Properties of Wheat Flour Dough for Bread making. *Int. J. Food Prop.* **2011**, *14*, 583–598. [CrossRef]
38. Shteinberg, T.S.; Shvedova, O.G.; Kandrokov, R.K.; Bolotov, V.I. Selection of color characteristics for the development of a method for identifying the admixture of soft wheat flour in durum wheat flour. *Bull. AGAU* **2017**, *12*, 175–181. (In Russian)
39. Mcguire, R.G. Reporting of objective color measurements. *HortScience* **1992**, *27*, 1254–1255. [CrossRef]
40. Sammalisto, S.; Laitinen, M.; Sontag-Strohm, T. Baking quality assessment of twenty whole grain oat cultivar samples. *Foods* **2021**, *10*, 2461. [CrossRef]
41. Efremova, E.N. Quality evaluation of bakery products. *Agrar. Bull. North Cauc.* **2016**, *3*, 20–24. (In Russian)
42. Nashat, S.; Abdullah, M.Z. In *Computer Vision Technology for Food Quality Evaluation*, Sun., D.-W., Ed.; Quality Evaluation of Bakery Products, 2nd ed.; Academic Press: Penang, Malaysia, 2016; pp. 525–589. [CrossRef]

foods

MDPI

Article

Elucidating the Nutritional Profile and Biochemical Characterization of High-Energy Nutritional Bar Formulated with Sukkari Date Paste and Mixed Nuts

Hassan Barakat [1,*], Hani A. Alfheeaid [1], Thamer Aljutaily [1], Raed Alayouni [1], Hend F. Alharbi [1] and Woroud A. Alsanei [2]

[1] Department of Food Science and Human Nutrition, College of Agriculture and Food, Qassim University, Buraydah 51452, Saudi Arabia; h.alfheeaid@qu.edu.sa (H.A.A.); thamer.aljutaily@qu.edu.sa (T.A.); r.alayouni@qu.edu.sa (R.A.); hf.alharbi@qu.edu.sa (H.F.A.)

[2] Department of Food and Nutrition, Faculty of Human Sciences and Design, King Abdulaziz University, Jeddah 21589, Saudi Arabia; walsanei@kau.edu.sa

* Correspondence: haa.mohamed@qu.edu.sa; Tel.: +966-547-141-277

Abstract

Growing health consciousness drives demand for convenient, nutrient-dense snacks. This study evaluates five Sukkari date-mixed-nut bar formulations (DNB1–DNB5; date/nut ratios 40:60–80:20) through comprehensive biochemical and nutritional analyses. Macronutrient profiling showed that higher date ratios increased moisture and carbohydrates, whereas higher nut ratios enhanced protein, fat, and caloric density. Mineral assays revealed progressive increases in calcium, phosphorus, magnesium, and trace elements as date content decreased. The assessment of phytochemicals and antioxidants demonstrated that total phenolics, flavonoids, and radical-scavenging activities peaked in nut-rich bars, declining by ~50% in date-rich bars, underscoring nuts' dominant antioxidant role. HPLC profiling identified catechol and vanillic acid as the major phenolics, with optimal release and retention at the 60:40 ratio (DNB3). Amino acid (AA) analysis confirmed positive correlations between nut content and total/essential AAs; DNB1–DNB2 achieved favorable essential-to-nonessential AA ratios (0.56–0.59) and higher protein quality indices. Fatty acid (FA) composition analysis revealed that oleic acid was identified as the major constituent across all formulations, coupled with optimal omega-6/omega-3 ratios. GC-MS analysis identified a total of 31 volatiles, mainly benzene derivatives and FA methyl esters. Results also revealed that notable variations attributed to different date/nut ratios significantly alter aroma profiles, with DNB3 yielding the most remarkable diversity of health-associated volatiles. Results from PCA and hierarchical clustering suggest that a single dominant dimension (PC1, 94.47% variance) governs compositional differences among the five date bar formulations, reflecting deliberate variation in ingredient proportions. The evidence suggests that DNB3's (60:40 Sukkari date to mixed nut ratio) delivers balanced macronutrients, robust antioxidants, and diverse bioactives, positioning it as a health-promoting functional snack, aligning with its suitability for athletes, clinical nutrition applications, and health-conscious populations. These findings support the commercial development of optimized date-nut bars as nutrient-dense functional snacks, and future work should focus on scale-up production, shelf-life stability, and assessing in vivo bioavailability.

Keywords: nutritional evaluation; high-energy bar; Sukkari Date; bioactive compounds; sustainable food systems; food supply

1. Introduction

High-energy nutritional bars have emerged as convenient, nutrient-dense snack formats tailored to meet the heightened dietary demands of athletes, shift workers, and individuals with increased energy requirements [1]. By developing nutritionally balanced snack bars with immediate-acting carbohydrates, muscle-protective proteins, and energy-dense lipids, these bars serve to enhance long-term nutritional health benefits and effective performance [2]. The incorporation of age-old ingredients with cutting-edge processing techniques of these products has been instrumental in improving both their functional attributes and consumers' acceptability [3]. In 2024, the global market for nutritional bars was valued at a remarkable growth, with an estimated value of approximately USD 7.4 billion and projected to reach USD 13.2 billion by 2034, with a notable compound annual growth rate of 6.1% [4]. This growth reflects rising health awareness and demand for convenient nutrition. Additionally, the functional food market is predicted to reach USD 441.66 billion by 2028, with nutritional bars involving protein, energy, meal-replacement, and specialty (e.g., gluten-free, organic, plant-based) formulations and constituting a key segment [3].

A recent study demonstrates that consumers are increasingly motivated by health benefits when selecting functional foods, with over 24.9% of male athletes preferring protein bars for quick energy and protein intake. This figure has doubled in three years [4]. Date fruits are recognized for their exceptional nutritional profile, comprising dietary fibers, natural sugars, micronutrients, and bioactive phytochemicals, including phenolic acids, polyphenols, and carotenoids that provide anti-inflammatory, antioxidant, and antimicrobial properties [5]. On the other hand, nuts are also a good source of high-quality proteins, essential fatty acids (FAs), vitamins, and minerals. This makes the date-nut combination a great base for nutrient-dense functional snack bars aimed at health-conscious customers [6].

Sukkari date (*Phoenix dactylifera* L.) is a premium cultivar widely grown in Saudi Arabia, prized for its soft texture, golden hue, and high natural sugar content [7,8]. Its carbohydrate nature, which is mostly glucose and fructose, makes it easy to get energy back after a meal quickly. Thus, the incorporation of Sukkari date paste into functional foods aligns with current trends toward a natural sweetener and binder for nutritious bar formulations [5,9]. Moreover, Sukkari dates contribute dietary fiber (4–8% *w*/*w*), which promotes gastrointestinal health and attenuates glycemic excursions [10]. Additionally, phenolic compounds in such dates significantly enhance their antioxidant capacity, thereby counteracting oxidative stress implicated in various chronic diseases [11–13].

Incorporation of mixed nuts like almonds, walnuts, and cashews to the mix increases the nutritious profile by adding high-quality plant proteins, unsaturated fatty acids, and a variety of micronutrients, including vitamin E, magnesium, and phosphorus [14]. Almonds and walnuts are especially high in monounsaturated and polyunsaturated fats, which are associated with decreasing the cardiovascular risk markers and improvement of blood lipid profiles [15,16]. Cashews and pistachios containing essential AAs and bioactive phytochemicals are related to enhancing the antioxidant and anti-inflammatory potential of different formulated bars [15,17]. This mix of date paste and mixed nuts has a lot of nutritional value, making it great for high-energy uses. Several studies have shown that date-based nutritious bars made with a variety of dietary items are good for your health and nutrition [5,9,10]. Similarly, Parn et al. [18] formulated novel fruit bars utilizing two popular date cultivars, including Sukkari, showing enhanced nutritional composition and consumer acceptability. Interestingly, mixed nuts provide unsaturated fats, plant proteins, and bioactive substances to the bars' health advantages. Mixed nuts maintain steady postprandial blood glucose levels compared to processed snacks and improve gut microbiome composition by increasing *Bifidobacterium longum* and reducing inflammation-associated species. [19,20]. Nonetheless, these individual components have undergone substantial examination; their synergistic effects in bar compositions, especially with premium date cultivars like Sukkari, remain inadequately defined. Several studies have shown that nutritious bars made with dates are good for your health and nutrition. Hong et al. [21] found that mixed nut consumption significantly reduced cardiovascular disease risk factors, including decreased triglycerides, total cholesterol, and LDL-cholesterol, while increasing adiponectin and reducing oxidative stress biomarkers [22]. These qualities make date-nut bars useful for sustained energy, cardiovascular health promotion, and metabolic control, especially for athletes, health-conscious customers, and groups who need efficient nutrition delivery [23,24].

Despite the documented nutritional benefits of date-based bars and mixed nuts separately, a critical knowledge gap exists regarding the comprehensive characterization of Sukkari date-mixed nut bars specifically. While Parn et al. [18] discovered enhanced nutritional value in Sukkari date bars, despite the absence of mixed nuts or AAs and FAs profiles. The previously described date-nut studies have concentrated on different types of dates or limited nutritional variables. This makes it hard to see their bioactive compounds, whole amino acid profile, and mineral content for use in functional foods. Furthermore, no research has thoroughly investigated the antioxidant activity of Sukkari date-mixed nut bars in relation to diverse phenolic profiles. Such a knowledge gap restricts the potential for developing a functional food for populations such as athletes and health-conscious consumers, who require high-energy and nutrient-dense products. Therefore, this study aims to develop high-energy bars formulated with varying ratios of Sukkari date paste and mixed nuts, focusing on their nutritional composition and biochemical characteristics, including FA and AA profiles, mineral quantification, phenolic compound identification, and antioxidant activity assessment.

2. Materials and Methods

2.1. Raw Ingredients

All ingredients, such as Sukari date paste and Sukkari date powder, were purchased from Al Emtyaz Dates Factory, Buraydah, Qassim Region, SA (https://www.alemtyaz.com.sa, accessed on 24 March 2024). At the same time, nuts, spices, and cow's milk powder were obtained from Damascus Winds supermarket, Buraydah, Qassim Region, SA. For each 1 kg nuts mix, dried Sukkari date powder (200 g), roasted almond (150 g), pistachio (100 g), whole oats (100 g), cow's milk powder (100 g), roasted cashew (75 g), roasted sesame seeds (75 g), walnuts (50), hulled pumpkin seeds (50 g), hulled sunflower seeds (50), chia seeds (30), dried cinnamon powder (10 g), and dried ginger powder (10 g) were mixed in a knife blender (Santos, VM0122E, Cleveland, OH, USA) at speed 3 for 60 sec min to prepare a homogenous nut mix and then kept at 4 ± 1 °C before further processing.

2.2. Formulation of High-Energy DNB

For each formula, the prepared date paste was thoroughly mixed with the predetermined nut mix in accordance with the ratios specified in Table 1 to create a uniform dough. The dense and sticky nature of date paste ensures effective incorporation of nuts and improves the cohesiveness of the bar matrix. Each formula was mechanically mixed using a mixer until the mixture resembled a homogeneous consistency mix. Immediately after mixing, the date-nut mix was portioned using a ready shaping frame and pressed systematically to guarantee uniform compaction and consistent thickness throughout each bar. The shaped bars (35 each) were then wrapped with polyethylene sheets and subjected to storage under cooling conditions of 4 ± 1 °C until further analysis; see Figure 1.

Table 1. Formulas of high-energy date-nuts-based (DNB) bars (g 100 g^{-1}).

Ingredients	DNB1	DNB2	DNB3	DNB4	DNB5
Sukkari date paste	40.00	50.00	60.00	70.00	80.00
dried Sukkari date powder	12.00	10.00	8.00	6.00	4.00
Roasted almond	9.00	7.50	6.00	4.50	3.00
Pistachio	6.00	5.00	4.00	3.00	2.00
Whole oats	6.00	5.00	4.00	3.00	2.00
Cow's milk powder	6.00	5.00	4.00	3.00	2.00
Roasted cashew	4.50	3.75	3.00	2.25	1.50
Roasted sesame seeds	4.50	3.75	3.00	2.25	1.50
Walnuts	3.00	2.50	2.00	1.50	1.00
Hulled pumpkin seeds	3.00	2.50	2.00	1.50	1.00
Hulled sunflower seeds	3.00	2.50	2.00	1.50	1.00
Chia seeds	1.80	1.50	1.20	0.90	0.60
Dried cinnamon powder	0.60	0.50	0.40	0.30	0.20
Dried ginger powder	0.60	0.50	0.40	0.30	0.20

DNB: date-nuts-based bars.

Figure 1. Formulas of high-energy DNB bars. The ingredients of formulas DNB1 to DNB5 were presented in Table 1.

2.3. Nutritional Composition and Mineral Content of Different DNBs

All DNB formulas underwent comprehensive chemical analyses to quantify moisture (AOAC 925.10), crude protein (AOAC 981.10), crude lipid (AOAC 963.15), ash (AOAC 942.05), dietary fiber (AOAC 992.16), available carbohydrate contents (by difference), and caloric value, following established AOAC procedures [25]. Mineral composition, specifically sodium and potassium levels, was assessed through flame photometry (Cambridge, UK) following AOAC 970.35. At the same time, atomic absorption spectroscopy was utilized to measure concentrations of calcium, magnesium, iron, copper, manganese, and zinc in accordance with AOAC protocols employed by a PerkinElmer AAnalyst™ 400 (Waltham, MA, USA) with appropriate hollow-cathode lamps and standard curves per AOAC 984.27 [25]. Furthermore, phosphorus content was determined using a standardized colorimetric method as outlined by Borah et al. [26], measuring absorbance at 880 nm using a UV–Vis spectrophotometer (Shimadzu UV-1800, Kyoto, Japan). All measurements were performed in triplicate, and calibration standards and quality controls were included in each analytical batch.

2.4. Phytochemical Analysis of Different DNBs

The total phenolic compounds (TPC) in DNBs were quantified using the Folin–Ciocalteu method, with results expressed as milligrams of gallic acid equivalents (mg GAE per 100 g), following the procedure outlined by Bettaieb et al. [27]. Total carotenoid (TC) content was measured colorimetrically according to a modified protocol [28]. Antioxidant capacity, assessed via DPPH radical scavenging activity (DPPH-RSA), involved a colorimetric assay using 2,2-diphenylpicrylhydrazyl (DPPH) radicals, calculating the percentage of radical inhibition and referencing a Trolox standard, with results reported as micromoles of Trolox equivalents per gram dry weight (μmol TE g^{-1}) [29]. Radical scavenging activity against ABTS radicals (ABTS-RSA) was tested using the method described by [30]. A 0.1 mL aliquot of extract was combined with 2.9 mL of ABTS solution, which was generated by incubating 7 mmol L^{-1} ABTS and 2.45 mmol L^{-1} potassium persulfate in distilled water, in the dark at room temperature for 12–16 h. Before use, this solution was diluted with ethanol to an absorbance of 0.70 ± 0.02 at 734 nm and equilibrated at 30 °C. Sample aliquots were then added, incubated at 30 °C for 20 min, and the absorbance reduction at 734 nm was recorded using a UV-Vis spectrophotometer (Shimadzu UV-1800, Kyoto, Japan). Antioxidant capacity was calculated as μmol Trolox equivalents per gram of sample (μmol TE g^{-1}) from a Trolox standard curve. Additionally, total flavonoid (TF) and total flavonol (TFL) contents were determined using methods by Barakat & Almundarij [31] and Kumaran & Karunakaran [32], respectively, with values expressed as milligrams of quercetin equivalents per gram (mg QE g^{-1}).

2.5. Quantification of Phenolic Compounds of Different DNBs

According to Schneider [33], phenolic compounds in different DNBs were analyzed using an Agilent 1260 Infinity HPLC system (Agilent Technologies, Palo Alto, CA, USA) equipped with a quaternary pump, autosampler, and a variable wavelength detector (VWD, Hewlett Packard 1050) set at 280 nm. The chromatographic separation was performed on a Kinetex® 1.7 μm EVO C18 column (150 mm × 4.6 mm, Phenomenex, Torrance, CA, USA) maintained at 30 °C. A ternary linear gradient elution was employed, consisting of (A) HPLC-grade water with 0.1% trifluoroacetic acid (TFA), (B) acetonitrile, and (C) HPLC-grade methanol. The gradient program was initiated at 94:3:3 (*v*/*v*) for phenolics and 98:1:1 (*v*/*v*) for flavonoids. Automated injection of 20 μL extract was used, with a flow rate of 1 mL min^{-1}, column temperature of 30 °C, and ambient laboratory temperature of 20 °C. Quantification of phenolic peaks was carried out in mg kg^{-1} using external calibration curves of appropriate standard solutions.

2.6. Determination of AA Profile in Different DNBs

The AA composition of various KBs enriched with GMB was analyzed using a Sykam AA analyzer (Sykam GmbH, Eresing, Bavaria, Germany). The system was configured with a solvent delivery unit (S 2100 quaternary pump), an autosampler (S 5200), an AA reaction module (S 4300) equipped with a dual-filter photometer operating at 440–570 nm with constant signal output and signal integration capability, and a refrigerated reagent organizer (S 4130). The analysis was performed using an LCA K06/Na column under gradient elution. Buffer A was prepared with 11.8 g trisodium citrate dihydrate, 6 g citric acid, 65 mL methanol, 6.5 mL of 32% HCl, and 0.5 g phenol per liter (pH 3.45, normality 0.12). In comparison, buffer B contained 19.6 g trisodium citrate dihydrate, 3.1 g NaOH, and 5 g boric acid per liter (pH 10.85, normality 0.20). The system operated at a flow rate of 1 mL min^{-1} with a programmed column temperature gradient between 57 and 74 °C. Detection was monitored at wavelengths of 440 and 570 nm. A standard stock solution (Sykam Type H, CAT No. S000029) containing 18 AAs at 2.5 μmol mL^{-1} (cystine at 1.25 μmol mL^{-1}) in 0.1 N HCl with 0.1% phenol was prepared. Standards were diluted (60 μL in 1.5 mL HPLC water), filtered (0.22 μm), and injected (100 μL). Samples (~1 g, weighed to four decimal places) underwent acid hydrolysis with 20 mL of 6N HCl at 110 °C for 16 h in sealed vessels. Hydrolysates were filtered, evaporated, reconstituted in 100 mL HPLC water, diluted 1:10, filtered, and injected (100 μL). AA concentrations were calculated using relative response factors from standards according to Cohen et al. [34]. Biological value and AA scores were determined following WHO guidelines (2007) [35] as described by Chavan et al. [36] using relevant equations.

2.7. Determination of the FA Profile in Different DNBs

According to Aldai et al. [37], the total FA components were converted into their corresponding methyl esters. These FA methyl esters (FAMEs), derived from the different date–nut bar (DNB) samples, were analyzed using gas–liquid chromatography (GLC) equipped with a flame ionization detector (FID). The chromatographic separation was performed under a programmed temperature gradient, starting at 100 °C and gradually increasing to 200 °C, with additional steps including a rise at 2 °C per minute until 230 °C, followed by a 10 min isothermal hold. The injection and detector ports were maintained at 250 °C and 300 °C, respectively. Data acquisition and peak integration were carried out using the Saturn GC Workstation Software, version 5.51.

2.8. Quantification of Volatile Components in Different DNBs

The chemical composition of NDB was performed using a Trace GC-TSQ mass spectrometer (Thermo Scientific, Austin, TX, USA) with a direct capillary column TG–5MS (30 m × 0.25 mm × 0.25 μm film thickness). The column oven temperature was initially held at 50 °C and then increased by 5 °C /min to 250 °C, held for 2 min, and then increased to the final temperature of 300 °C by 30 °C /min and held for 2 min. The injector and MS transfer line temperatures were kept at 270 and 260 °C, respectively; helium was used as a carrier gas at a constant flow rate of 1 mL/min. The solvent delay was 4 min, and diluted samples of 1 μL were injected automatically using Autosampler AS1300 coupled with GC in the split mode. EI mass spectra were collected at 70 eV ionization voltages over the range of m/z 50–650 in full scan mode. The ion source temperature was set at 200 °C. The components were identified by comparison of their mass spectra with those of the WILEY 09 and NIST 14 mass spectral databases.

Methylation Method

H_2SO_4-methanol 2% (v/v) was added to a vial containing 10 mg of the sample that had previously been weighed. The vial was heated at 80 °C with occasional shaking. Afterwards, 0.25 mL of the neutralized aqueous solution (sodium hydroxide 1 M) was added, and it was smoothly shaken. Allowed to cool at room temperature, then after neutralization, 1 mL KCl 59.64 g L^{-1} was added and shaken vigorously, then 1 mL hexane was added and shaken vigorously, then the upper organic layer was taken for GC-MS analysis following the methodology of Hewavitharana et al. [38].

2.9. Statistical Analysis

Statistical evaluation was performed using one-way ANOVA in SPSS version 22 (IBM Corp., Armonk, NY, USA, 2013) under a completely randomized design. Post hoc multiple comparisons were conducted using Duncan's test at a $p < 0.05$ significance level according to Steele et al. [39]; all analyses were performed in triplicate. Additionally, principal component analysis (PCA) was performed to identify underlying patterns and relationships within the dataset. The PCA was performed on the 5 date bar samples (DNB1-DNB5) across 27 nutritional and bioactive variables. The data were standardized using z-score normalization before analysis to account for different scales and units across variables.

3. Results

3.1. Nutritional Composition of Formulated High-Energy DNBs

The nutritional composition of the five Sukkari DNBs (DNB1–DNB5) varied significantly ($p < 0.05$) with changing date-to-nut ratios (Table 2). Moisture content rose considerably from 9.68% in DNB1 to 13.43% in DNB5, reflecting increased date paste (DNB1 vs. DNB5, $p < 0.05$). Conversely, protein significantly declined from 10.89% to 5.567% (DNB1 vs. DNB5, $p < 0.05$), and total fat significantly decreased from 18.66% to 6.62% ($p < 0.05$). Ash content showed a modest but significant reduction (3.16% to 2.92%; DNB1 vs. DNB5, $p < 0.05$). Dietary fiber remained relatively constant (5.74–6.23%; $p > 0.05$). Total carbohydrates significantly increased from 57.62% in DNB1 to 71.47% in DNB5 ($p < 0.05$), while energy value decreased significantly from 417.01 kcal in DNB1 to 344.74 kcal in DNB5 ($p < 0.05$).

Table 2. Nutritional composition of different formulated high-energy DNBs (mean ± SE), $n = 3$.

Composition *	High-Energy DNB Formulas				
	DNB1	DNB2	DNB3	DNB4	DNB5
Moisture	9.68 ± 0.14 ^{a}	10.61 ± 0.16 ^{b}	11.55 ± 0.15 ^{c}	12.49 ± 0.21 ^{d}	13.43 ± 0.18 ^{e}
Protein	10.89 ± 0.16 ^{a}	9.56 ± 0.15 ^{b}	8.23 ± 0.11 ^{c}	6.90 ± 0.11 ^{d}	5.56 ± 0.07 ^{e}
Total fat	18.66 ± 0.27 ^{a}	15.65 ± 0.24 ^{b}	12.64 ± 0.17 ^{c}	9.63 ± 0.16 ^{d}	6.62 ± 0.09 ^{e}
Ash	3.16 ± 0.05 ^{a}	3.11 ± 0.05 ^{ab}	3.04 ± 0.04 ^{b}	2.98 ± 0.05 ^{bc}	2.92 ± 0.04 ^{c}
Dietary Fiber	6.23 ± 0.09 ^{a}	6.11 ± 0.09 ^{a}	5.99 ± 0.08 ^{a}	5.87 ± 0.1 ^{a}	5.74 ± 0.08 ^{a}
Total carbohydrates	57.62 ± 0.82 ^{a}	61.08 ± 0.94 ^{b}	64.54 ± 0.86 ^{c}	68.01 ± 1.13 ^{d}	71.47 ± 0.95 ^{e}
Energy (kcal)	417.01 ± 5.96 ^{a}	398.95 ± 5.70 ^{b}	380.88 ± 5.44 ^{c}	362.81 ± 5.18 ^{d}	344.74 ± 4.92 ^{e}

*: Presented on 100 g wet weight, SE: Standard error, ^{a,b,c,d,e}: No significant difference ($p > 0.05$) between any two means within the same row with the same superscripted letters.

3.2. Mineral Content of Formulated High-Energy DNBs

The mineral analysis of the five formulated bars (DNB1–DNB5) revealed significant differences ($p < 0.05$) across all measured elements with changing date-to-nut ratios (Table 3). Calcium content showed the most pronounced variation, decreasing from 422.12 mg 100 g^{-1} in DNB1 to 241.06 mg 100 g^{-1} in DNB5, with statistically significant differences between all formulations ($p < 0.05$). Potassium exhibited a similar declining trend from 1247.08 mg 100 g^{-1} (DNB1) to 1079.54 mg 100 g^{-1} (DNB5), though differences between DNB2, DNB3, and DNB4 were not significant ($p > 0.05$). Phosphorus content decreased substantially from 785.95 to 430.77 mg 100 g^{-1} across the formulation range, with significant differences between most treatments. Magnesium followed a similar pattern, declining from 326.52 mg 100 g^{-1} (DNB1) to 195.66 mg 100 g^{-1} (DNB5), while sodium decreased from 57.54 to 35.97 mg 100 g^{-1}. Among trace elements, manganese showed significant reductions from 3.57 mg 100 g^{-1} (DNB1) to 1.95 mg 100 g^{-1} (DNB5), with all formulations differing significantly ($p < 0.05$). Copper content declined from 1.67 to 0.91 mg 100 g^{-1}, though some intermediate formulations showed non-significant differences. Iron decreased from 6.97 to 4.02 mg 100 g^{-1}, with DNB4 and DNB5 showing similar values ($p > 0.05$). Zinc content reduced from 5.46 mg 100 g^{-1} (DNB1) to 2.91 mg 100 g^{-1} (DNB5), while selenium decreased from 21.89 µ100 g^{-1} to 11.84 µ100 g^{-1}, with significant differences between most formulations ($p > 0.05$).

Table 3. Mineral content (mg 100 g^{-1}) * of different formulated high-energy DNBs (mean ± SE), n = 3.

Minerals and Trace Elements (mg kg^{-1})	High-Energy DNB Formulas				
	DNB1	DNB2	DNB3	DNB4	DNB5
Calcium	422.12 ± 7.04 ^{a}	360.1 ± 6.55 ^{b}	324.88 ± 6.5 ^{c}	296.37 ± 5.39 ^{d}	241.06 ± 4.02 ^{e}
Sodium	57.54 ± 0.96 ^{a}	49.95 ± 0.91 ^{b}	45.87 ± 0.92 ^{c}	42.68 ± 0.78 ^{d}	35.97 ± 0.6 ^{e}
Potassium	1247.0 ± 20.78 ^{a}	1165.9 ± 21.2 ^{b}	1147.59 ± 22.95 ^{b}	1145.02 ± 20.82 ^{b}	1079.54 ± 17.99 ^{c}
Phosphorus	785.95 ± 13.1 ^{a}	665.46 ± 12.1 ^{b}	595.68 ± 11.91 ^{c}	538.58 ± 9.79 ^{d}	430.77 ± 7.18 ^{e}
Magnesium	326.52 ± 5.44 ^{a}	281.1 ± 5.11 ^{b}	256.01 ± 5.12 ^{c}	235.99 ± 4.29 ^{d}	195.66 ± 3.26 ^{e}
Manganese	3.57 ± 0.06 ^{a}	3.02 ± 0.05 ^{b}	2.70 ± 0.09 ^{c}	2.44 ± 0.11 ^{d}	1.95 ± 0.03 ^{e}
Copper	1.67 ± 0.07 ^{a}	1.41 ± 0.03 ^{ab}	1.26 ± 0.11 ^{b}	1.14 ± 0.12 ^{bc}	0.91 ± 0.08 ^{c}
Iron	6.97 ± 0.12 ^{a}	5.96 ± 0.11 ^{b}	5.38 ± 0.11 ^{c}	4.92 ± 0.09 ^{d}	4.02 ± 0.07 ^{e}
Zinc	5.46 ± 0.09 ^{a}	4.60 ± 0.08 ^{b}	4.09 ± 0.08 ^{c}	3.68 ± 0.07 ^{d}	2.91 ± 0.05 ^{e}
Selenium (μg kg^{-1})	21.89 ± 0.36 ^{a}	18.49 ± 0.34 ^{b}	16.51 ± 0.33 ^{c}	14.88 ± 0.27 ^{d}	11.84 ± 0.21 ^{e}

*: Presented on wet weight, SE: Standard error, ^{a,b,c,d,e}: No significant difference (p > 0.05) between any two means within the same row with the same superscripted letters.

3.3. Antioxidant Activities of Formulated High-Energy DNBs

The antioxidant activity and phytochemical content of the five formulated high-energy bars (DNB1–DNB5) demonstrated significant variations (p < 0.05) corresponding to different date-to-nut ratios (Table 4). Total phenolic content (TPC) decreased progressively from 37.98 mg GAE g^{-1} in DNB1 to 18.17 mg GAE g^{-1} in DNB5, with all formulations showing statistically significant differences (p < 0.05). DPPH radical scavenging activity (DPPH-RSA) followed a similar declining trend, ranging from 48.29 μmol TE g^{-1} in DNB1 to 23.09 μmol TE g^{-1} in DNB5, with significant differences between all treatments. ABTS radical scavenging activity (ABTS-RSA) exhibited the highest values in DNB1 (62.76 μmol TE g^{-1}) and progressively decreased to 27.71 μmol TE g^{-1} in DNB5, with statistical significance across all formulations (p < 0.05). Total flavonoids (TF) content declined from 26.01 mg QE g^{-1} (DNB1) to 10.91 mg QE g^{-1} (DNB5). In comparison, total flavonols (TFL) decreased from 20.80 mg QE g^{-1} to 7.75 mg QE g^{-1}, both showing significant differences between most formulations (p < 0.05).

Table 4. Antioxidant activity, total phenolic content, total flavonoids, and total flavonols in different formulated high-energy DNBs (mean ± SE), n = 3.

Phytochemical Analysis *	High-Energy DNB Formulas				
	DNB1	DNB2	DNB3	DNB4	DNB5
TPC [mg GAE g^{-1}]	37.98 ± 2.37 ^{a}	33.77 ± 3.52 ^{ab}	25.98 ± 1.35 ^{bc}	22.08 ± 3.37 ^{bc}	18.17 ± 2.24 ^{cd}
DPPH-RSA [μmol of TE g^{-1}]	48.29 ± 6.69 ^{a}	40.53 ± 5.62 ^{ab}	29.88 ± 3.40 ^{c}	29.26 ± 2.49 ^{cd}	23.09 ± 4.31 ^{d}
ABTS-RSA [μmol of TE g^{-1}]	62.76 ± 8.91 ^{a}	48.65 ± 5.75 ^{ab}	38.84 ± 6.52 ^{bc}	36.57 ± 4.61 ^{c}	27.71 ± 2.37 ^{d}
TF [mg QE g^{-1}]	26.01 ± 3.37 ^{a}	20.26 ± 3.31 ^{ab}	15.33 ± 2.2 ^{b}	12.14 ± 1.2 ^{c}	10.91 ± 0.91 ^{d}
TFL [mg QE g^{-1}]	20.80 ± 1.73 ^{a}	15.19 ± 2.33 ^{b}	9.96 ± 1.53 ^{c}	9.11 ± 2.15 ^{c}	7.75 ± 2.21 ^{c}

*: Presented on wet weight, SE: Standard error, ^{a,b,c,d}: No significant difference (p > 0.05) between any two means within the same column with the same superscripted letters.

3.4. Phenolic Profile of Formulated High-Energy DNBs

The HPLC analysis of phenolic acids and flavonoids in formulated Sukkari bars (Table 5) revealed that catechol was the predominant phenolic acid across all formulations, with concentrations of 4183.11, 4640.26, and 3676.47 mg kg^{-1} in DNB1, DNB3, and DNB5, respectively. Vanillic acid was the second most abundant, peaking at 528.95 mg kg^{-1} in DNB3 and decreasing to 418.57 mg kg^{-1} in DNB5, indicating that a 60:40 date-to-nut ratio optimizes retention of this compound. Gallic acid and *p*-hydroxybenzoic acid also followed a similar pattern, with the highest levels in DNB3 (15.56 and 89.09 mg kg^{-1}, respectively). Phenolic acids such as syringic and *p*-coumaric acids were found only in DNB3 (5.31 mg kg^{-1}) and at trace levels in DNB1 and DNB5. Flavonoid profiling revealed quercetin as the major flavonol, with a concentration of 2.79 mg kg^{-1} in DNB1, 0.66 mg kg^{-1} in DNB3, and increased to 5.82 mg kg^{-1} in DNB5. Apigenin and rutin were the most abundant in DNB1 (1.48 and 2.11 mg kg^{-1}, respectively) but declined with increasing date ratio, indicating that nuts contribute significantly to these compounds. Kaempferol and myricetin were detected at low levels (<0.6 mg kg^{-1}) across all treatments. Overall, the 60:40 date-to-nut formulation (DNB3) accomplished the highest total phenolic acids (≈5282 mg kg^{-1}) and substantial flavonoid levels, balancing date-derived antioxidants and nut-derived flavonols.

Table 5. The phenolic acids and flavonoids content (mg kg^{-1}) * in formulated DNBs.

RT	Name	DNB1	DNB3	DNB5
		Phenolic acids (mg kg^{-1})		
2.97	Gallic acid	10.44	15.56	10.73
3.43	Catechol	4183.11	4640.26	3676.47
3.89	*p*-hydroxybenzoic acid	76.03	89.09	75.66
5.33	Vanillic acid	495.79	528.95	418.57
6.73	Syringic acid	0	5.31	0
6.92	*p*-coumaric	2.38	0	2.87
9.22	O-cumaric	3.92	3.22	0
		Flavonoids (mg kg^{-1})		
9.64	Rutin	2.11	1.35	0
10.85	Myricetin	0.59	0.25	0
12.1	Quercetin	2.79	0.66	5.82
14.04	Kaempferol	0.14	0.39	0
14.55	Apigenin	1.48	0.69	0

* Presented means are the average of duplicate analyses.

3.5. AA Composition of Different DNBs

High-energy DNB formulas were analyzed for their contents of AAs, and the data are shown in Table 6. The AAs analysis of the five formulated Sukkari-DNBs (DNB1–DNB5) revealed substantial compositional differences corresponding to varying date-to-nut ratios. The total amount of AAs was down from 8.47 g 100 g^{-1} in DNB1 to 5.29 g 100 g^{-1} in DNB5. The amounts for DNB2, DNB3, and DNB4 were 7.99, 8.02, and 6.26 g 100 g^{-1}, respectively. Leucine was always the most common EAAs in all formulations, with amounts ranging from 0.33 g 100 g^{-1} in DNB5 to 0.57 g 100 g^{-1} in DNB1. Valine came in third, with levels ranging from 0.30 to 0.43 g 100 g^{-1}, while phenylalanine came in second, with levels ranging from 0.31 to 0.46 g 100 g^{-1}. There was a big difference in the amount of lysine, an important amino acid for nutrition, from 0.19 g 100 g^{-1} in DNB5 to 0.33 g 100 g^{-1} in DNB1. Methionine had the lowest levels of EAAs, with amounts ranging from 0.09 g 100 g^{-1} (DNB4) to 0.14 g 100 g^{-1} (DNB1). The total amount of EAAs went down from 3.03 g 100 g^{-1} in DNB1 to 1.97 g 100 g^{-1} in DNB5. In all formulations, glutamic acid was the most common NEAAs, with the highest content in DNB1 (1.92 g 100 g^{-1}) and the lowest in DNB5 (1.08 g 100 g^{-1}). The second most common NEAA was aspartic acid, which ranged from 0.504 g 100 g^{-1} (DNB5) to 0.84 g 100 g^{-1} (DNB1). Proline levels ranged from 0.42 g 100 g^{-1} (DNB4) to 0.69 g 100 g^{-1} (DNB1). Arginine, on the other hand, had an opposite effect on nut content, peaking at 0.71 g 100 g^{-1} in DNB2 and dropping to 0.34 g 100 g^{-1} in DNB5. The total amount of NEAAs varied from 3.32 g 100 g^{-1} (DNB5) to 5.45 g 100 g^{-1} (DNB1). The EAAs/NEAAs ratios, on the other hand, ranged from 0.46 for DNB3 to 0.59 for DNB5. DNB1, DNB2, and DNB4 had intermediate values of 0.56, 0.59, and 0.56, respectively. These ratios reflect the changing protein quality characteristics across different formulations.

Table 6. AA composition (g 100 g^{-1} bar) * of different formulated high-energy DNBs.

AAs	DNB1	DNB2	DNB3	DNB4	DNB5
		Essential AAs (EAAs)			
Leucine	0.57	0.55	0.50	0.42	0.33
Lysine	0.33	0.32	0.28	0.24	0.19
Valine	0.43	0.42	0.35	0.32	0.30
Methionine	0.14	0.13	0.10	0.09	0.11
Histidine	0.35	0.32	0.28	0.26	0.22
Threonine	0.29	0.27	0.24	0.22	0.17
Phenylalanine	0.45	0.46	0.396	0.36	0.31
Isoleucine	0.25	0.25	0.21	0.18	0.13
Cystine	0.22	0.23	0.1769	0.16	0.20
		Non-Essential AAs (NEAAs)			
Aspartic Acid	0.84	0.79	0.72	0.64	0.50
Glutamic Acid	1.92	1.80	1.64	1.46	1.08
Serine	0.43	0.41	0.37	0.33	0.26
Glycine	0.43	0.40	0.37	0.33	0.25
Arginine	0.64	0.71	0.62	0.46	0.34
Alanine	0.41	0.38	0.36	0.32	0.25
Tyrosine	0.08	0.12	0.87	0.07	0.09
Proline	0.69	0.43	0.56	0.42	0.56
Essential AAs	3.03	2.95	2.52	2.24	1.97
Non-Essential AAs	5.45	5.04	5.50	4.03	3.32
EAAs/NEAAs ratio	0.56	0.59	0.46	0.56	0.59
Total AAs	8.47	7.99	8.02	6.26	5.29

*: Presented means are the average of duplicate analyses.

Table 7 displays the AAs%, computed biological efficiency, essential AA index (EAAI), and requirement index of different prepared bars. The five DNB bars exhibited apparent formulation-dependent differences in protein quality metrics. DNB1 had the most total AAs (8.48 g 100 g^{-1}) and total essential AAs (3.03 g 100 g^{-1}). DNB5 had the least (5.29 and 1.97 g 100 g^{-1}, respectively). The highest levels of branched-chain AAs were in DNB2, where they reached 152.71 mg g^{-1} protein. The highest levels of basic AAs were in the same formulation, where they reached 168.73 mg g^{-1} protein. The biological value (BV) went from 43.06 (DNB1) to 52.05 (DNB5), with DNB2 at 50.44. The EAAI went from 53.90 (DNB4) to 56.27 (DNB2). DNB3 had the most total conditional AAs (526.57), while DNB4 had the least (462.32). Age-specific requirement indices for infants, preschoolers, schoolchildren, and adults exceeded 100 for all formulations, with DNB2 consistently showing the highest indices (120.55–150.67), indicating superior capacity to meet age-dependent AA needs.

Table 7. The AAs% and calculated biological efficiency, essential AA index, estimated protein efficiency ratio, and requirement index of different age groups.

Parameters	DNB1	DNB2	DNB3	DNB4	DNB5
Total BCAAs (mg g^{-1} protein)	148.07	152.71	131.33	145.75	144.80
Total BAAs (mg g^{-1} protein)	155.50	168.73	147.42	154.05	141.78
Total Aromatic AA (mg g^{-1} protein)	62.29	73.48	157.89	67.53	75.05
Total uncharged polar AAs (mg g^{-1} protein)	445.61	452.50	500.00	458.49	433.08
Total Conditional AA (mg g^{-1} protein)	469.21	462.39	526.57	462.32	475.80
BV	43.06	50.44	45.84	45.09	52.05
EAAI	54.09	56.27	53.58	53.90	55.92
Requirement index (infants)	115.89	120.55	114.79	115.47	119.80
Requirement index (preschool child)	125.86	130.93	124.67	125.41	130.11
Requirement index (schoolchild)	137.73	143.27	136.42	137.23	142.38
Requirement index (adult)	144.84	150.67	143.47	144.32	149.73

BAAs: Basic AAs, BV: Calculated biological value, EAAI: Essential AA index.

3.6. FA Composition of Formulated High-Energy DNBs

The FA composition analysis revealed distinct patterns across the three selected DNB formulations (Table 8). Oleic acid (C18:1 n9) was the predominant FA, decreasing from 44.94% in DNB1 to 40.51% in DNB5, while palmitic acid (C16:0) increased substantially from 15.06% to 20.26%. Total saturated FAs increased progressively from 24.05% in DNB1 to 32.92% in DNB5, accompanied by corresponding decreases in monounsaturated FAs (45.63% to 41.38%) and polyunsaturated FAs (30.31% to 25.70%). Linoleic acid (C18:2 n6) content decreased from 27.38% in DNB1 to 23.12% in DNB5, while α-linolenic acid (C18:3 n3) remained relatively stable (2.90–2.52%), resulting in omega-6 to omega-3 ratios ranging from 9.0:1 to 9.5:1 across formulations.

Table 8. FA composition (g 100 g^{-1}) * of high-energy DNBs.

	DNB1	DNB3	DNB5
Saturated FAs	%	%	%
Butyric (C4:0)	0.19	0.24	0.32
Caproic (C6:0)	0.14	0.18	0.28
Caprylic (C8:0)	0.09	0.15	0.19
Capric (C10:0)	0.24	0.29	0.57
Lauric (C12:0)	0.39	0.53	0.77
Myristic (C14:0)	1.85	2.18	3.16
Myristoleic (C14:0)	0.06	0.21	0.11
Pentadecanoic (C15:0)	0.22	0.26	0.38
Palmitic (C16:0)	15.06	16.39	20.26
Margaric acid (C17:0)	0.17	0.19	0.24
Stearic (C18:0)	5.33	5.44	6.42
Arachidic (C20:0)	0.24	0.22	0.22
Behenic (C22:0)	0.13	0.12	0.11
Monounsaturated FAs			
cis-10-pentadecenoic acid (C15:1)	0.03	0.04	0.06
Palmitoleic (C16:1)	0.51	0.62	0.62
cis-10-Heptadecenoic acid (C17:1)	0.09	0.08	0.08
Oleic (C18:1 n9)	44.94	43.76	40.51
Polyunsaturated FAs			
Arachidonic acid (20:4 n6)	0.03	0.04	0.06
Linoleic (C18:2 n6)	27.38	26.12	23.12
Linolenic (C18:3 n3)	2.9	2.92	2.52

*: Presented means are the average of duplicate analyses.

3.7. GC-MS Volatile Profile Analysis of Formulated High-Energy DNBs

The GC-MS analysis of the three selected formulated Sukkari DNBs (DNB1, DNB3, and DNB5) with varying ratios of Sukkari date paste and nut mixture (40:60%) in DNB1 and (80–20%) in DNB5 revealed a complex volatile organic compound (VOC) profile comprising 31 distinct compounds (Table 9). The identified compounds belonged to several chemical classes, with benzene derivatives, FA methyl esters, and organic acids being the predominant groups. For benzene derivatives and aromatic compounds, the most significant findings were the detection of numerous benzene derivatives, which constituted the major portion of the volatile profile. Compounds such as benzene (1-butylhexyl)-, benzene (1-propylheptyl)-, benzene (1-ethyldecyl)-, and benzene (1-methylnonyl)- were identified across all formulations. These compounds are typically generated during thermal processing of nuts and thus closely correlate with the distinctive nutty and roasted flavors in nut-based food products.

Table 9. GC-MS volatile profile analysis of different formulated DNBs.

Name	RT	DNB1	DNB3	DNB5
Benzene, (1-butylhexyl)-	18.06	1.68	1.13	1.71
Benzene, (1-propylheptyl)-	18.26	1.38	0.91	1.45
Benzene, (1-ethyloctyl)-	18.68	1.16	0.74	1.23
Benzene, (1-methylnonyl)-	19.52	1.05	-	1.08
Benzene, (1-pentylhexyl)-	20.24	1.67	0.99	1.82
Benzene, (1-butylheptyl)-	20.31	4.31	2.53	4.63
Benzene, (1-propyloctyl)-	20.53	3.1	1.82	3.4
Benzene, (1-ethylnonyl)-	20.98	2.59	1.51	2.76
Benzene, (1-methyldecyl)-	21.79	2.46	1.54	2.55
Benzene, (1-pentylheptyl)-	22.38	4.65	3.82	4.77
Benzene, (1-butyloctyl)-	22.47	4.95	3.15	5.37
Benzene, (1-propylnonyl)-	22.73	3.44	2.08	3.65
Benzene, (1-ethyldecyl)-	23.17	2.72	1.58	2.92
Benzene, (1-methylundecyl)-	23.97	2.39	1.39	2.78
Benzene, (1-pentyloctyl)-	24.44	3.74	2.29	4.01
Benzene, (1-butylnonyl)-	24.57	2.92	1.61	2.99
Benzene, (1-propylheptadecyl)-	24.83	2.18	1.06	2
Benzene, (1-ethylundecyl)-	25.29	1.78	0.85	1.56
7-Hexadecenoic acid, methyl ester, (Z)-	25.88	0.00	0.48	0.00
Benzene, (1-methyldodecyl)-	26.06	1.73	1.07	1.68
Hexadecanoic acid, methyl ester	26.4	7.74	12.6	9.07
17-Octadecynoic acid	27.63	-	-	0.38
9,12-Octadecadienoic acid (Z, Z)-, methyl ester	29.42	12.8	16.06	6.65
9-Octadecenoic acid, methyl ester,(E)-	29.59	25.31	30.08	19.21
Methyl stearate	30.15	3.29	6.52	4.97
Hexadecanoic acid, 2-hydroxy-1-(hydroxymethyl)ethyl ester	32.08	-	0.62	0.72
9-Octadecenoic acid (Z)-	32.19	-	-	1.08
9 12-octadecadienoic acid (Z)- 2-hydroxy-1-(hydroxymethyl)ethyl ester	34.78	-	0.59	0.58
9,12-Octadecadienoyl chloride, (Z, Z)-	34.94	0.84	2.99	3.59
Mono(2-ethylhexyl) phthalate	36.62	0.38	-	1
1,2-Benzenedicarboxylic acid	39.81	-	-	0.14

- Not detected.

Several FA methyl esters were detected, including 7-hexadecenoic acid methyl ester and 17-octadecenoic acid, which are essential volatile compounds in nuts and dates. The 7-hexadecenoic acid methyl ester, identified in peaks 19 and 22, is a monounsaturated FA derivatives that contributes to the characteristic aroma profile of date-based products. The analysis revealed the presence of various organic acids and their derivatives, including hexadecanoic acid (palmitic acid) and 9,12-octadecadienoic acid compounds. Hexadecanoic acid (peak 21, 26) is commonly found in nuts and dates and contributes to the fatty, waxy notes in the aroma profile. Notably, the volatile profile showed significant variation among the five formulations. DNB3 (60% date paste, 40% nut mixture) exhibited unique characteristics with several compounds, as shown in Table 9.

Interestingly, the principal component analysis (PCA) of all measured variables is clearly presented for DNB1–DNB5 in Figure 2. The first two principal components explained 98.18% of total variance, with PC1 accounting for 94.47% and PC2 for 3.71%, indicating a dominant pattern of systematic variation among formulations. Samples showed clear separation along PC1 in a gradient pattern from DNB1 to DNB5. Hierarchical clustering grouped DNB1-DNB2 as nutrient-dense formulations and DNB4-DNB5 as lower nutrient-density samples (Figure 3). Minerals (Se, Zn, P, Mn, Ca, Cu, Fe, Mg, Na), macronutrients (protein, total fat, energy, dietary fiber), and bioactive compounds (TPC, DPPH-RSA, ABTS-RSA, TF, TFL) showed strong positive associations with PC1. In contrast, moisture and total carbohydrates were negatively associated. EAAs, MUFAs, and PUFAs also contributed positively to PC1. The heatmap with hierarchical clustering revealed tight co-variation among minerals and antioxidant compounds, indicating coherent nutritional patterns across formulations. The systematic gradient reflects a compositional trade-off between nutrient density and straightforward carbohydrate content, with DNB1 representing the most nutrient-dense formulation (Figure 3).

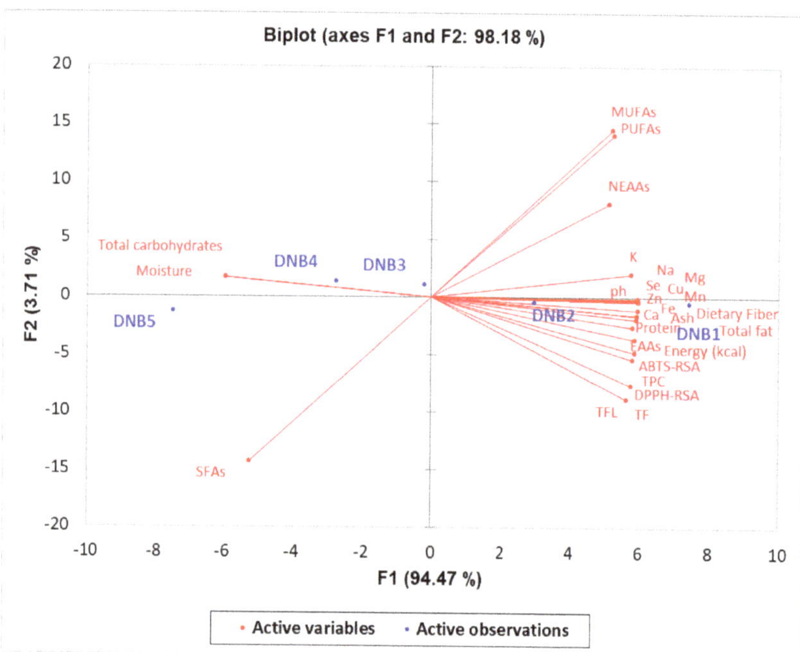

Figure 2. PCA biplot showing the relationships between samples and variables in the first two principal components.

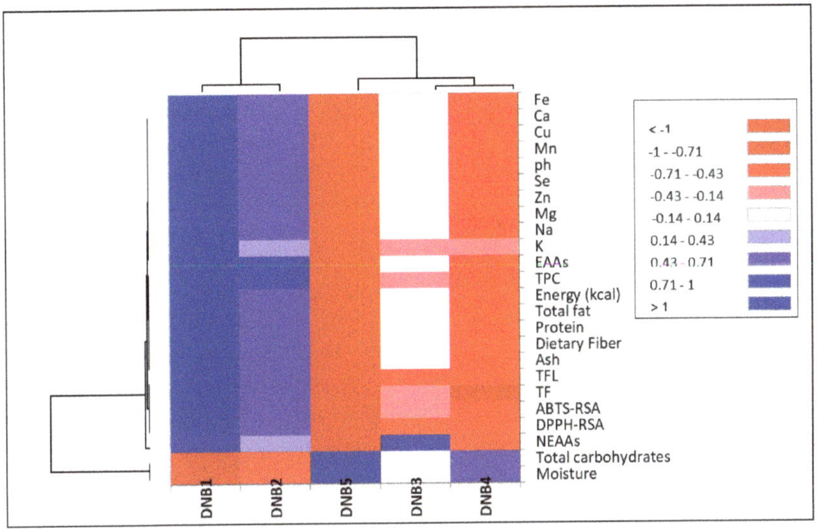

Figure 3. Heatmap with hierarchical clustering showing standardized values of nutritional and bioactive components across five date bar formulations.

4. Discussion

The global nutritional bar market is undergoing rapid expansion, driven by growing health awareness and consumer demand for convenient and nutrient-dense snacks. Fruit-based snack bars are the healthiest, with natural sugars, vitamins, minerals, and other bio-nutrients to fulfill consumers' daily nutritional needs [40]. Date-based bars have materialized as a promising subcategory. The date palm market is predicted to grow, and the global nuts market, supplying essential proteins and healthy fats, is estimated to rise. Date fruits had anti-inflammatory, anti-tumor, antihypertensive, anti-hypercholesterolemia, and antimicrobial properties [41–43]. Sukkari dates contribute dietary fiber, natural sugars, micronutrients, and bioactive compounds (phenolic acids, polyphenols, and carotenoids) associated with anti-inflammatory, antioxidant, and antimicrobial effects [5]. Mixed nuts provide high-quality proteins, essential FAs, vitamins, and minerals [44]. Together, these ingredients form a synergistic matrix ideally suited for developing functional, high-nutrition snack bars that fulfill modern consumer preferences for healthful, on-the-go foods.

The inverse relationship between date content and both protein and fat levels reflects the low protein (2.5–4.7%) and fat (<1%) of Sukkari dates compared to nuts, which contribute 15–25% protein and 40–60% fat [44,45]. DNB1 (60% nuts) demonstrated the highest protein and lipid content, supporting its capability as a protein-rich, energy-dense snack [46]. In contrast, DNB5 (80% dates) presented higher moisture and carbohydrate contents, delivering natural sugars and moisture maintenance characteristic of date-based formulations [47]. The consistent dietary fiber across formulations (≈6%) highlights the complementary contributions of date pulp and oat/nut matrix to fiber content [48]. The significant increase in total carbohydrates and corresponding decrease in caloric density with higher date ratios align with the dilution of lipid energy and highlight the trade-off between quick-release carbohydrates and sustained energy from fats [49,50]. Ash content reduction suggests minor mineral dilution by carbohydrate-rich dates, although all formulations remain good mineral sources. Statistically, each nutritional parameter differed significantly across formulations ($p < 0.05$), except dietary fiber, indicating that formulation adjustments can precisely tailor macronutrient profiles to nutritional targets. Ultimately, DNB1–DNB3 balance protein, fat, and energy optimally for sports and meal-replacement applications, while DNB4–DNB5 prioritize carbohydrate energy and palatability for daytime snacking.

The decline in mineral content with increasing date proportion reflects the fundamental compositional differences between dates and nuts, where the mineral profile of blended nuts surpasses that of date paste [44,51]. The substantial calcium reduction (43% from DNB1 to DNB5) aligns with reported values for nuts, particularly almonds and pistachios, compared to date fruits [52,53]. This trend has significant dietary implications, as calcium deficiency is prevalent globally, and nut-rich formulations could contribute substantially to daily calcium requirements [54]. The trend of declining potassium content across formulations; however, its overall high concentration remained markedly, reflecting the natural mineral profile of the constituents of both dates and nuts [14,47]. These levels support cardiovascular health and blood pressure regulation, particularly relevant for active populations and individuals with hypertension risk [40]. The phosphorus reduction (45% decrease) underscores nuts' superior phosphorus density, with implications for energy metabolism, bone health, and cellular function [14]. Magnesium content patterns reflect the remarkable magnesium density of nuts compared to dates, with DNB1 providing levels that could significantly contribute to daily magnesium requirements [53,55]. Magnesium deficiency is associated with metabolic disorders, muscle dysfunction, and cardiovascular disease; therefore, the increased magnesium content observed in these formulations offers notable health advantages [56]. Magnesium protects against oxidative stress by stabilizing and activating antioxidant enzymes (e.g., SOD, catalase, glutathione reductase) and supporting glutathione synthesis and mitochondrial function, thus reducing reactive oxygen species and lipid peroxidation [57].

The trace element profiles demonstrate nuts' superiority as sources of bioavailable minerals necessary for enzymatic functions and antioxidant defense systems. The zinc content reduction from DNB1 to DNB5 reflects the high zinc density in nuts compared to dates, with associations for immune function, wound healing, and protein synthesis [14,55]. Similarly, copper and manganese reductions align with nuts' excellent trace element profiles, where these minerals serve as cofactors for antioxidant enzymes and metabolic processes [53]. Iron content, while declining with increased date proportion, remained significant across all formulations, contributing to iron intake requirements, particularly important for women and athletes [58]. The selenium content, though reduced in higher date formulations, remained within ranges that could support antioxidant defense systems and thyroid function [59]. The statistical analysis reveals that formulation adjustments can precisely control mineral profiles, with most minerals showing significant differences between extreme formulations (DNB1 vs. DNB5). The non-significant differences between some intermediate formulations (DNB2, DNB3, DNB4) for certain minerals suggest gradual transitions that allow fine-tuning of mineral content to meet specific nutritional objectives. The high mineral density inherent in DNB1 and DNB2 provides a suitable platform for development into functional foods aimed at improving mineral status in at-risk populations and athletes [9]. The mineral profiles support the bars' potential as convenient sources of essential minerals often inadequate in modern diets, particularly calcium, magnesium, and trace elements [54,56]. Higher date formulas (DNB4, DNB5), while providing lower mineral density, offer adequate mineral content combined with natural sugars, fiber, and bioactive compounds characteristic of dates, making them suitable for general snacking applications where mineral density is less critical than palatability and energy provision [40].

The fact that the antioxidant capacity and phytochemical content go down as the date proportion goes up shows that mixed nuts are better at fighting free radicals than date paste, which goes against what you might think based on dates' reputation as antioxidant-rich fruits [5,47]. The large drop in TPC (52% from DNB1 to DNB5) is in keeping with what has been said about the phenolic content of nuts, especially walnuts, almonds, and pistachios, when compared to fresh dates [14,60]. This finding suggests that phenolic compounds in the nut mix, including ellagic acid, gallic acid, and catechins derived from almonds and walnuts, are instrumental in enhancing the bars' oxidative defense properties [44]. The DPPH and ABTS radical scavenging activities were substantially linked to the TPC values. This is in line with other research that found a favorable link between phenolic content and antioxidant capacity in nut-based products [61,62]. The higher ABTS-RSA results than DPPH-RSA values for all formulations show that ABTS assays can find a wider range of antioxidants since they evaluate both hydrophilic and hydrophobic antioxidant activity. DPPH, on the other hand, mostly finds hydrophobic antioxidants [62]. This distinction is especially significant in nut-based formulations that incorporate lipophilic antioxidants such as tocopherols and tocotrienols [14].

The flavonoid and flavonol content patterns reinforce the nuts' contribution to phytochemical density, with compounds such as quercetin, catechins, and proanthocyanidins being abundant in nuts [44]. The 62% reduction in TF content from DNB1 to DNB5 indicates that nuts serve as primary flavonoid sources in these formulations, contributing compounds with established cardiovascular and neuroprotective benefits [14]. While dates contain flavonoids, including quercetin, rutin, and luteolin, their concentrations (1.74–3.39 mg catechin equivalent 100 g^{-1}) are substantially lower than those found in nuts [60,61]. The enhanced antioxidant activity in DNB1 and DNB2 formulations highlights their potential role as functional foods for managing oxidative stress-related conditions, particularly among athletes and health-conscious populations [63]. The phenolic compounds identified in similar date-nut products, including gallic acid, ferulic acid, and catechins, have demonstrated anti-inflammatory, cardioprotective, and neuroprotective effects in clinical studies [5,61].

The declining antioxidant trends with increased date content reflect the trade-off between palatability and functional properties, where higher date formulations (DNB4, DNB5) prioritize natural sweetness and energy density over antioxidant capacity [46]. However, even the lowest values (DNB5) maintain meaningful antioxidant activity comparable to conventional snack bars, suggesting adequate functional benefits across all formulations [64].

The phenolic and flavonoid profiles of the formulated Sukkari date–nut bars underscore the intricate interplay between date pulp and nut matrices in determining antioxidant composition and potential bioactivity. Catechol emerged as the most abundant phenolic, reflecting the high catechol derivative content inherent to Sukkari dates and corroborating previous reports on the dominant role of catechol in date fruit antioxidant potential [47]. The peak catechol and vanillic acid levels observed in DNB3 (60% date) suggest that moderate date incorporation enhances phenolic extractability, likely due to optimal matrix porosity facilitating solvent penetration and phenolic release [65]. Results indicate that increased date content in DNB3 correlates with higher vanillic acid concentrations, a potent radical scavenger, compared to DNB5 (80% date), leading to phenolic-cell wall interactions, thus limiting phenolic bioavailability [49]. Gallic and *p*-hydroxybenzoic acids followed similar trends, reinforcing the notion that phenolic recovery is maximized at intermediate pulp ratios. The restricted detection of syringic and *p*-coumaric acids primarily in DNB3 further highlights the importance of formulation balance: these compounds, which possess anti-inflammatory and cardioprotective effects, were virtually absent in DNB1 and DNB5 formulations, suggesting that a 60:40 date-to-nut ratio fosters a synergistic matrix environment conducive to their liberation [10]. The overall total phenolic acid content peaked in DNB3, outperforming both DNB1 and DNB5 and underscoring the critical role of matrix structure in phenolic bioaccessibility [66].

Flavonoid analysis revealed quercetin as the principal flavonol, with its highest level in DNB5 aligning with reports that date peels are a rich quercetin source [66,67]. Conversely, apigenin and rutin peaked in DNB1, indicating that the nut mixture—composed of almond, pistachio, and other nuts—contributes substantially to these compounds, consistent with the high rutin content reported in almond and pistachio [68,69]. During extraction, a reduction in apigenin and rutin levels with increasing date ratio suggests competitive interactions between date phenolics and nut flavonoids, where increased date sugar concentrations potentially influence solvent polarity and flavonoid solubility [70]. Kaempferol and myricetin remained minor constituents across all formulations, in agreement with their low prevalence in both date and nut matrices [71].

From a functional perspective, the rich ensemble of phenolic acids and flavonoids confers robust antioxidant, anti-inflammatory, and cardioprotective properties to the date–nut bars. Catechol and gallic acid exhibit strong hydrogen-donating capacity, mitigating oxidative stress implicated in chronic diseases [47], while vanillic and *p*-coumaric acids have demonstrated endothelial-protective and anti-hyperglycemic effects in vivo [49,72]. Quercetin, rutin, and apigenin synergistically modulate inflammatory pathways and inhibit platelet aggregation, thereby promoting cardiovascular health in nut-enriched diets [69,73]. The DNB3 achieves an optimal balance, delivering high phenolic acid content alongside substantial flavonoid levels, thus maximizing the bar's functional potential without overemphasizing either date or nut contributions. The phenolic acid and flavonoid profiles of Sukkari date–nut bars are highly formulation-dependent, with moderate date incorporation (60%) fostering superior release and retention of key bioactives.

The AA profile of the formulated Sukkari date-nut bars demonstrates the significant impact of ingredient ratios on protein quality and nutritional value. The progressive decline in total AA content with increasing date proportion directly reflects the fundamental difference in protein content between dates and nuts, confirming previous reports on the protein-diluting effect of fruit-based ingredients in mixed formulations [44,45]. Concerning protein quality assessment and EAA adequacy, the dominance of leucine across all formulations aligns with the typical AA pattern in date-nut products, where leucine represents approximately 10-12% of total protein content [46,48]. This abundance is nutritionally significant, as leucine serves as a key regulator of muscle protein synthesis and metabolic processes [74]. The high lysine content in nut-rich formulations (DNB1, DNB2) effectively compensates a critical limitation of lysine in plant-based dietary proteins, which often serve as the first limiting AAs [75]. When evaluated against the FAO/WHO/UNU [35] AA scoring pattern, DNB1 demonstrated superior protein quality potential. The leucine content (0.57 g 100 g^{-1}) substantially exceeds the recommended 70 mg g^{-1} protein for preschool children, while lysine levels (0.334 g 100 g^{-1}) approach the 58 mg g^{-1} protein requirement (FAO/WHO/UNU, 2007). However, the methionine + cysteine combination (0.361 g 100 g^{-1} total) remains potentially limiting relative to the 25 mg g^{-1} protein recommendation, consistent with typical sulfur AA limitations in plant-based proteins [76].

Regarding EAAs/NEAAs ratios and nutritional implications, the EAAs/NEAAs ratios approach the ideal value of 0.6 recommended for high-quality proteins, with DNB2 and DNB5 achieving ratios closest to this target (FAO/WHO/UNU, 2007). Interestingly, the higher ratios in date-rich formulations (DNB4, DNB5) result from proportionally greater reductions in NEAAs rather than EAAs preservation, suggesting differential AA stability during processing or inherent compositional differences between date and nut proteins [77]. For functionality, high glutamic acid content across all formulations plays a dual role in flavor enhancement via umami taste and potential health promotion [5]. The substantial arginine levels, particularly in DNB2, provide cardiovascular benefits through nitric oxide pathway enhancement. At the same time, the balanced essential AA profile supports various physiological functions beyond basic protein synthesis requirements [45]. Also, the AA profile reflects not only ingredient composition but also potential matrix interactions that may influence protein digestibility and bioavailability. The fiber-rich date matrix may create physical barriers affecting protein accessibility. At the same time, the lipid content from nuts could enhance the absorption of fat-soluble vitamins and modify protein digestion kinetics [46]. These factors suggest that AA bioavailability may vary among formulations independent of total content.

Based on comprehensive AA analysis, DNB1 emerges as the optimal formulation for protein quality, offering the highest total AA content, a superior EAAs profile, and a favorable EAAs/NEAAs ratio. This formulation successfully balances the nutritional benefits of dates—including natural sugars, fiber, and bioactive compounds—with the superior protein quality characteristics of mixed nuts [77]. DNB2 represents a viable alternative for applications requiring moderate protein content while maintaining sensory acceptability. In contrast, formulations with higher date content (DNB4, DNB5) are better suited for applications prioritizing carbohydrate energy and natural sweetness over protein density. The results support the feasibility of developing personalized nutritional products with flexible ingredient ratios to meet specific individual dietary and preference parameters [44].

The protein quality profiles demonstrate that a balanced date-to-nuts ratio enhances AAs adequacy and utilization efficiency in functional snack bars. DNB2's high BV and EAAI reflect optimal alignment with FAO/WHO reference patterns for essential AAs. At the same time, its superior requirement indices across all age groups underscore its potential to support growth and maintenance (FAO/WHO/UNU, 2007). DNB3's elevated conditional AAs suggest enhanced protein utilization under increased physiological demands, although its BV and EAAI are slightly lower. Overall, DNB2 emerges as the best formulation for maximizing protein quality, combining the natural sweetness, fiber, and bioactives of dates with the superior AA profile of mixed nuts to deliver a nutritionally robust snack suitable for diverse age groups [78,79].

The FA profiles demonstrate that DNB1 provides the most favorable nutritional composition from a cardiovascular health perspective, featuring the highest oleic acid content and the lowest saturated FA proportion. The predominance of oleic acid across all formulations aligns with Mediterranean diet principles. It supports cardiovascular benefits, as oleic acid consumption has been linked to improved HDL/LDL cholesterol ratios and reduced inflammatory markers [80]. The progressive increase in saturated FAs, particularly palmitic and myristic acids, from DNB1 to DNB5 may lessen the overall cardioprotective potential of higher-nut formulations, as excessive palmitic acid intake has been associated with increased cardiovascular disease risk compared to oleic acid-rich diets [81]. While the omega-6 to omega-3 ratios remain elevated compared to optimal recommendations (1–4:1), they fall within acceptable ranges for functional foods and reflect the nutritional limitations of plant-based ingredients in providing long-chain omega-3 FAs.

The GC-MS analysis of Sukkari date-nut bars revealed 31 volatiles whose profiles varied markedly with date-to-nut ratios. Benzene derivatives, originating from roasted nuts [82], dominated the aroma and imparted characteristic nutty notes. At the same time, FA methyl esters (e.g., 7-hexadecenoic and octadecenoic acid esters) reflected lipid breakdown and contributed fruity, sweet nuances [82,83]. Formulation DNB3 (60% date, 40% nuts) exhibited notably lower levels of several volatiles, indicating that ingredient ratio strongly influences volatile release and retention [46,84]. These compounds not only define sensory quality but also signal the presence of bioactive FAs and antioxidants inherent to dates and nuts, underscoring the bars' nutritional value [71,83]. The consistency of volatile profiles across storage suggests good product stability and safety [46,85]. Optimizing date-nut ratios based on volatile profiles can therefore tailor flavor while preserving functional and healthful properties, providing a robust foundation for quality control and consumer-driven product development. The GC-MS volatile profiling analysis demonstrates that formulated Sukkari date-nut bars represent a promising category of functional food products with complex volatile profiles that contribute to both sensory appeal and potential health benefits [5]. The optimization of formulation ratios based on volatile compound profiles offers opportunities for developing customized products that meet diverse consumer preferences while maintaining nutritional integrity and food safety standards [9]. PCA and hierarchical clustering revealed that a single dominant dimension (PC1, 94.47% variance) governs compositional differences among the five date bar formulations, reflecting deliberate variation in ingredient proportions. Samples formed a transparent gradient from nutrient-dense (DNB1–DNB2) to carbohydrate-rich (DNB4–DNB5) profiles, corroborated by clustering patterns. Strong positive loadings of minerals, proteins, and antioxidant compounds versus negative loadings of moisture and carbohydrates indicate a trade-off between nutrient density and simple carbohydrates, in line with previous food formulation studies. This multivariate approach underscores its utility for guiding formulation optimization in functional food development [9,40].

In summary, higher date-to-nut ratios shift bars from protein- and fat-rich (DNB1–DNB3) toward carbohydrate-rich formulations (DNB4–DNB5), altering mineral density, antioxidant capacity, and protein quality. DNB1–DNB2 suit sports-nutrition, while DNB4–DNB5 favor general snacking. Limitations include in vitro digestibility and no sensory data; future work should assess in vivo bioavailability, consumer acceptability, and bioactive stabilization.

5. Conclusions

The study demonstrates that high-energy bars formulated with Sukkari date paste and mixed nuts yield distinct nutritional and functional profiles that can be tailored by adjusting date-to-nut ratios: nut-rich formulations optimize protein content, essential AA balance, mineral density, and antioxidant capacity, while date-rich variants enhance natural carbohydrate provision and palatability. A 60:40 date-to-nut ratio strikes an optimal balance, maximizing phenolic recovery, antioxidant activity, and sensory appeal, positioning such bars as versatile functional foods for athletes, health-conscious consumers, and populations with specific dietary needs. Future work should focus on clinical trials to validate metabolic and performance benefits, process optimization for industrial scale-up and shelf-life stability, sensory evaluation across demographics, and the incorporation of additional bioactives or probiotics. Sustainability assessments of ingredient sourcing and the development of targeted formulations such as low-glycemic, mineral-fortified, or age-specific bars will further enhance the commercial and public health impact of these nutritionally engineered snack products.

Author Contributions: Conceptualization, H.B. and H.A.A.; methodology, H.B., T.A. and H.F.A.; investigation, H.B. and H.A.A.; data curation, H.B. and T.A.; formal analysis, H.B., W.A.A. and R.A.; visualization, H.B. and R.A.; writing—original draft preparation, H.B., W.A.A. and H.F.A.; review and editing, H.B., W.A.A. and H.A.A. All authors have read and agreed to the published version of the manuscript.

Funding: The researchers would like to thank the Deanship of Graduate Studies and Scientific Research at Qassim University for financial support (QU-APC-2025).

Institutional Review Board Statement: Not applicable.

Informed Consent Statement: Not applicable.

Data Availability Statement: The original contributions presented in this study are included in the article. Further inquiries can be directed to the corresponding author.

Conflicts of Interest: The authors declare no conflicts of interest.

References

1. AlJaloudi, R.; Al-Dabbas, M.M.; Hamad, H.J.; Amara, R.A.; Al-Bashabsheh, Z.; Abughoush, M.; Choudhury, I.H.; Al-Nawasrah, B.a.A.; Iqbal, S. Development and Characterization of High-Energy Protein Bars with Enhanced Antioxidant, Chemical, Nutritional, Physical, and Sensory Properties. *Foods* **2024**, *13*, 259. [CrossRef]
2. Carreiro, A.L.; Dhillon, J.; Gordon, S.; Higgins, K.A.; Jacobs, A.G.; McArthur, B.M.; Redan, B.W.; Rivera, R.L.; Schmidt, L.R.; Mattes, R.D. The Macronutrients, Appetite, and Energy Intake. *Annu. Rev. Nutr.* **2016**, *36*, 73–103. [CrossRef] [PubMed]
3. Fekete, M.; Lehoczki, A.; Kryczyk-Poprawa, A.; Zábó, V.; Varga, J.T.; Bálint, M.; Fazekas-Pongor, V.; Csípő, T.; Rząsa-Duran, E.; Varga, P. Functional Foods in Modern Nutrition Science: Mechanisms, Evidence, and Public Health Implications. *Nutrients* **2025**, *17*, 2153. [CrossRef] [PubMed]
4. Pulidindi, K. Nutritional Bars Market—By Product, by Category, by Distribution Channel—Global Forecast, 2025–2034 (Report Id: Gmi5644). 2025. Available online: https://www.gminsights.com/industry-analysis/nutritional-bars-market (accessed on 13 September 2025).
5. Barakat, H.; Alfheeaid, H.A. Date Palm Fruit (*Phoenix dactylifera*) and Its Promising Potential in Developing Functional Energy Bars: Review of Chemical, Nutritional, Functional, and Sensory Attributes. *Nutrients* **2023**, *15*, 2134. [CrossRef]
6. Nazzaro, C.; Uliano, A.; Lerro, M.; Stanco, M. From Claims to Choices: How Health Information Shapes Consumer Decisions in the Functional Food Market. *Foods* **2025**, *14*, 699. [CrossRef]
7. Al-Farsi, M.; Alasalvar, C.; Morris, A.; Baron, M.; Shahidi, F. Comparison of Antioxidant Activity, Anthocyanins, Carotenoids, and Phenolics of Three Native Fresh and Sun-Dried Date (*Phoenix dactylifera* L.) Varieties Grown in Oman. *J. Agric. Food Chem.* **2005**, *53*, 7592–7599. [CrossRef]
8. Al-Farsi, M.; Alasalvar, C.; Al-Abid, M.; Al-Shoaily, K.; Al-Amry, M.; Al-Rawahy, F. Compositional and Functional Characteristics of Dates, Syrups, and Their by-Products. *Food Chem.* **2007**, *104*, 943–947. [CrossRef]
9. Alfheeaid, H.A.; Barakat, H.; Althwab, S.A.; Musa, K.H.; Malkova, D. Nutritional and Physicochemical Characteristics of Innovative High Energy and Protein Fruit- and Date-Based Bars. *Foods* **2023**, *12*, 2777. [CrossRef]
10. Barakat, H.; Almutairi, A.S. The Organoleptic and Nutritional Characteristics of Innovative High-Fiber Khalas Date-Based Bar. *Ital. J. Food Sci.* **2024**, *36*, 13. [CrossRef]
11. Majid, A.; Naz, F.; Bhatti, S.; Phull, A.-R. Phenolic Profile and Antioxidant Activities of Three Date Seeds Varieties (*Phoenix dactylifera* L.) of Pakistan. *Explor. Res. Hypothesis Med.* **2023**, *8*, 195–201. [CrossRef]
12. Tassoult, M.; Kati, D.E.; Fernández-Prior, M.Á.; Bermúdez-Oria, A.; Fernandez-Bolanos, J.; Rodríguez-Gutiérrez, G. Antioxidant Capacity and Phenolic and Sugar Profiles of Date Fruits Extracts from Six Different Algerian Cultivars as Influenced by Ripening Stages and Extraction Systems. *Foods* **2021**, *10*, 503. [CrossRef]
13. Nawaz, A.; Li, E.; Khalifa, I.; Walayat, N.; Liu, J.; Ahsan, H.M.; Irshad, S.; Barakat, H.; Lorenzo, J.M.; Pateiro, M.; et al. Effect of Structurally Different Pectin on Dough Rheology, Structure, Pasting and Water Distribution Properties of Partially Meat-Based Sugar Snap Cookies. *Foods* **2021**, *10*, 2692. [CrossRef] [PubMed]
14. Ros, E. Health Benefits of Nut Consumption. *Nutrients* **2010**, *2*, 652–682. [CrossRef] [PubMed]
15. Olas, B. The Cardioprotective Properties of Selected Nuts: Their Functional Ingredients and Molecular Mechanisms. *Foods* **2024**, *13*, 242. [CrossRef]
16. Singar, S.; Kadyan, S.; Patoine, C.; Park, G.; Arjmandi, B.; Nagpal, R. The Effects of Almond Consumption on Cardiovascular Health and Gut Microbiome: A Comprehensive Review. *Nutrients* **2024**, *16*, 1964. [CrossRef] [PubMed]
17. Aslam, N.; Hassan, S.A.; Mehak, F.; Zia, S.; Bhat, Z.F.; Yıkmış, S.; Aadil, R.M. Exploring the Potential of Cashew Waste for Food and Health Applications- a Review. *Future Foods* **2024**, *9*, 100319. [CrossRef]
18. Parn, O.J.; Bhat, R.; Yeoh, T.; Al-Hassan, A. Development of Novel Fruit Bars by Utilizing Date Paste. *Food Biosci.* **2015**, *9*, 20–27. [CrossRef]
19. Snelson, M.; Biesiekierski, J.R.; Chen, S.; Sultan, N.; Cardoso, B.R. The Effects of Nut Intake on Gut Microbiome Composition and Gut Function in Adults: A Systematic Review and Meta-Analysis. *Adv. Nutr.* **2025**, *16*, 100465. [CrossRef]
20. Mederle, A.L.; Dima, M.; Stoicescu, E.R.; Căpăstraru, B.F.; Levai, C.M.; Hațegan, O.A.; Maghiari, A.L. Impact of Gut Microbiome Interventions on Glucose and Lipid Metabolism in Metabolic Diseases: A Systematic Review and Meta-Analysis. *Life* **2024**, *14*, 1485. [CrossRef]
21. Hong, M.Y.; Groven, S.; Marx, A.; Rasmussen, C.; Beidler, J. Anti-Inflammatory, Antioxidant, and Hypolipidemic Effects of Mixed Nuts in Atherogenic Diet-Fed Rats. *Molecules* **2018**, *23*, 3126. [CrossRef]

22. Park, G.; Johnson, K.; Miller, K.; Kadyan, S.; Singar, S.; Patoine, C.; Hao, F.; Lee, Y.; Patterson, A.D.; Arjmandi, B. Almond Snacking Modulates Gut Microbiome and Metabolome in Association with Improved Cardiometabolic and Inflammatory Markers. *NPJ Sci. Food* **2025**, *9*, 35. [CrossRef]
23. Alu'datt, M.H.; Rababah, T.; Tranchant, C.C.; Al-u'datt, D.; Gammoh, S.; Alrosan, M.; Bani-Melhem, K.; Aldughpassi, A.; Alkandari, D.; AbuJalban, D. Date Palm (*Phoenix dactylifera*) Bioactive Constituents and Their Applications as Natural Multifunctional Ingredients in Health—Promoting Foods and Nutraceuticals: A Comprehensive Review. *Compr. Rev. Food Sci. Food Saf.* **2025**, *24*, e70084. [CrossRef] [PubMed]
24. Vignesh, A.; Amal, T.C.; Sarvalingam, A.; Vasanth, K. A Review on the Influence of Nutraceuticals and Functional Foods on Health. *Food Chem. Adv.* **2024**, *5*, 100749. [CrossRef]
25. AOAC. *Official Methods of Analysis of the Aoac*, 17th ed.; Association of Official Analytical Chemists: Rockville, MD, USA, 2000.
26. Borah, S.; Baruah, A.; Das, A.; Borah, J. Determination of Mineral Content in Commonly Consumed Leafy Vegetables. *Food Anal. Methods* **2009**, *2*, 226–230. [CrossRef]
27. Bettaieb, I.; Bourgou, S.; Wannes, W.A.; Hamrouni, I.; Limam, F.; Marzouk, B. Essential Oils, Phenolics, and Antioxidant Activities of Different Parts of Cumin (*Cuminum cyminum* L.). *J. Agric. Food Chem.* **2010**, *58*, 10410–10418. [CrossRef] [PubMed]
28. Khalifa, I.; Barakat, H.; El-Mansy, H.; Soliman, S. Optimizing Bioactive Substances Extraction Procedures from Guava, Olive and Potato Processing Wastes and Evaluating Their Antioxidant Capacity. *J. Food Chem. Nanotechnol.* **2016**, *2*, 170–177. [CrossRef]
29. Zhang, D.; Hamauzu, Y. Phenolics, Ascorbic Acid, Carotenoids and Antioxidant Activity of Broccoli and Their Changes During Conventional and Microwave Cooking. *Food Chem.* **2004**, *88*, 503–509. [CrossRef]
30. Almundarij, T.I.; Alharbi, Y.M.; Abdel-Rahman, H.A.; Barakat, H. Antioxidant Activity, Phenolic Profile, and Nephroprotective Potential of Anastatica Hierochuntica Ethanolic and Aqueous Extracts against Ccl4-Induced Nephrotoxicity in Rats. *Nutrients* **2021**, *13*, 2973. [CrossRef]
31. Barakat, H.; Almundarij, T.I. Phenolic Compounds and Hepatoprotective Potential of *Anastatica Hierochuntica* Ethanolic and Aqueous Extracts against Ccl4-Induced Hepatotoxicity in Rats. *Tradit. Chin. Med.* **2020**, *40*, 947.
32. Kumaran, A.; Karunakaran, R.J. In Vitro Antioxidant Activities of Methanol Extracts of Five Phyllanthus Species from India. *LWT-Food Sci. Technol.* **2007**, *40*, 344–352. [CrossRef]
33. Schneider, S. Quality Analysis of Extra Virgin Olive Oils—Part 6 Nutritive Benefi Ts—Phenolic Compounds in Virgin Olive Oil. Agilent Technology Application Note 2016. Available online: https://www.agilent.com/cs/library/applications/5991-3801EN.pdf?srsltid=AfmBOooDCe5tYCzUiGVHbzNcfo0MsWZ3sCxKzyl8T-9CSY9Y8swms0Xp (accessed on 9 April 2024).
34. Cohen, S.A.; Meys, M.; Travin, T.L. *The Pico Tag Method a Manual of Advanced Techniques for Amino Acid Analysis*; Waters Chromatography Division: Milford, MA, USA, 1989.
35. WHO/FAO/UNU. *Protein and Amino Acid Requirements in Human Nutrition*; 0512-3054 (Print); World Health Organization: Geneva, Switzerland, 2007; pp. 1–265.
36. Chavan, U.D.; McKenzie, D.B.; Shahidi, F. Protein Classification of Beach Pea (*Lathyrus maritimus* L.). *Food Chem.* **2001**, *75*, 145–153. [CrossRef]
37. Aldai, N.; Osoro, K.; Barron, L.; Nájera, A. Gas–Liquid Chromatographic Method for Analysing Complex Mixtures of Fatty Acids Including Conjugated Linoleic Acids (Cis9trans11 and Trans10cis12 Isomers) and Long-Chain (N-3 or N-6) Polyunsaturated Fatty Acids: Application to the Intramuscular Fat of Beef Meat. *J. Chromatogr. A* **2006**, *1110*, 133–139. [PubMed]
38. Hewavitharana, G.G.; Perera, D.N.; Navaratne, S.B.; Wickramasinghe, I. Extraction Methods of Fat from Food Samples and Preparation of Fatty Acid Methyl Esters for Gas Chromatography: A Review. *Arab. J. Chem.* **2020**, *13*, 6865–6875. [CrossRef]
39. Steel, R.G. *Pinciples and Procedures of Statistics a Biometrical Approach*, 3rd ed.; McGraw-Hill: Boston, MA, USA, 1997.
40. Ayad, A.A.; Williams, L.L.; Gad El-Rab, D.A.; Ayivi, R.; Colleran, H.L.; Aljaloud, S.; Ibrahim, S.A. A Review of the Chemical Composition, Nutritional and Health Benefits of Dates for Their Potential Use in Energy Nutrition Bars for Athletes. *Cogent Food Agric.* **2020**, *6*, 1809309. [CrossRef]
41. Al-Dashti, Y.A.; Holt, R.R.; Keen, C.L.; Hackman, R.M. Date Palm Fruit (*Phoenix dactylifera*): Effects on Vascular Health and Future Research Directions. *Int. J. Mol. Sci* **2021**, *22*, 4665. [CrossRef]
42. Al-Sayyed, H.F.; Abu-Qatouseh, L.F.; Malkawy, M.; Al-Wawi, S.; Al Kafaween, M. Extracts of Jordanian Date Palm Fruit (*Phoenix dactylifera* L.) Inhibit Human Mammary Adenocarcinoma (Mcf-7) Cells in Vitro by Inducing Cell Viability. *Curr. Res. Nutr. Food Sci.* **2021**, *9*, 423–430. [CrossRef]
43. Al-Zeiny, S.S.M.; Alyaqubi, K.J.; Abbas, D.A.H. In Vitro: Anticancer Effect of Oily and Methanolic Extracts of Al-Zahdi (*Phoenix dactylifera* L.) from Dry Dates and Leaves on Amn3, Hela and Ref Cancer Cell Cultures. *Kufa J. Vet. Med. Sci.* **2022**, *13*, 1–12. [CrossRef]
44. Gonçalves, B.; Pinto, T.; Aires, A.; Morais, M.C.; Bacelar, E.; Anjos, R.; Ferreira-Cardoso, J.; Oliveira, I.; Vilela, A.; Cosme, F. Composition of Nuts and Their Potential Health Benefits—An Overview. *Foods* **2023**, *12*, 942. [CrossRef]
45. Assirey, E.A.R. Nutritional Composition of Fruit of 10 Date Palm (*Phoenix dactylifera* L.) Cultivars Grown in Saudi Arabia. *J. Taibah Univ. Sci.* **2015**, *9*, 75–79. [CrossRef]

46. Eid, W.A.M.; Azab, D.E.-S.H.; Negm, S.H. Characterization of a Novel Date Energy Bar Fortified with Moringa Oleifera Leaves Powder. *J. Future Foods* **2025**, *5*, 266–275. [CrossRef]
47. Al-Farsi, M.A.; Lee, C.Y. Optimization of Phenolics and Dietary Fibre Extraction from Date Seeds. *Food Chem.* **2008**, *108*, 977–985. [CrossRef]
48. Bouaziz, M.A.; Besbes, S.; Blecker, C.; Wathelet, B.; Deroanne, C.; Attia, H. Protein and Amino Acid Profiles of Tunisian Deglet Nour and Allig Date Palm Fruit Seeds. *Fruits* **2008**, *63*, 37–43. [CrossRef]
49. Vayalil, P.K. Antioxidant and Antimutagenic Properties of Aqueous Extract of Date Fruit (*Phoenix dactylifera* L. Arecaceae). *J. Agric. Food Chem.* **2002**, *50*, 610–617. [CrossRef] [PubMed]
50. Vayalil, P.K. Date Fruits (*Phoenix dactylifera* Linn): An Emerging Medicinal Food. *Crit. Rev. Food Sci. Nutr.* **2012**, *52*, 249–271. [CrossRef] [PubMed]
51. Markiewicz-Żukowska, R.; Puścion-Jakubik, A.; Grabia, M.; Perkowski, J.; Nowakowski, P.; Bielecka, J.; Soroczyńska, J.; Kańgowski, G.; Bołtryk, J.M.; Socha, K. Nuts as a Dietary Enrichment with Selected Minerals—Content Assessment Supported by Chemometric Analysis. *Foods* **2022**, *11*, 3152. [CrossRef] [PubMed]
52. Dghaim, R.; Hammami, Z.; Al Ghali, R.; Smail, L.; Haroun, D. The Mineral Composition of Date Palm Fruits (*Phoenix dactylifera* L.) under Low to High Salinity Irrigation. *Molecules* **2021**, *26*, 7361. [CrossRef] [PubMed]
53. Venkatachalam, M.; Sathe, S.K. Chemical Composition of Selected Edible Nut Seeds. *J. Agric. Food Chem.* **2006**, *54*, 4705–4714. [CrossRef]
54. Weaver, C.M.; Dwyer, J.; Fulgoni, V.L., III; King, J.C.; Leveille, G.A.; MacDonald, R.S.; Ordovas, J.; Schnakenberg, D. Processed Foods: Contributions to Nutrition. *Am. J. Clin. Nutr.* **2014**, *99*, 1525–1542. [CrossRef]
55. Al-Shahib, W.; Marshall, R.J. The Fruit of the Date Palm: Its Possible Use as the Best Food for the Future? *Int. J. Food Sci. Nutr.* **2003**, *54*, 247–259. [CrossRef]
56. Volpe, S.L. Magnesium in Disease Prevention and Overall Health. *Adv. Nutr.* **2013**, *4*, 378S–383S. [CrossRef]
57. Zheltova, A.A.; Kharitonova, M.V.; Iezhitsa, I.N.; Spasov, A.A. Magnesium Deficiency and Oxidative Stress: An Update. *BioMedicine* **2016**, *6*, 20. [CrossRef]
58. Hurrell, R.; Egli, I. Iron Bioavailability and Dietary Reference Values. *Am. J. Clin. Nutr.* **2010**, *91*, 1461S–1467S. [CrossRef]
59. Rayman, M.P. Selenium and Human Health. *Lancet* **2012**, *379*, 1256–1268. [CrossRef]
60. Mohamed, R.M.; Fageer, A.S.; Eltayeb, M.M.; Mohamed Ahmed, I.A. Chemical Composition, Antioxidant Capacity, and Mineral Extractability of Sudanese Date Palm (*Phoenix dactylifera* L.) Fruits. *Food Sci. Nutr.* **2014**, *2*, 478–489. [CrossRef] [PubMed]
61. Al-Shwyeh, H.A. Date Palm (*Phoenix dactylifera* L.) Fruit as Potential Antioxidant and Antimicrobial Agents. *J. Pharm. Bioallied Sci.* **2019**, *11*, 1–11. [CrossRef] [PubMed]
62. Youn, U.-Y.; Kim, R.-H.; Kim, G.-N.; Lee, S.-C. Antioxidant and Anti-Adipogenic Activities of the Nuts of *Castanopsis cuspidata* Var. Thunbergii. *Food Sci. Biotechnol.* **2017**, *26*, 1407–1414. [CrossRef]
63. Abdelbaky, A.S.; Tammam, M.A.; Ali, M.Y.; Sharaky, M.; Selim, K.; Semida, W.M.; Abd El-Mageed, T.A.; Ramadan, M.F.; Oraby, H.F.; Diab, Y.M. Antioxidant and Anticancer Assessment and Phytochemical Investigation of Three Varieties of Date Fruits. *Metabolites* **2023**, *13*, 816. [CrossRef] [PubMed]
64. Abuelgassim, A.O.; Eltayeb, M.A.; Ataya, F.S. Palm Date (*Phoenix dactylifera*) Seeds: A Rich Source of Antioxidant and Antibacterial Activities. *Czech J. Food Sci.* **2020**, *38*, 171–178. [CrossRef]
65. Staveckienė, J.; Kulaitienė, J.; Levickienė, D.; Vaitkevičienė, N.; Vaštakaitė-Kairienė, V. The Effect of Ripening Stages on the Accumulation of Polyphenols and Antioxidant Activity of the Fruit Extracts of Solanum Species. *Plants* **2023**, *12*, 2672. [CrossRef]
66. Mrabet, A.; Jiménez-Araujo, A.; Fernández-Bolaños, J.; Rubio-Senent, F.; Lama-Muñoz, A.; Sindic, M.; Rodríguez-Gutiérrez, G. Antioxidant Phenolic Extracts Obtained from Secondary Tunisian Date Varieties (*Phoenix dactylifera* L.) by Hydrothermal Treatments. *Food Chem.* **2016**, *196*, 917–924. [CrossRef]
67. Chaari, A.; Abdellatif, B.; Nabi, F.; Khan, R.H. Date Palm (*Phoenix dactylifera* L.) Fruit's Polyphenols as Potential Inhibitors for Human Amylin Fibril Formation and Toxicity in Type 2 Diabetes. *Int. J. Biol. Macromol.* **2020**, *164*, 1794–1808. [CrossRef] [PubMed]
68. Nguyen, T.H.D.; Vu, D.C.; Alvarez, S.; Nguyen, K.D.; Nguyen, T.L.T.; Tuan, N.N.; Minh-Nguyet, N.T.; Tam, L.N.; Ho, T.L.; Vo, X.T. Comparative Examination of Phytonutrients and Antioxidant Activity of Commonly Consumed Nuts and Seeds Grown in Vietnam. *Horticulturae* **2022**, *8*, 521. [CrossRef]
69. Ghazzawi, H.A.; Al-Ismail, K. A Comprehensive Study on the Effect of Roasting and Frying on Fatty Acids Profiles and Antioxidant Capacity of Almonds, Pine, Cashew, and Pistachio. *J. Food Qual.* **2017**, *2017*, 9038257. [CrossRef]
70. Otles, S.; Selek, I. Phenolic Compounds and Antioxidant Activities of Chestnut (*Castanea sativa* Mill.) Fruits. *Qual. Assur. Saf. Crops Foods* **2012**, *4*, 199–205. [CrossRef]
71. Saafi, E.B.; El Arem, A.; Chahdoura, H.; Flamini, G.; Lachheb, B.; Ferchichi, A.; Hammami, M.; Achour, L. Nutritional Properties, Aromatic Compounds and in Vitro Antioxidant Activity of Ten Date Palm Fruit (*Phoenix dactylifera* L.) Varieties Grown in Tunisia. *Braz. J. Pharm. Sci.* **2022**, *58*, e18871. [CrossRef]

72. Kong, C.-S.; Jeong, C.-H.; Choi, J.-S.; Kim, K.-J.; Jeong, J.-W. Antiangiogenic Effects of P-Coumaric Acid in Human Endothelial Cells. *Phytother. Res.* **2013**, *27*, 317–323. [CrossRef] [PubMed]
73. Bettaieb, I.; Kilani, A.; Ben Othman, K.; Benabderrahim, M.A.; Elfalleh, W. Phenolic Profile, Sugar Composition, and Antioxidant Capacities of Some Common Date Palm (*Phoenix dactylifera* L.) Cultivars as a Potential Nutraceutical and Functional Food Ingredients. *J. Food Qual.* **2023**, *2023*, 2474900. [CrossRef]
74. Wolfe, R.R.; Church, D.D.; Ferrando, A.A.; Moughan, P.J. Consideration of the Role of Protein Quality in Determining Dietary Protein Recommendations. *Front. Nutr.* **2024**, *11*, 1389664. [CrossRef]
75. Gorissen, S.H.M.; Crombag, J.J.R.; Senden, J.M.G.; Waterval, W.A.H.; Bierau, J.; Verdijk, L.B.; van Loon, L.J.C. Protein Content and Amino Acid Composition of Commercially Available Plant-Based Protein Isolates. *Amino Acids* **2018**, *50*, 1685–1695. [CrossRef] [PubMed]
76. Millward, D.J. Amino Acid Scoring Patterns for Protein Quality Assessment. *Br. J. Nutr.* **2012**, *108*, S31–S43. [CrossRef]
77. Ibrahim, A.S.; Sukor, R.; Anwar, F.; Murugesu, S.; Selamat, J.; Raseetha, S. Nutritional, Nutraceutical Attributes, Microbiological and Chemical Safety of Different Varieties of Dates—A Review. *Future Foods* **2024**, *10*, 100421. [CrossRef]
78. Adhikari, S.; Schop, M.; de Boer, I.J.M.; Huppertz, T. Protein Quality in Perspective: A Review of Protein Quality Metrics and Their Applications. *Nutrients* **2022**, *14*, 947. [CrossRef]
79. Chapman, D.G.; Castillo, R.; Campbell, J.A. Evaluation of Protein in Foods. I. A Method for the Determination of Protein Efficiency Ratios. *Can. J. Biochem. Physiol.* **1959**, *37*, 679–686. [CrossRef]
80. Yajima, K.; Chiba, S.; Park, I.; Ogata, H.; Kayaba, M.; Ishihara, A.; Tanaka, Y.; Simeng, Z.; Jaehoon, S.; Katakura, M.; et al. Dietary Palmitic Acid to Oleic Acid Ratio Modulates Energy Metabolism and Biological Rhythms in Young Healthy Japanese Males. *Br. J. Nutr.* **2024**, *131*, 447–460. [CrossRef]
81. Alsuhebani, H.; Sakr, S.S.; Elkashef, H.; Algheshairy, R.M.; Alfheeaid, H.A.; Algeffari, M.; Alharbi, H.F. Novel High Protein-Energy Balls Formulated with Date Paste Enriched with Samh Seeds Powder and/or Different Milk Protein Origins: Effect on Protein Digestibility in Vitro and Glycemic Response in Young Adults. *Front. Nutr.* **2025**, *12*, 1538441. [CrossRef]
82. Valdés García, A.; Sánchez Romero, R.; Juan Polo, A.; Prats Moya, S.; Maestre Pérez, S.E.; Beltrán Sanahuja, A. Volatile Profile of Nuts, Key Odorants and Analytical Methods for Quantification. *Foods* **2021**, *10*, 1611. [CrossRef]
83. Flowers, J.M.; Hazzouri, K.M.; Lemansour, A.; Capote, T.; Gros-Balthazard, M.; Ferrand, S.; Lebrun, M.; Amiri, K.M.A.; Purugganan, M.D. Patterns of Volatile Diversity Yield Insights into the Genetics and Biochemistry of the Date Palm Fruit Volatilome. *Front. Plant Sci.* **2022**, *13*, 853651. [CrossRef] [PubMed]
84. Zheng, A.-R.; Wei, C.-K.; Liu, D.-H.; Thakur, K.; Zhang, J.-G.; Wei, Z.-J. Gc-Ms and Gc×Gc-Tof-Ms Analysis of Roasted/Broth Flavors Produced by Maillard Reaction System of Cysteine-Xylose-Glutamate. *Curr. Res. Food Sci.* **2023**, *6*, 100445. [CrossRef] [PubMed]
85. Nadeem, M.; Salim-ur-Rehman; Muhammad Anjum, F.; Murtaza, M.A.; Mueen-ud-Din, G. Development, Characterization, and Optimization of Protein Level in Date Bars Using Response Surface Methodology. *Sci. World J.* **2012**, *2012*, 518702. [CrossRef] [PubMed]

foods

MDPI

Article

Lemon Dietary Fibre-Based Powder as a Promising Ingredient for the Food Industry: Enhancing *Mortadella* Nutritional Quality

Daniela Magalhães [1], Cristina V. Rodrigues [1], Carmen Botella-Martinez [2], Nuria Muñoz-Tebar [2], José Angel Pérez-Álvarez [2], Manuel Viuda-Martos [2], Paula Teixeira [1] and Manuela Pintado [1,*]

[1] CBQF—Centro de Biotecnologia e Química Fina—Laboratório Associado, Escola Superior de Biotecnologia, Universidade Católica Portuguesa, Rua Diogo Botelho 1327, 4169-005 Porto, Portugal; dmagalhaes@ucp.pt (D.M.); civrodrigues@ucp.pt (C.V.R.); pcteixeira@ucp.pt (P.T.)

[2] IPOA Research Group, Institute on Agri-Food and Agri-Environmental Research and Innovation (CIAGRO-UMH), Miguel Hernandez University of Elche, Carretera de Beniel, km 3.2, 03312 Orihuela, Alicante, Spain; c.botella@umh.es (C.B.-M.); nmunoz@umh.es (N.M.-T.); ja.perez@umh.es (J.A.P.-Á.); mviuda@umh.es (M.V.-M.)

* Correspondence: mpintado@ucp.pt

Abstract: Lemon co-products are valuable due to their high dietary fibre, making them significant for valorisation. This research aimed to characterise an innovative lemon dietary fibre (LDF) obtained through integrated extraction (of essential oil, phenolic compounds (PCs), and pectin) by evaluating its chemical, physicochemical, structural, techno-functional, total phenolic content, and antioxidant and antibacterial properties. The effects of incorporating LDF (3% and 6%) into *mortadella*, a bologna-type sausage, on chemical, physicochemical, technological, and sensory properties were analysed. LDF exhibited a total dietary fibre content of 85.79%, mainly insoluble (52.55%). Hesperidin (89.97–894.44 mg/100 g DW) and eriocitrin (68.75–146.35 mg/100 g DW) were the major free PCs. The major bound PCs were vanillin (5.90–9.16 mg/100 g DW) and apigenin-7-O-glucoside (8.82 mg/100 g DW). This functional ingredient demonstrated antioxidant and antibacterial activity. LDF significantly influenced *mortadella*'s colour, texture, and mineral composition. Higher levels of LDF result in a paler colour and increased hardness and contribute to reducing sodium levels of the final product. It also decreased residual nitrite levels, although this reduction was followed by a slight increase in lipid oxidation, which remained below the rancidity threshold ($\geq$1.0), ensuring acceptable product quality. Sensory evaluation revealed positive feedback, favouring the 3% LDF formulation.

Keywords: dietary fibre; lemon co-products; bioactive properties; sensorial analysis; functional meat foods; nutritional value

1. Introduction

The search for innovative and sustainable ingredients remains a key focus in the constantly evolving food industry. Since consumer preferences have been shifting towards healthier and environmentally conscious choices, the valorisation of lemon co-products addresses waste management concerns and offers a valuable source of functional ingredients [1]. Citrus juice industries generate a significant amount of co-products, which are mainly used for animal feed. These wastes account for 50% of the original whole fruit and mainly comprise peels (albedo and flavedo), almost one-fourth of the whole fruit mass, seeds, and fruit pulp after juice and essential oil extraction. However, because of their high fibre content, they can be used as a good source of dietary fibre (DF) [2,3]. DF is a complex mixture of polysaccharides, which can be classified into two categories based on water solubility: soluble (SDF) and insoluble dietary fibre (IDF). Soluble dietary fibre includes gums, pectin, glucans, and some biological and synthetic polysaccharides, while IDFs include cellulose, hemicellulose, lignin, and non-cellulosic polysaccharides [4,5].

Lemon fibres are known for their exceptional quality, attributed to their low caloric content and abundant fibre content, leading to favourable functional characteristics, including efficient colonic fermentability [6,7]. Moreover, citrus fruits have better quality than other sources of dietary fibre due to the presence of bioactive compounds, such as phenolic compounds, with antioxidant and antimicrobial properties. These bioactive compounds may confer higher health-promoting effects than the dietary fibre itself [4,8,9]. Dietary fibre may exhibit antioxidant effects due to the presence of phenolic compounds attached through hydrogen bonds to the polysaccharides, known as bound phenolic compounds (BP). Free phenolic compounds (FP) are not bound to the fibre matrix and can be extracted with water and organic solvents, whereas BP requires extraction by acid or alkaline hydrolysis [10]. Free phenolic compounds can be absorbed in the upper parts of the gastrointestinal tract due to their free form. In contrast, BP must be released from the fibre, a process that partially can occur during gastrointestinal digestion (depending on the preprocessing) but predominantly occurs in the colon after the dietary fibre undergoes fermentation. Consequently, dietary fibre bound to (poly)phenols might have an antioxidant effect associated with the presence of these compounds [11]. For this reason, lemon co-products exhibit remarkable antioxidant potential among citrus fruits, making them an optimal choice for dietary strategies to prevent cardiovascular diseases, among others [12].

Dietary fibres offer a multitude of advantages beyond their nutritional and bioactive value, encompassing a spectrum of functional and technological attributes. These include water-holding capacity (WHC), oil-holding capacity (OHC), swelling capacity (SWC), viscosity modulation, gel-forming ability, textural enhancement, and chelating capabilities [13]. DF consumption plays an important role in decreasing intestinal transit time and increasing stool bulk, reducing blood total and/or LDL cholesterol levels and consequently reducing the risk factors for cardiovascular diseases, reducing post-prandial blood glucose and/or insulin levels and consequently managing diabetes by improving insulin sensitivity, decreasing body weight gain, and being fermentable by colonic microflora with all the benefits of positive gut microbiota modulation [14–16]. DF intake in industrialised nations is currently estimated to be less than 25 g per person/day. However, recommendations from the European Food Safety Authority (EFSA) and the World Health Organization (WHO) on adequate dietary fibre intake differ worldwide and by age group, but 25–30 g or more per day is widely recommended for adults [17,18]. Recommended intakes were defined as those necessary for maintaining normal laxation and cardiovascular health. However, emerging evidence points to benefits extending well beyond these by modulation of the gut microbiota [19]. The insufficient intake of DF in the population's diet is often related to various health disorders such as colon and cardiovascular diseases, obesity, and cancer [20]. This trend has prompted research into incorporating natural fibres into food products, including processed meats, to improve their nutritional value while maintaining their sensory appeal. Among the diverse sources of fibre available, those extracted from citrus fruits stand out most due to their functional, technological, and bioactive properties [21–23].

Meat is an excellent source of high-quality protein, offering a well balanced ration of essential amino acids and having a high biological value. Additionally, it is an important source of micronutrients such as selenium, iron, magnesium, potassium, and sodium. Despite its nutritional benefits, meat has a notable drawback as it is deficient in dietary fibre [24]. The era of globalisation, combined with rapid urbanisation, has significantly influenced consumer behaviour and demand for many convenient processed meat products [25,26]. However, meat products lack the minimum contents of DF to meet the recommended daily fibre intake. *Mortadella*, a cooked-cured sausage, is often associated with high fat content and an unfavourable nutritional image. To address these concerns, incorporating potentially functional ingredients into *mortadella* formulations, such as lemon dietary fibre (LDF), emerges as a promising strategy to develop a healthier meat product that not only improves its nutritional profile but also aligns with evolving consumer demands for healthier food options. Additionally, the addition of DF can contribute to reducing fat and nitrite residual levels, substituting commercial preservatives, creating new textures, and improving stability during processing and storage, thereby increasing shelf-life [27–31]. According to the European Union (EU) regulation (EC) No 1924/2006, a claim that a product is "high in fibre" or "source of fibre" can be made when the product contains at least 6 g/100 g or 3 g/100 g fibre, respectively. These guidelines provide a framework for developing functional foods and represent an important benchmark to meet the expectations of consumers who are increasingly focused on the health and nutritional quality of their food choices.

Commercial citrus fibres have been noted for their water-binding capacity in low-fat frankfurters [31] and as a fat replacer in low-fat beef hamburgers [30]. Additionally, the incorporation of lemon albedo recovered from commercial lemons into bologna sausages has been studied for its fibre content (22–30%) [32]. Furthermore, the incorporation of fruit-derived dietary fibre powders into meat products has gained increasing attention as a strategy to enhance both nutritional and functional properties, while helping to balance protein and fibre content [33,34]. Insoluble dietary fibres recovered from fruits, such as kiwi pomace, have been studied and incorporated into meatballs, showing improvements in swelling capacity as well as water, oil, and fat holding capacities. Additionally, meatballs reformulated with 3% kiwi pomace demonstrated high overall acceptability, making them a suitable option for consumers [35].

In contrast, our research is innovative because it has developed a lemon dietary fibre-based powder (LDF) that boasts a notably higher fibre. Unlike previous studies that used commercial fibre/lemons, our research recovers natural fibre from lemon co-products, promoting sustainability and resource efficiency. With a focus on the development of innovative, clean-label, and sustainable food products, this research aims to develop an LDF from lemon co-products, evaluating the chemical, physicochemical, tecno-functional, and food preservation properties of the obtained powder. Furthermore, this study aims to explore the use of LDF in *mortadella* formulations, assessing the effects on the physicochemical, technological, and sensory properties of the product. The fibre concentrations used in the formulations were established following the regulatory claims of 3% and 6%, allowing a comparative analysis between products classified as "source of fibre" and "high in fibre". In this way, the goal is to contribute to the development of innovative meat products that meet the growing demand for healthier, more nutritious foods without compromising the quality and taste that consumers expect.

2. Materials and Methods

2.1. Chemicals and Reagents

Acetonitrile and methanol (HPLC-gradient grade) were purchased from VWR Chemicals (Rosny-sous-Bois-cedex, France). Ethanol 96% was purchased from LabChem (Santo Antão do Tojal, Portugal). Hydrogen peroxide 33% (*w*/*v*), nitric acid 69% (*v*/*v*) (analytical grade), trichloroacetic acid, thiobarbituric acid, zinc acetate 2-hydrate, and potassium hexacyanoferrate (II) 3-hydrate for analysis were purchased from PanReac AppliChem (Darmstadt, Germany). Griess-Ilosvays nitrite reagent for microbiology (contains acetic acid) was purchased from Merck-Millipore (Darmstadt, Germany). Folin–

Ciocalteu's reagent and hydrochloric acid 37% (*v*/*v*) (HCl) were purchased from Merck (Algés, Portugal). The 2-azinobis-3-ethylbenzothiazoline-6-sulphonic acid (ABTS), 2,2′-azo-bis-(2-methylpropionamidine)-dihydrochloride (AAPH), 2,2-diphenyl-1-picrylhydrazyl (DPPH), fluorescein, gallic acid, sodium carbonate (Na_2CO_3), trifluoroacetic acid (TFA), and trolox (97%) were purchased from Sigma-Aldrich (Sintra, Portugal). The standards (HPLC grade; ≥95% purity) vanillin, eriocitrin, p-coumaric acid, ferulic acid, apigenin-7-O-glucoside, hesperidin, rutin, luteolin-7-O-glucoside, and hesperetin were purchased from Extrasynthese (Genay, France).

2.2. Lemon Dietary Fibre (LDF)-Based Powder Preparation

The lemon co-products (*Eureka* cultivar) were collected from Frutas Tereso company (Algarve, Portugal). The lemons used for fibre extraction were cultivated using pest control substances, and organic fertilisers approved for organic farming were also applied. The fertilisation includes the application of bludiamond classic 5-3-10 and bludiamond azo N-32. Foliar fertilisation is carried out using a combination of products, including nutriman N-24, sergomil, urea 46% yara, sergomax L90, feedser, verdezin extra, servalac 10-5-30, bioalgae, power stop, and red KSC, all aimed at enhancing nutrient uptake and improving plant vigour. In addition, phytosanitary treatments are performed to protect crops against pests and diseases, using products such as nutriman N-24, peters professional 20-20-20, ninja zeon, verdezin extra, urea 46% yara, repulser +3, and envidor.

The trees were grown on clayey-sandy soil and irrigated using a drip (drop-by-drop) irrigation system. The lemons were produced under an integrated production system, with non-mobilised soil and grassing between tree rows (inter-row grass cover).

After the integrative extraction of the lemon co-products, bioactive fractions, hydrolat, essential oils (EOs), phenolic compounds (PCs), and pectin, a dietary fibre-rich biomass was obtained and carefully dried in an oven at 55 °C for 48 h (Figure 1a). Following the drying phase, the biomass underwent grinding and granulometric separation, producing a finely powdered substance with a particle size below 100 μm. This process resulted in a lemon dietary fibre-based powder, as illustrated in Figure 1b.

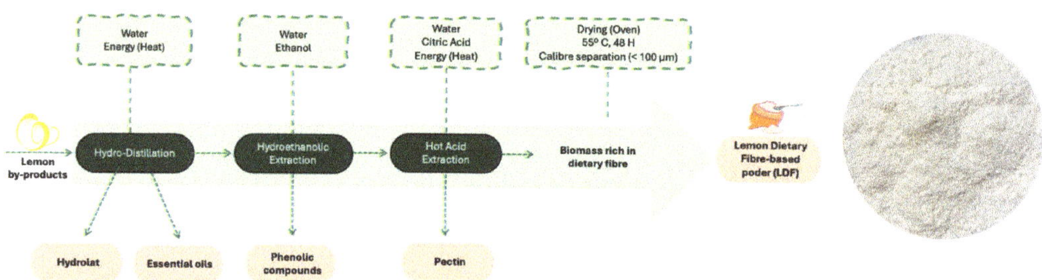

(a) (b)

Figure 1. (**a**) Process flow chart: recovery of lemon dietary fibre (LDF)-based powder; (**b**) final LDF-based powder (*Eureka* cultivar).

2.3. Chemical Composition Analysis

2.3.1. Proximate Composition

Dry matter (DM), moisture, ash, protein, fat, carbohydrates, and dietary fibre were determined in triplicate according to standard procedures using the AOAC methodology [36]. The results were expressed as g/100 g of product (dry weight).

2.3.2. Mineral Composition

Minerals were quantified after mineralisation of the lyophilised samples (0.5 g) with 67% nitric acid and 33% hydrogen peroxide by a microwave system using Inductively Coupled Plasma Mass Spectrometry (ICPMS-2030-Shimadzu, Kyoto, Japan). The final value was the average of three reads. Mineral content was expressed in mg/100 g of product.

2.3.3. Cellulose, Hemicellulose, and Lignin

For the determination of cellulose (as glucose), hemicellulose (as arabinose, mannose, galactose, and xylose), and lignin (soluble and insoluble) in lemon dietary fibre-based powder, the methodology of Ribeiro et al. (2020) was followed [37]. LDF samples were submitted to a two-step sequential acid hydrolysis, and further determination/quantification of structural carbohydrates was achieved using High-Performance Liquid Chromatography (HPLC) [37]. The insoluble lignin content was calculated gravimetrically after hydrolysis residue filtration, and soluble lignin was estimated by ultraviolet (UV) spectrophotometry at 340 nm [38]. The results were expressed in g/100 g DW.

2.4. Physicochemical Properties

2.4.1. pH and Water Activity (a_w)

The pH was measured using a calibrated pH metre GLP21 (Crison, Barcelona, Spain) on a 10% (*w*/*v*) aqueous suspension for 2 min. The water activity (a_w) was determined using a LabSwift-aw Novasina Hygrometer (Lachen, Switzerland) at 25 °C.

2.4.2. Colour

The colour was measured using a CM-700 Minolta spectrophotocolourimeter (Minolta, Osaka, Japan) in CIELAB mode under CIE Standard Illuminant D65 and an observation angle of 10°. The lemon dietary fibre-based powder was placed in a Petri dish for colour measurements, and six readings were performed at room temperature (25 °C). The following CIELAB coordinates were determined: lightness (L*), redness (a* ± red/green), and yellowness (b* ± yellow/blue). From these coordinates, psychophysical magnitudes chrome (C*) and hue (h°) were calculated using Equations (1) and (2), respectively. Colour measurements were made according to the American Meat Science Association (AMSA) guidelines for colour assessment [39].

$$C^* = \sqrt{a^{*2} + b^{*2}} \quad (1)$$

$$h^\circ = \text{arctg}\ (b^*/a^*) \quad (2)$$

2.5. Techno-Functional Properties

The water-holding capacity (WHC), oil-holding capacity (OHC), and swelling capacity (SWC) were determined according to the methodology described by Fernández-López et al., 2015 [21]. The analysis was performed in triplicate.

2.5.1. Gelling Capacity and Precipitate in the Oily Phase

Gelling capacity (GC) and precipitate in the oily phase (POP) are defined as the capacity of the sample to absorb fat and water in a matrix, forming a gel. The analysis was made in triplicate. Distilled water and sunflower oil (20 g of each one) were mixed, heated, and centrifugated. Two phases, the gel phase and precipitate in the oily phase, were identified in the tubes, and the relation between these two phases concerning the total volume was calculated. The GC and POP were calculated according to Equations (3) and (4), respectively.

$$\text{GC} = \frac{\textit{Gel layer volume}}{\textit{Total sample volume}} \times 100 \quad (3)$$

$$\text{POP} = \frac{Precipitate\ layer\ volume}{Total\ sample\ volume} \times 100 \quad (4)$$

2.5.2. Emulsifying Activity and Emulsion Stability

Emulsifying activity (EA) and emulsion stability (ES) were carried out following the methods described by Chau and Huang (2003) [40]. The analysis was made in triplicate. The EA and ES results were calculated according to Equations (5) and (6), respectively.

$$\text{EA} = \frac{Emulsified\ layer\ volume}{Total\ sample\ volume} \times 100 \quad (5)$$

$$\text{ES} = \frac{Remaining\ emulsified\ layer\ volume}{Original\ emulsion\ volume} \times 100 \quad (6)$$

2.6. Structural Characterisation

2.6.1. Fourier Transform Infrared Spectroscopy (FT-IR)

Infrared spectra were obtained by analysing the sample with the Spectrum 100 FT-IR Spectrometer (Perkin Elmer, Waltham, MA, USA), with an attachment of total attenuated internal reflection (ATR) in the spectral region between 500 cm^{-1} and 4000 cm^{-1}. The resolution used was 2.0 cm^{-1}, and each spectrum is the average of 100 scans in transmittance (%T) mode.

2.6.2. Scanning Electron Microscopy (SEM)

The morphology of the LDF samples was evaluated using a Phenom XL G2 (Thermo Fischer Scientific, Waltham, MA, USA) scanning electron microscope (SEM). Sample powder was placed over double-sided adhesive carbon tape (NEM tape; Nisshin, Nisshin-shi, Japan), which covered the observation pins and were coated with gold/palladium on a sputter coater (Polaron, Bretzfeld, Germany). An acceleration voltage of 20 kV in high-vacuum mode was used; all images were obtained using the secondary electron detector and are representative of the morphology of the sample.

2.7. Bioactive Characterisation

2.7.1. Free and Bound Phenolic Compound Extraction

Free phenolic (FP) compound extracts from LDF were prepared as described by Tinh et al. (2021) with some modifications [41]. LDF was extracted with water (H_2O) and compared with methanol (MeOH). It was then homogenised using an Ultra-Turrax digital apparatus (IKA T18, Wilmington, NC, USA) at 20,000 rpm for 1 min. Then, the mixture was sonicated using an ultrasonic bath for 60 min and left under continuous stirring (200 rpm) at room temperature (25 °C) overnight. Lastly, the mixture was centrifuged at 10,000 rpm at 4 °C for 15 min, and the supernatants were collected and filtered using a 0.45 µm filter (Orange Scientific, Braine-l'Alleud, Belgium).

Bound phenolic (BP) compound extracts from LDF were determined with alkaline and acid hydrolyses according to the method described by Ribeiro et al. (2020) with some modifications [37]. Briefly, the free phenolics solid residue was hydrolysed for 4 h with 20 mL of 4 M NaOH in an orbital shaker at 250 rpm. After that, the mixture was acidified to pH 2.0 using HCl and centrifuged (8000 rpm, 15 min at 4 °C); the supernatant was extracted 3 times with ethyl acetate (15 mL). This extract was Alkaline Bound Phenolics (ALK-BP). The pellet obtained after alkaline hydrolyses was submitted to an acid hydrolyse with 5 mL of 2M HCl for 1 h at 85 °C. After that, the mixture was acidified to pH 2.0 using NaOH and centrifuged (8000 rpm, 15 min at 4 °C); the supernatant was extracted 3 times with ethyl acetate (15 mL). This extract was Acid Bound Phenolics (ACD-BP). The ethyl acetate of the extracts was evaporated to dryness using a rotary vacuum evaporator at 40 °C. The resulting residue was subsequently dissolved in methanol or water to a final volume of 5 mL. The extracts obtained were stored at −80 °C until analysis.

2.7.2. Total Phenolic Content (TPC)

The TPC of the LDF extracts (in MeOH and H_2O) was determined by the Folin–Ciocalteu method followed by Vilas-boas et al. method (2020) with some modifications [42]. Gallic acid was used as a standard for the calibration curve (0.025–0.200 mg/mL), and the results were expressed as milligrams equivalent of gallic acid per gram of dry weight (mg GAE/g DW). The analyses were performed in triplicate.

2.7.3. Phenolic Compound Quantification by HPLC-DAD

The quantitative profiling of phenolic compounds (PCs) in LDF extract was conducted using a Waters Alliance e2695 separation module system interfaced with a photodiode array UV/Vis detector 2998 (PDA 190–600 nm; Waters, Mildford, MA, USA). The separation took place in a reversed-phase C18 column (ZORBAX Eclipse XDB-C18, 80 A°; 4.6 × 250 mm; 5 µm; Agilent, Santa Clara, CA, USA) at 25 °C. The mobile phase, the gradient elution, the injection volume, and the flow rate were prepared according to [42]. Data acquisition and analysis were carried out using Software Empower 3. Detection was performed at 280, 320, and 350 nm, and phenolic compound identification was achieved by comparing the retention time and absorbance spectra with pure standards. Quantification was conducted using calibration curve interpolation, and the results were expressed as milligrams per 100 g of dry weight (mg/100 g DW). The analyses were performed in triplicate.

2.7.4. Antioxidant Capacity

ABTS Assay

The free radical scavenging capacity of the LDF extract, described in Section 2.6.1, was evaluated using the ABTS (2,2′-azino-bis-(3-ethylbenzothiazoline-6-sulfonic acid) radical cation decolourisation assay [43]. The results were expressed as micromoles of Trolox equivalent per gram of extract (µmol TE/g). The analyses were performed in triplicate.

DPPH Assay

The electron donation capacity of the extracts was measured by bleaching the purple-coloured solution of 1,1-diphenyl-2-picrylhydrazyl radical (DPPH) according to the method of Bondet et al. (1997) with some modifications [42,44]. Trolox was used as a standard to prepare a calibration curve (0.025–0.175 µmol/mL). The results were expressed as micromoles of Trolox equivalent per millilitre of extract (µmol TE/mL). The analyses were performed in triplicate.

Oxygen Radical Absorbance Capacity Assay (ORAC)

The ORAC assay, which belongs to the hydrogen atom transfer (HAT)-based assay family, gives a measure of species competing for peroxyl radicals and was carried out following the method established by Dávalos et al. (2004) [45]. Each plate experiment also included a calibration curve (1–8 µM of Trolox). The excitation wavelength was set at 485 nm and the emission wavelength at 528 nm [46]. The software used was the Multidetection plate reader (Synergy H1, Winooski, VT, USA) controlled by the Gen5 Biotek software version 3.04. All analyses are conducted in triplicate. Antioxidant curves (fluorescence versus time) were normalised to the blank curve corresponding to the same assay by multiplying the original data by fluorescenceblank, t = 0/fluorescencesample, t = 0. Employing the normalised curves, the area under the fluorescence decay curve (AUC) was calculated according to Equation (7).

$$AUC = \sum_{i=1}^{i=n-1} \left(\frac{f_i + f_{i+1}}{2}\right) * (t_{i+1} - t_i) \quad (7)$$

where f_i is the fluorescence at reading I and t_i is the time (minutes) at reading i.

2.7.5. Antibacterial Activity

Bacterial Culture

The following bacterial strains were used: *Escherichia coli* (ATCC (American Type Culture Collection) 25922) and *Salmonella enterica* (ATCC 13076), *Bacillus cereus* (NCTC (National Collection of Type Cultures) 2599), and *Staphylococcus aureus* (ATCC 25923). These bacteria were cultured in Mueller–Hinton broth (MHB) and incubated at 37 °C overnight. After the incubation, the inoculum concentration for testing was ca. 10^8 CFU/mL.

Determination of Bacterial Inhibition

The bacterial inhibition of LDF was tested for 3% fibre concentration (0.349 g of LDF powder diluted in 10 mL of MHB). These samples were inoculated with bacteria at 10^6 CFU/mL. The inoculated 24-well plates (24-well plates; Orange Scientific 4430300N) were placed in an orbital shaker at 40 rpm for 24 h at 37 °C overnight. After 24 h, serial dilutions (from −1 to −6) were performed, and 20 µL of each dilution was plated on Muller–Hinton Agar Petri Plates. Furthermore, the same procedure was carried out for a negative control (MHB only) and a positive control (bacteria without the LDF sample) to validate the experiment. All tests were conducted in triplicate. The results were expressed as a log reduction in viable cell counts and percentage reduction (%). The log reduction was calculated according to Equation (8).

$$\text{Log reduction} = \log_{10}(A) - \log_{10}(B) \quad (8)$$

where A is the number of viable microorganisms before treatment, and B is the number of viable microorganisms after treatments.

2.8. Sausage Elaboration Process

Mortadellas were manufactured according to a traditional formula (only the meat percentage adds up to 100% while the percentage of the other ingredients is related to the meat): 70% lean pork meat and 30% pork backfat; 15% water (ice, *w*/*w*), 3% potato starch (*w*/*w*), 1.5% sodium chloride (*w*/*w*), 500 mg/kg sodium ascorbate, 300 mg/kg sodium tripolyphosphate, 150 mg/kg sodium nitrite, and 0.3% spices. This original mixture was used as a control, while lemon dietary fibre (LDF) (3% and 6%) was added to the other samples. The LDF was obtained after the extraction of (essential oils, phenolic compounds, and pectin), resulting after extraction into a biomass composed mainly of fibre. To achieve fibre incorporation rates of 3% and 6% in the *mortadellas*, 3.49 g and 6.99 g of LDF per 100 g of *mortadellas*, respectively, were added, accounting for the 85% of fibre present in the lemon dietary fibre-based powder.

The products were prepared in the IPOA (Innovación de Productos Alimentarios) research group pilot plant and followed industrial processing techniques. The frozen raw material of animal origin, except pork backfat, was transferred to the cutter (Tecator 1094 Homogenizer, Tekator, Höganäs, Sweden) with sodium chloride to extract salt-soluble proteins; after comminution, the other ingredients and additives were added. The pork backfat, previously cut into 10 × 10 × 10 cm cubes, was then added. After homogenisation, the mixture was stuffed into a Fibran-Pack (Fibran, Girona, Spain) artificial casing 100 × 150 mm long, clipped at both ends (Polyclip system/Niedecker, Germany), and cooked in a water bath. The sausages were kept in the bath until the coldest point (the geometric centre of each *mortadella*, corresponding to the thickest part of the product, reached 72 °C). A thermocouple probe (Omega Engineering, Inc., Stamford, CT, USA) positioned in the geometric centre of the *mortadella* was used to monitor product temperature. The sausages were immediately chilled on ice when the endpoint temperature was achieved. After reaching 20 °C, the product was transferred to the laboratory in ice-filled insulated boxes. The *mortadella* sausages (1000 g each) were stored at 4 °C until analysis for no longer than two weeks. Three replications of this elaboration process were performed on different days (Figure 2).

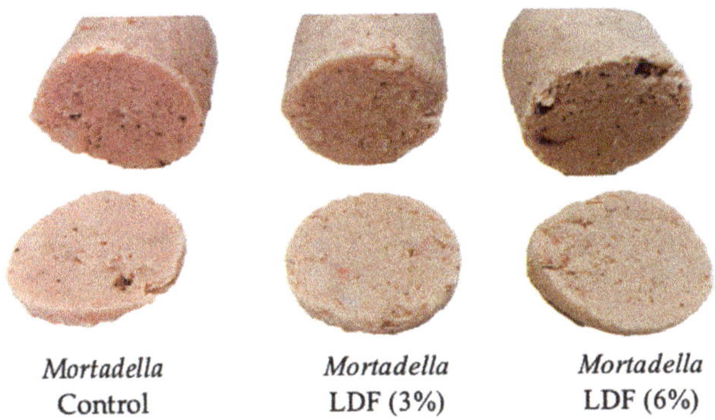

Figure 2. Reformulated *mortadellas* with LDF at 3 and 6%, and control.

2.9. Emulsion Stability of Meat

The emulsion stability of meat batters (before cooking) was assessed by measuring the total expressible fluid (TEF) following the method of Botella-Martínez et al. (2021) with slight modifications [47]. TEF represents the proportion of fluid separated from the emulsion, serving as an indicator of stability. A higher %TEF represents reduced stability as more fluid is released, whereas lower %TEF values indicate greater emulsion stability with minimal fluid separation.

The samples were initially centrifugated at 3000 rpm for 1 min. Subsequently, they were heated in a water bath at 70 °C for 30 min and then cooled to room temperature. Next, they were centrifuged again for 3 min at 3000 rpm. The samples were left standing upside down to allow the expressible fluid (fat and water) to flow into filter paper. All determinations were conducted in triplicate for each sample. The results are expressed in grams of total fluid expelled per 100 g of sample and were calculated using the following expression (Equation (9)):

$$\%TEF = \frac{Weight\ of\ tube\ with\ sample - Weight\ of\ tube\ with\ pellet}{Weight\ of\ sample} \times 100 \quad (9)$$

2.10. Chemical Analysis

2.10.1. Proximate Composition

Moisture, ash, lipid, and protein were determined using the corresponding AOAC methods [48]. Moisture content (g water/100 g sample) was determined by drying a 3 g sample at 105 °C to constant weight. Ash content was measured after incineration at 550 °C for 2 h (g ash/100 g sample). Lipid content (g lipids/100 g sample) was determined by weight loss after a 6-cycle extraction with petroleum ether in a Soxhlet apparatus. Protein (g protein/100 g sample) was analysed according to the Kjeldahl method, with a nitrogen to crude protein conversion factor of 6.25.

2.10.2. Minerals Composition

Minerals were quantified after mineralisation of the lyophilised raw samples (0.5 g) with 67% nitric acid and 33% hydrogen peroxide using a microwave system. Analysis was performed using Inductively Coupled Plasma Mass Spectrometry (ICP-MS-2030-Shimadzu, Kyoto, Japan). Final values represent the average of three replicates, expressed in mg/100 g of DW.

2.11. Technological Properties

2.11.1. pH and Water Activity (a_w)

The pH of *mortadellas* was measured directly using a pH metre (Model 507, Crison, Barcelona, Spain) equipped with a Crison combination electrode probe (Cat. no. 52, Crison, Barcelona, Spain). Measurements were performed three times, changing the position of the electrode insertion. The water activity (a_w) was assessed with a LabSwift Novasina Hygrometer (Novasina; Axair Ltd., Pfaeffikon, Switzerland) at 25 °C.

2.11.2. Colour

Colour measurements were conducted using the CIELab colour space with a CM-2600 d colourimeter (Minolta Camera Co., Osaka, Japan) using illuminant D65, an observer angle of 10°, SCI mode, 11 mm aperture for illumination, and 8 mm for measurement. The samples were placed in Petri dishes for measurement, and the following colour parameters were recorded: lightness (L*), redness (a* ± red-green), and yellowness (b* ± yellow-blue). From these coordinates, hue (h* = tan − 1 b*/a*) and chroma (C* = (a*2 + b*2) 1/2) were calculated. Nine determinations per sample were carried out [(three measures on the same inner face of three slices (2 cm height)].

2.11.3. Textural Properties

Texture profile analysis (TPA) was conducted using a TA-XT2 Texture analyser (Stable Micro Systems, Surrey, England). *Mortadella* samples were removed from the casing, cut into cubes (1 × 1 × 1 cm), and subjected to a 2-cycle compression test. All texture analyses were performed on chilled (4 °C) samples. Each sample was subjected to a two-cycle compression to 75% of its original height with a speed of 1 mm/s, and at 20–25 °C, the corresponding force–time deformation curves were obtained. The texture profile, evaluated according to Bourne (1978) [49], included hardness (N), springiness (mm), cohesiveness, and chewiness (N × mm). Each sample was analysed in six replicates.

2.12. Residual Nitrite Level

The residual nitrite level (mg $NaNO_2$/kg sample) was determined according to standards ISO/DIS 2918.26 [50].

2.13. Lipid Oxidation

Lipid oxidation was evaluated in triplicate using the 2-thiobarbituric acid (TBA) test following the recommendations of Buege and Aust (1978) [51]. TBAR values were calculated from a malonaldehyde (MDA) standard curve and expressed as mg MDA/kg sample. Briefly, the extracts were prepared by mixing the sample with TBA and trichloroacetic acid solutions, followed by heating and centrifugation. The absorbance of the resulting supernatant was then measured at 532 nm.

2.14. Sensory Evaluation

A sensory panel of 45 untrained participants (20 males, 25 females) aged 18–60 years was recruited from the staff and students at Miguel Hernández University to evaluate the three *mortadella* formulations. The protocols for sensory analysis were accepted by the Project Evaluation Office of the Miguel Hernández University (OEP, UMH, Elche, Alicante, Spain). This analysis was operated under white, fluorescent lights in individual booths. Three pieces (ca. 1.0 cm^2), one from each bath, were cut from the *mortadellas* and served at room temperature. Unsalted crackers and mineral water (room temperature) were provided to clean the palate between samples. A 9-point hedonic scale (1: dislike extremely and 9: like extremely) was used to evaluate the following attributes: global appearance, colour appearance, general quality, hardness, homogeneity, general flavour, acid taste, and bitter taste.

2.15. Statistical Analysis

Conventional statistical methods were used to calculate means and standard deviations and are shown in corresponding tables and figures. Data were evaluated by one-way analysis of variance (ANOVA), and if statistically significant differences were found, a Tukey post hoc test was performed at a 5% significance level ($p < 0.05$). For the *mortadella* characterisation, three samples were obtained from each of the three formulations during each elaboration process (3). ANOVA was applied for each parameter with one-factor treatment and three levels (control, LDF (3%), and LDF (6%)). The statistical analyses were made using GraphPad Prism Software (version 8).

3. Results and Discussion

3.1. LDF Analysis

3.1.1. Proximate Composition

The proximate composition analysis of lemon dietary fibre-based powder (LDF) recovered from lemon co-products (*Eureka* variety) is outlined in Table 1. LDF is primarily composed of total dietary fibre (TDF), which accounts for approximately 85.8% of its composition. Within the TDF fraction, insoluble dietary fibre (IDF) makes up 52.6%, while soluble fibre (SDF) represents 33.2%. Additionally, LDF is predominantly composed of carbohydrates (90.5%), protein (4.6%), and ash (1.2%). The composition analysis of LDF regarding structural carbohydrates was determined by HPLC and revealed the presence of cellulose (9.6%), hemicellulose (6.8%), soluble lignin (9.9%), and insoluble lignin (5.6%).

Table 1. Proximate composition (g/100 g sample) of lemon dietary fibre-based powder.

	LDF
Moisture (%)	3.8 ± 0.0
Protein (%)	4.6 ± 0.0
Lipids (%)	0.0 ± 0.0
Ash (%)	1.2 ± 0.0
Carbohydrates (%)	90.5 ± 0.0
Total Dietary Fibre (TDF) (%)	85.8 ± 1.1
Insoluble Dietary Fibre (IDF) (%)	52.6 ± 0.7
Soluble Dietary Fibre (SDF) (%)	33.2 ± 0.8
Cellulose (as glucose)	9.6 ± 0.9
Hemicellulose	6.8 ± 0.4
Soluble Lignin	9.9 ± 0.4
Insoluble Lignin	5.6 ± 0.4

The values are expressed as g/100 g DW (dry weight), where n = 3.

Lemon co-products typically contain high levels of dietary fibre, which vary depending on factors such as cultivar, variety, and origin. Generally, they have higher concentrations of insoluble fibre compared to soluble fibre. For instance, lemons with minimal albedo (the white part of the peel) inherently have a lower fibre content. According to research by Nempeque et al. (2021), dried lemon (*Citrus latifolia*) peels contained 92.86% of dietary fibre [52]. Wang et al. (2015) assessed the fibre content of dried lemon peels (*Citrus limon* (L.) Burm.f.) sourced from commercial orchards in China [53]. These peels contained

64.07% TDF, with IDF accounting for 50.32% and SDF representing 12.89%. Similarly, a research group reported comparable values when producing dietary fibre from lemon peel waste, with SDF and IDF representing approximately 19.55% and 80.56%, respectively [54]. A recent study by Xiao et al. (2024) evaluated the dietary fibre content of lemon peels (sourced from China) after extracting lemon oil and juice [55]. The resulting powder contained 73.66% TDF, comprising 22.05% SDF and 51.58% ISF. Additionally, the composition included 26.70% cellulose, 12.43% hemicellulose, and 5.39% lignin.

3.1.2. Mineral Composition

The mineral quantitative profile of lemon dietary fibre-based powder is shown in Table 2. The most abundant mineral was calcium (Ca), 425.96 ± 4.54 mg/100 g, followed in descending order by phosphorus (P), 184.60 ± 7.27 mg/100 g, and potassium (K), 77.92 ± 6.53 mg/100 g. The mineral profile observed in the LDF may be influenced by the agro-technical practices employed during cultivation. The use of organic fertilisers, clayey-sandy soil, and drip irrigation can influence and affect the nutritional composition of fruits and promote mineral uptake. Therefore, it is plausible that the growing conditions in this study contributed to the mineral content of the lemon fibre [56,57]. Minerals can be divided into macro-minerals, such as potassium, calcium, and magnesium, and micro-minerals, such as iron, zinc, copper, and manganese. Macro-minerals play a vital role and are required by the human body in large quantities [58]. Essential minerals, which the body cannot produce, must be obtained from the diet. Poor-quality diets can result in deficiencies, with calcium, phosphorus, and potassium being a significant concern [59]. Furthermore, WHO recommends a daily intake of 1000 mg of calcium daily [60], while EFSA advises a phosphorus consumption of 700 mg/per day for young adults [61].

Table 2. Minerals composition (mg/100 g; n = 3) of lemon dietary fibre-based powder.

	LDF (mg/100 g)
Calcium	425.96 ± 4.54
Phosphorus	184.60 ± 7.27
Potassium	77.92 ± 6.53
Magnesium	15.31 ± 1.60
Sodium	9.92 ± 1.33
Iron	1.28 ± 0.46
Copper	0.62 ± 0.01
Zinc	0.58 ± 0.27
Manganese	0.17 ± 0.00

Minerals have been documented in the literature as having important physiological roles. Czech et al. (2020) reported that lemon peel from the *Citrus limon*, Interdonato cultivar (Turkey), showed more macro- and micro-minerals than lemon pulp and the whole fruit [62]. The main macro-minerals detected in lemon peel were potassium (127.00 mg/100 g), calcium (31.80 mg/100 g), and phosphorus (23.90 mg/100 g). Regarding micro-minerals, lemon peel had substantially lower quantities, with the principal constituents being iron (0.34 mg/100 g) and zinc (0.28 mg/100 g). Trace element levels in fruit may be influenced by various factors, including the soil's mineral composition, the quality of irrigation water, weather conditions, and agricultural practices such as fertiliser usage and fruit variety. However, detailed information on the minerals present in lemon co-products remains limited. Furthermore, Lario et al. (2004) formulated a fibre-rich powder but did not analyse its mineral content [12]. Therefore, our study is particularly relevant, as we have successfully developed a powder that is not only rich in fibre but also contains essential minerals crucial to promote health and well-being.

3.2. Physicochemical Properties

pH, a_w, and Colour

Table 3 shows the physicochemical properties (pH, a_w, and CIELAB colour coordinates) of the LDF. The pH is one of the most critical parameters affecting their processing and storage quality. The LDF exhibited an acidic pH of 3.27. This value is consistent with previous studies. For instance, Fernández-López et al. (2015) [21] reported pH values for lemon fibres ranging from 3.84 to 4.70, while Lario et al. (2004) [12] observed pH values between 3.83 and 3.98 for dried lemon fibre. Water activity (a_w) indicates available water for biological functions. Lowering a_w reduces microbial growth. The limit value of aw for microbial growth is around 0.6, and below this, food is microbiologically safe [63]. The water activity in LDF is 0.424. Bakshi and Ananthanarayan (2022) [64] reported a similar aw value for lemon peel powder (0.45). In contrast, other studies reported even lower aw values (0.13–0.21) for lemon fibres, likely due to their lower moisture content [12].

Table 3. Physicochemical [pH, water activity (a_w) (n = 3), and colour coordinates (n = 6) (L*, lightness; a*, red/green coordinate; b*, yellow/blue coordinate; C*, chrome; h°, hue)] of lemon dietary fibre-based powder.

	LDF
pH	3.27 ± 0.01
a_w	0.424 ± 0.02
L^*	77.91 ± 1.35
a^*	0.69 ± 0.08
b^*	13.59 ± 0.38
C^*	13.74 ± 0.37
h°	86.89 ± 0.43

Colour is one of the most important quality parameters in food ingredients/products. L* represents lightness from black to white on a scale of 0 to 100 and correlates with the moisture content; L* values increase as moisture content increases. L* value in LDF was 77.91, which follows other authors' reports (64.91–78.53) [12]. The a* values represent (red (+)/green (−)), and the b* values indicate (yellow (+)/blue (−)) components that define the colour chromaticity. LDF showed higher yellow components (b* = 13.59) than red components (a* = 0.69), which is characteristic of lemon fruits and their co-products. C* (Chroma), which means the intensity or purity of the colour. A higher C* value suggests a more vibrant or saturated colour, while a lower value indicates a more muted or desaturated colour. Therefore, the LDF (C* = 13.74) indicates how far the colour is from being saturated. On the other hand, h° denotes the hue angle, representing the specific colour direction on the CIELAB colour wheel. In this case, 86.89° in LDF is close to the yellow-green area.

3.3. Structural Characterisation Analysis

3.3.1. FT-IR Spectra

Fourier Transform Infrared Spectroscopy (FT-IR) is a widely used analytical technique for detecting and quantifying chemical bonding species and functional groups within a sample. It serves as a valuable tool for evaluating proteins, fibres, and other substances. The FT-IR spectrum of LDF is presented in Figure 3. According to the infrared spectra analysis, LDF had a characteristic peak near 3400 cm^{-1}, due to the stretching vibration of O–H, which is the characteristic band of cellulose. The absorption peak at 2924 cm^{-1} represents the stretching vibration peak of aliphatic saturated C–H in cellulose and hemicellulose [65]. Absorption peaks at 1742 cm^{-1} and 1646 cm^{-1} are attributed to the C=O stretching vibrations. In addition, the characteristic bending or tensile vibration peak of lignin is at 1518 and 1272 cm^{-1} [66]. The characteristic peak near 1020 cm^{-1} was attributed to the stretching vibration of the C-O-C glycosidic bond and C-O-H side group, which indicates the existence of a pyran ring. This peak is a strong indicator of the polysaccharide content in plant-based fibres.

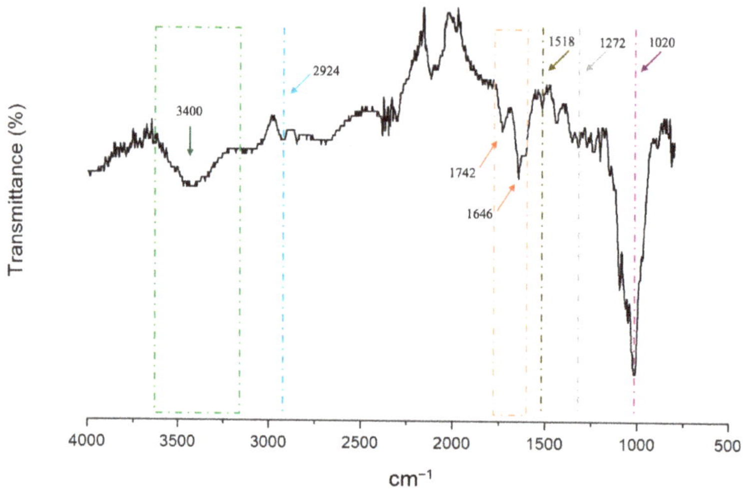

Figure 3. FT-IR spectra in the 4000–500 cm^{-1} region of the lemon dietary fibre-based powder.

3.3.2. SEM

The microscopic surface morphology of lemon dietary fibre-based powder was analysed by scanning electron microscope (SEM) (Figure 4). The microstructural analysis of LDF revealed important structural characteristics that directly influence its physicochemical and functional properties in food applications. At a magnification of 1000 times, the particles exhibited distinct sizes, being fragmented and with irregular shapes, ranging from small fragments to larger structures, which indicates a heterogeneous distribution. This variability can be attributed to the composition of the fibre but especially to the mechanical processing, such as milling, which typically produces particles of differing dimensions. At magnifications of 2000 and 2500 times, the overall morphology appeared flaky. The microstructure of citrus fibre plays a critical role in determining its physicochemical properties, such as water-holding capacity and cation exchange capacity. Citrus fibres with a large surface area offer more binding sites for water molecules through hydrogen bonding and/or dipole interactions. This property enhances their functionality, making them valuable as texturisers, stabilisers, and moisture retainers in food formulations [67].

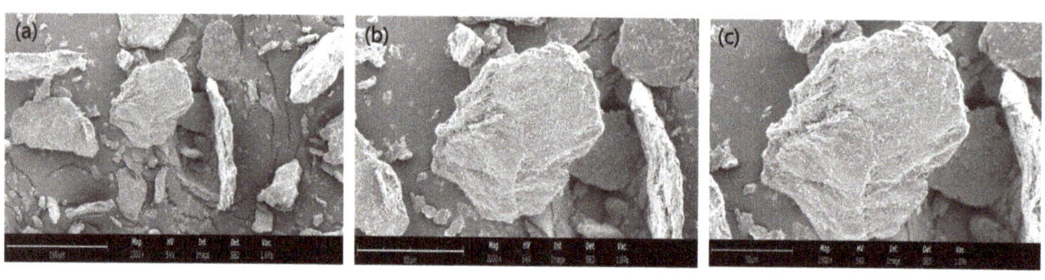

Figure 4. Scanning electron microscopy images (SEM) of lemon dietary fibre (LDF). (**a**) LDF: 1000× magnification and scale bar of 150 μm; (**b**) LDF:2000× magnification and scale bar of 80 μm; (**c**) LDF: 2500× magnification and scale bar of 50 μm.

3.4. Techno-Functional Properties

Figure 5 shows the techno-functional properties of the lemon dietary fibre-based powder (LDF). The hydration properties, evaluated as their water-holding capacity (WHC) and swelling capacity (SWC), are beneficial in determining their optimal usage levels in foods as they contribute to desirable texture properties. The WHC of fibre relies on its processing, chemical composition, and physical structure, particularly influenced by its soluble dietary fibre content. Dry milling alters the physical structure of lemon fibre, compacting it and reducing its water-holding capacity by increasing fibre density through broken pores. However, the prominent WHC (8.49 ± 0.24 g water/g) and SWC (9.57 ± 0.04 mL/g) of LDF indicate its potential as a functional ingredient in food applications. It can help to mitigate syneresis, modify the texture and viscosity, and contribute to the calorie reduction in foods. Oil-holding capacity (OHC) is another important technical property of dietary fibre, which plays an important role in preventing fat loss during food processing, and LDF showed an OHC of 1.55 ± 0.13 g oil/g. In addition, LDF showed a gelling capacity (GC) of 67%, precipitation in the oil phase (POP) of 30%, emulsifying ability (EA) of 75%, and emulsion stability (EE) of 82%. These properties suggest that a significant portion of LDF interacts with the oil phase, enhancing emulsion stability. As a result, LDF is highly effective in forming and maintaining stable emulsions, making it a valuable ingredient for improving texture in meat products. Its ability to retain moisture and stabilise oil–water mixtures contributes to a more consistent and desirable texture in processed meats. The existing research has been mainly focused on evaluating the WHC, OHC, and SWC. A study conducted by Núñez-Gómez et al. (2024) [68] reported a SWC of 9.9 mL/g, a WHC of 1.1 g water/g, and an OHC of 9.3 g oil/g in lemon peels. These results differ from our findings, as their study demonstrated a higher fat absorption capacity and a lower water retention capacity in lemon peels. In contrast, Huang et al. (2021) [69] observed functional property values in untreated citrus peels that are more in line with our results, reporting an SWC of 8 mL/g, a WHC of 8 g/g, and an OHC of 2 g/g. Similarly, a recent study by Xiao et al. (2024) [55] also presented values comparable to our findings, further supporting the consistency across different studies. These previous studies have focused on evaluating the techno-functional properties of citrus peels. The novelty of this research lies in being the first to evaluate these properties in lemon dietary fibre obtained through a biorefinery extraction approach.

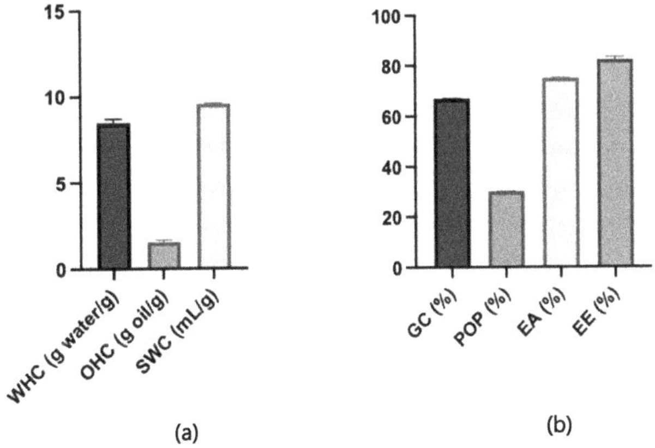

Figure 5. Techno-functional properties: (**a**) (WHC, water-holding capacity (g water/g); OHC, oil-holding capacity (g oil/g); SWC, swelling capacity (mL/g)); (**b**) (GC, gelling capacity (%); POP, precipitate in oily phase (%); EA, emulsion ability (%); EE, emulsion stability (%)) of lemon dietary fibre-based powder. Data are presented as mean ± SD, where n = 3.

3.5. *Bioactive Properties*

3.5.1. Total Phenolic Content

The total phenolic content (TPC) evaluated and compared across three types of extracts, free phenolic extracts (FP), bound phenolic extracts obtained via alkaline hydrolysis (BP-ALK), and bound phenolic extracts obtained via acid hydrolysis (BP-ACD), are illustrated in Figure 6. Furthermore, the differences in TPC among these extracts were analysed using two solvents: methanol (MeOH) and water (H_2O). The results revealed that FP extracts consistently exhibited the highest TPC levels, irrespective of the solvent used. Notably, the use of MeOH as a solvent resulted in a significantly higher TPC concentration in FP extracts compared to water ($p < 0.05$). In contrast, both BP-ALK and BP-ACD extracts displayed lower TPC levels, with the solvent type having no significant effect on their TPC concentrations ($p > 0.05$). A study conducted by Al-Qassabi et al. (2018) evaluated the TPC in imported lemon peels using six different extraction methods [70]. The study found that methanol extracts exhibited the highest TPC (7.75 mg/g), while water extracts had the lowest (5.34 mg/g). These results are consistent with ours and further support the efficacy of methanol as a solvent for extracting PCs from lemon co-products.

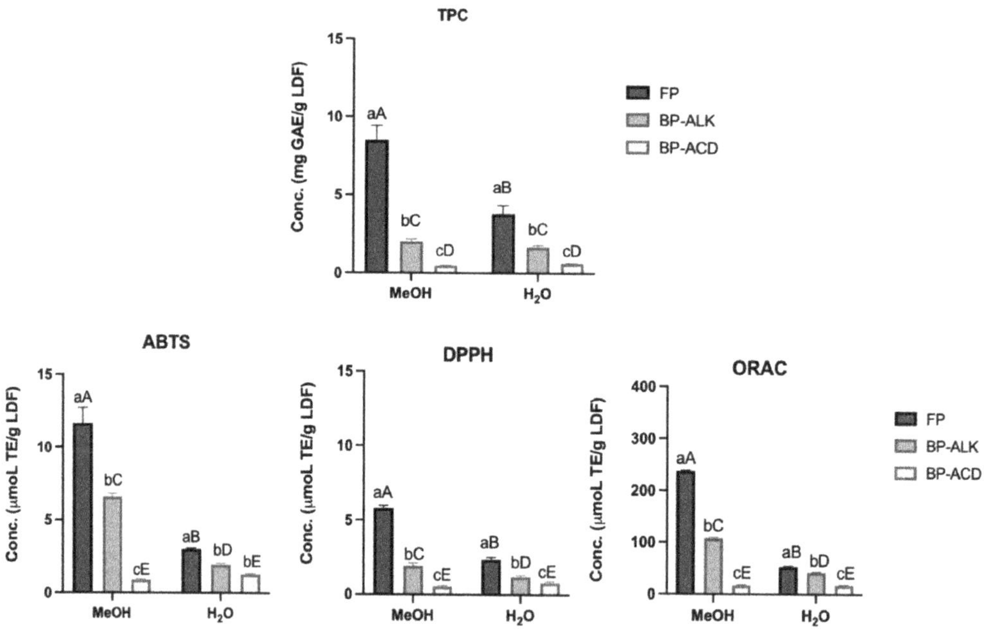

Figure 6. Total phenolic content (TPC) and antioxidant activity (ABTS, DPPH, and ORAC) of LDF in free phenolic extracts (FP), bound phenolic extracts obtained by alkaline hydrolysis (BP-ALK), and by acid hydrolysis (BP-ACD). Different lowercase letters indicate significant differences between extraction types (FP, BP-ALK, and BP-ACD), while different uppercase letters indicate significant differences between solvents (MeOH and H_2O) ($p < 0.05$), as determined by Tukey's Multiple Range Test. Data are presented as mean ± SD, where n = 3.

3.5.2. Phenolic Compound Quantification by HPLC-DAD

The quantification of phenolic compounds by High-Performance Liquid Chromatography with Diode Array Detection (HPLC-DAD) in FP, BP-ALK, and BP-ACD extracts is presented in Table 4. Hesperidin and eriocitrin were identified and quantified in free phenolic extracts as the main phenolic compounds. Both compounds were found in significantly higher concentrations when methanol was used as the solvent compared to water

($p < 0.05$). For hesperidin, the concentration was 894 mg/100 g DW with MeOH, compared to 89.97 mg/100 g DW with H_2O. Similarly, eriocitrin was measured at 146 mg/100 g DW MeOH, versus 68.85 mg/100 g DW with H_2O. Regarding the non-extractable fractions, normally called "bound phenolic compounds", most phenolic compounds were identified and quantified using the first alkaline hydrolysis extraction (BP-ALK). This is likely due to the fact that alkaline hydrolysis effectively breaks ester bonds that link phenolic compounds to cell wall components such as lignin and hemicellulose. As a result, these bound phenolics are released into the extractable fraction, leading to a higher recovery and improved quantification [71]. Six phenolic compounds were quantified in this fraction, ordered by concentration as follows: vanillin > apigenin-7-O-glucoside > luteololin-7-O-glucoside > p-coumaric acid > ferulic acid > rutin. However, when water was used as the extraction solvent, only three of them were detected (vanillin, p-coumaric, and ferulic acid). This highlights the significant influence of the extraction solvent on the concentrations of these phenolic compounds and their bioactivities. In the acid hydrolysis extraction (BP-ACD), only hesperetin was quantified, with concentrations ranging from 5.30 to 6.00 mg/100 g DW ($p < 0.05$).

Table 4. Phenolic compound quantification by HPLC-DAD of LDF in free phenolic extracts (FP), bound phenolic extracts obtained by alkaline hydrolysis (BP-ALK), and by acid hydrolysis (BP-ACD).

Phenolic Compound	RT (min)	Free Phenolics (FP)		Bound Phenolics (BP-ALK)		Bound Phenolics (BP-ACD)	
		MeOH	H_2O	MeOH	H_2O	MeOH	H_2O
		(mg/100 g DW)					
Vanillin	22.02			9.16 ± 2.14 a	5.90 ± 1.80 b		
Eriocitrin	23.42	146.35 ± 0.81 a	68.75 ± 4.23 b				
p-coumaric acid	23.99			7.06 ± 0.33 a	5.47 ± 1.70 b		
Ferulic acid	25.80			5.43 ± 0.55 a	3.93 ± 0.19 b		
Apigenin-7-O-Glucoside	29.02			8.82 ± 1.84			
Hesperidin	29.24	894.44 ± 4.63 a	89.97 ± 1.39 b				
Rutin	29.55			3.41 ± 0.22			
Luteolin-7-O-Glucoside	30.27			7.43 ± 1.93			
Hesperetin	48.11					5.30 ± 0.21 a	6.00 ± 0.19 a

Different lowercase letters in same row indicate significant differences ($p < 0.05$), as determined by Tukey's Multiple Range Test. Data are presented as mean ± SD, where n = 3.

According to Durmus and Kilic-Akyilmaz (2023), who investigated the phenolic compounds in extractable and non-extractable fractions of lemon peel, ten phenolic compounds were identified [72]. These included caffeic acid, p-coumaric acid, ferulic acid, o-coumaric acid, rutin, hesperidin, quercetin-3-O-glucoside, quercitrin, hesperetin, and apigenin. Among these, hesperetin was detected in the highest concentrations, with 219.9 mg/100 g DW in extractable phenolics. In the non-extractable phenolics, hesperetin concentrations varied between 103.4 and 423.5 mg/100 g DW. Hesperidin was the second most abundant compound, with 136.3 mg/100 g DW in extractable phenolics and concentrations ranging from 107.7 to 269.4 mg/100 g DW in non-extractable phenolics, depending on the extraction methodology. Notably, the ultrasound-enzyme-assisted extraction method yielded the highest concentrations of these non-extractable phenolics. Additionally, it is important to emphasise that our results show a lower hesperetin content compared to those reported in crude lemon peels. This reduction can be attributed to the integrative biorefinery extraction process and the applied temperatures, which caused the degradation of certain phenolic compounds, thereby explaining the lower observed values.

3.5.3. Antioxidant Activity

The food industry is increasingly adopting plant-based ingredients and extracts in response to concerns about the potential adverse effects of synthetic antioxidants. Among these natural sources, lemon co-products are distinguished for their significant antioxidant properties, offering both biological and economic advantages. The antioxidant activity of the FP, BP-ALK, and BP-ACD extracts from LDF was assessed using two solvents (MeOH and H_2O) across three different chemical assays: ABTS, DPPH, and ORAC (Figure 6). It was observed that in all antioxidant assays, FP exhibited the highest activity, followed by BP-ALK and BP-ACD. These findings align with the TPC analysis results, suggesting that the higher antioxidant activity of FP is attributed to its greater TPC. However, it is noteworthy that bound phenolic compounds, especially BP-ALK, also demonstrated significant antioxidant activity, highlighting the potential of fibre-bound phenolics as valuable contributors to antioxidant activity. Among the assays, LDF showed higher activity against ABTS radicals compared to DPPH radicals. This can be explained by the fact that ABTS is effective for both lipophilic and hydrophilic compounds, while DPPH is more selective for the former. The ORAC method, which measures antioxidant potential under conditions that mimic cellular processes due to its similarity to physiological conditions, recorded higher antioxidant activities compared to both ABTS and DPPH. Additionally, statistically significant differences were observed between the two solvents for FP and BP-ALK ($p < 0.05$), while no significant differences were detected for BP-ACD ($p > 0.05$). A study conducted by Zappia et al. (2023) evaluated the antioxidant profile of a phenolic extract from lemon by-products obtained after extracting juice and essential oils using a 50% hydroalcoholic mixture as the solvent [73]. Their findings are consistent with ours, as they also observed higher antioxidant activity in the ABTS assay (19.42 μmol TE/g) compared to the DPPH assay (8.25 μmol TE/g). Furthermore, our study is particularly innovative and significant, as it investigates the antioxidant activity within the lemon dietary fibre, following an integrated extraction process that includes essential oils, phenolic compound-rich extract, and pectin. This approach underscores the potential of LDF as a valuable source of antioxidant activity.

3.5.4. Antibacterial Activity

The antibacterial activity of lemon dietary fibre is presented in Table 5. As shown in Figure 6, LDF demonstrates a high water absorption capacity. When tested at a 6% concentration, the bacterial growth medium was completely absorbed, leading to the inhibition of bacterial growth. This result is likely due to the absence of a suitable growth medium rather than the intrinsic antibacterial activity of LDF. To address this issue, the antibacterial activity of LDF was evaluated at a lower concentration (3%), where sufficient medium was available for bacterial growth. The results showed that LDF exhibited the highest antibacterial activity against Gram-positive bacteria, particularly *B. cereus*, followed by *S. aureus*, which demonstrated a 0.3 log reduction in growth ($p < 0.05$). In contrast, the antibacterial activity against Gram-negative bacteria was much lower, with *E. coli* showing a 0.1-log reduction in growth ($p < 0.05$), while *S. enterica* was not significantly affected ($p > 0.05$). These findings suggest that LDF exhibits a low antibacterial effect, with a relatively greater impact on Gram-positive bacteria. Presentato et al. (2020) reported stronger antibacterial activity using pectin derived from lemon peel waste [74]. At a concentration of 3 mg/mL, a 1.5-log unit reduction for *S. aureus* was observed compared to the control, decreasing from 8.3 log CFU/mL in the control to 6.8 log CFU/mL with lemon pectin.

Table 5. Antibacterial activity of lemon dietary fibre-based powder.

		Log_{10} (CFU/mL)	Log Reduction
Gram-positive	*S. aureus* Control	9.5 ± 0.0 [a]	-
	S. aureus LDF (3%)	9.2 ± 0.0 [b]	0.3 ± 0.1
	B. cereus Control	9.8 ± 0.0 [a]	-
	B. cereus LDF (3%)	Total inhibition [b]	
Gram-negative	*E. coli* Control	9.6 ± 0.0 [a]	-
	E. coli LDF (3%)	9.5 ± 0.1 [b]	0.1 ± 0.0
	S. enterica Control	9.9 ± 0.0 [a]	-
	S. enterica LDF (3%)	9.9 ± 0.0 [a]	0.0 ± 0.0

Values followed by the same letter within the same column are not significantly different ($p > 0.05$), according to the *t*-test (non-parametric test) ($p > 0.05$). Data are presented as the mean values of replications ± SD, where n = 3.

3.6. Sausage Analysis

3.6.1. Emulsion Stability

The emulsion stability results presented in Figure 7 are expressed as the percentage of total expressible fluid (%TEF). This metric indicates the proportion of fluid that separates from the emulsion, serving as a measure of stability. A higher %TEF suggests lower stability, as more fluid is released, whereas lower %TEF values signify greater emulsion stability, with minimal fluid separation. The LDF (6%) sample exhibited a lower %TEF value (0.008%) than the other samples, indicating good emulsion stability. However, the differences in values are not significant ($p > 0.05$), probably due to the heterogeneous texture of the *mortadella*. Consequently, concentrations above 3% are required to significantly reduce fluid release. Other researchers have also assessed this parameter in meat products. Jeong and Han (2019) demonstrated that increasing the concentration of *Opuntia* fruit powder significantly reduced water and fat loss in sausages ($p < 0.05$) [75]. The group containing 10% of this powder exhibited the lowest water and fat loss, highlighting its effectiveness in improving sausage quality. Zhao et al. (2018) evaluated the effects of regenerated cellulose (0.4, 0.8, and 1.2%) on the emulsion stability of pork sausages [76]. The cellulose showed a high water-binding capacity, thereby enhancing emulsion stability.

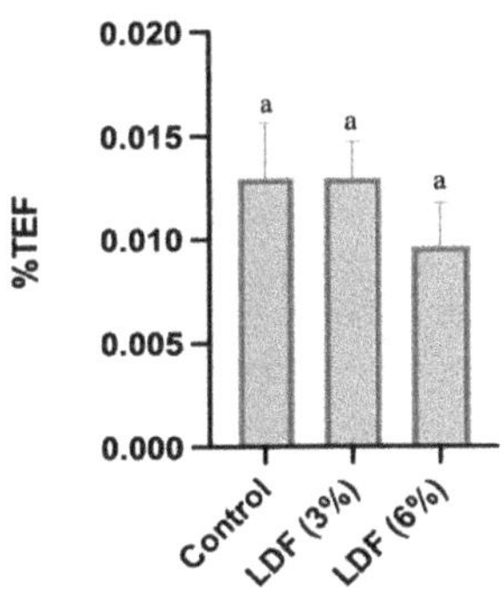

Figure 7. Effect of incorporation of lemon dietary fibre (3 and 6%) on emulsion stability of *mortadellas*. Results are expressed as % total expressible fluid (%TEF). For each parameter, results followed by same case letter are not significantly different according to Tukey's HSD post hoc test ($p > 0.05$). Data are presented as mean values of replications ± SD, where n = 3.

3.6.2. Chemical Composition

Table 6 shows the proximate composition of *mortadellas* formulated with 3% and 6% lemon dietary fibre. The addition of 6% lemon dietary fibre significantly ($p < 0.05$) reduced the moisture to 62.15%, compared to 66.48% in the control sample. In contrast, the levels of fat, protein, and ash did not exhibit significant differences ($p > 0.05$). Diao et al. (2024) recently investigated the effects of incorporating citrus fibres recovered from mandarin peels at concentrations of 1%, 2%, and 4% into frankfurter sausages [77]. The study revealed that significant differences in moisture and fat content were observed only in samples with 4% citrus fibre compared to the control. Ash content exhibited statistically significant differences in formulations with 2% and 4% citrus fibre relative to the control ($p < 0.05$), while protein content did not show any significant changes ($p > 0.05$).

Table 6. Chemical composition of *mortadellas* formulated with lemon dietary fibre (3% and 6%).

Formulation	Moisture (%)	Fat (%)	Protein (%)	Ash (%)
Control	66.48 ± 0.00 [a]	11.94 ± 0.05 [a]	14.75 ± 0.44 [a]	3.78 ± 0.00 [a]
LDF (3%)	63.65 ± 0.02 [ab]	11.75 ± 0.57 [a]	14.88 ± 0.53 [a]	3.46 ± 0.00 [a]
LDF (6%)	62.15 ± 0.00 [b]	11.32 ± 0.26 [a]	15.15 ± 1.26 [a]	3.53 ± 0.00 [a]

Control: *mortadella* without lemon by-products; LDF 3%: *mortadella* formulated with LDF at 3%; LDF 6%: *mortadella* formulated with LDF at 6%. Values followed by same letter within same column are not significantly different ($p > 0.05$), according to Tukey's HSD post hoc test ($p > 0.05$). Data are presented as mean values of replications ± SD, where n = 3.

3.6.3. Mineral Composition

The mineral composition of the *mortadella* samples is illustrated in the form of a heat map in Figure 8, which provides a detailed visualisation of each mineral's distribution and relative abundance across the samples. The mineral composition of the *mortadella* samples included zinc (Zn), phosphorus (P), sodium (Na), manganese (Mn), magnesium (Mg), potassium (K), iron (Fe), copper (Cu), and calcium (Ca). Among these, the highest concentration was for P, followed by Na and K. Statistically significant differences were observed in the levels of P and K across all samples ($p < 0.05$). However, Na exhibited significant differences between the control sample and those containing 3% and 6% of LDF but not between the two LDF treatments (3% and 6%). This indicates that while P and K varied consistently among all samples, Na content was specifically influenced by the presence and concentration of LDF. The incorporation of LDF into *mortadellas* significantly reduced Na levels ($p < 0.05$), highlighting its potential as a viable strategy for Na reduction. This is particularly important given the critical role of Na salts as additives in food processing, where they contribute to flavour enhancement, texture improvement, and food preservation. However, excessive Na intake is a global health concern associated with cardiovascular and cerebrovascular diseases [78]. Worldwide, sodium salt intake exceeds the maximum recommended level by the WHO (2012) (salt not exceeding 5 g/day, sodium not exceeding 2 g/day) [79]. In Europe, the average sodium salt intake is about 8–12 g/day. Efforts to reduce sodium levels face several challenges, including a decline in both their textural properties and characteristic flavour. Consequently, current research has focused on achieving sodium reduction without compromising saltiness or texture quality.

This study demonstrates that lemon dietary fibre can serve as a natural, clean-label alternative for lowering sodium levels in processed meat products. This approach addresses consumer demand for healthier options without compromising product quality, making it a promising solution in the ongoing pursuit of a balanced and healthy diet. Recent research by Powell et al. (2019) evaluated the effects of replacing sodium tripolyphosphate with citrus fibre (0.5%, 0.75%, and 1%) in a bologna sausage [29]. The results showed that this replacement did not significantly affect most physical, chemical, or sensory characteristics of the product during refrigerated storage.

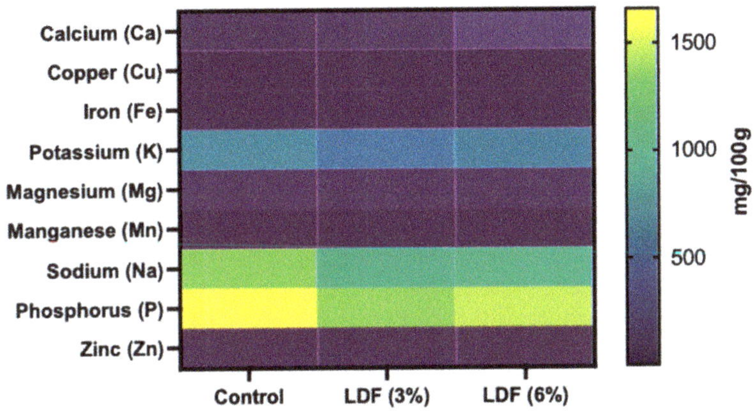

Figure 8. Heat map illustrating impact of incorporating lemon dietary fibre (3% and 6%) on mineral composition of *mortadellas*. Data are presented as mean values of replications ± SD, where n = 3.

3.6.4. Physicochemical Parameters

pH and Water Activity

The pH and water activity (a_w) values for the *mortadella* samples are presented in Table 7. The results indicate that the incorporation of LDF had a significant impact on the pH of the samples ($p < 0.05$). The natural acidity of the lemon fraction LDF contributed to a reduction in pH, resulting in more acidic *mortadella* compared to the control. Among samples, LDF at 6% had the lowest pH value of 5.36, followed by LDF at 3% with a pH of 5.66, while the control sample had the highest pH at 6.02. Although in this study we did not incorporate the whole lemon peel flour, these findings are in accordance with those reported by Ibrahim et al. (2018), who incorporated 1 and 2% dried lemon peel in beef patties and observed a reduction in pH values to 5.80 and 5.71, respectively, compared to the control sample, which had a pH of 6.01 [80]. This demonstrates the acidifying effect of lemon peel, consistent with its expected acidity. In contrast, a_w did not show significant differences among the samples ($p > 0.05$). This indicates that the addition of LDF, irrespective of its concentration, did not significantly alter the aw values of the *mortadella* formulations, probably due to the high water content present in the product. However, a study conducted by Viuda-Martos et al. (2010) reduced the aw to 0.87 by adding 1% of orange dietary fibre, compared to the control sample, which had an aw of 0.89, attributed because the *mortadellas* in this study showed a lower moisture content (61.27–61.32%) [22].

Table 7. pH and aw values of *mortadellas* formulated with lemon dietary fibre (3% and 6%).

Formulation	pH	a_w
Control	6.02 ± 0.02 [a]	0.965 ± 0.00 [a]
LDF (3%)	5.66 ± 0.03 [b]	0.965 ± 0.00 [a]
LDF (6%)	5.36 ± 0.03 [c]	0.963 ± 0.00 [a]

Control: *mortadella* without lemon by-products; LDF 3%: *mortadella* formulated with LDF at 3%; LDF 6%: *mortadella* formulated with LDF at 6%. Values followed by same letter within same column are not significantly different ($p > 0.05$), according to Tukey's HSD post hoc test ($p > 0.05$). Data are presented as mean values of replications ± SD, where n = 3.

Colour

Table 8 shows instrumental colour evaluation for *mortadellas* formulated with lemon dietary fibre. For lightness (L*) values, the treatment effects were statistically significant ($p < 0.05$), with higher L* values observed in samples containing higher concentrations of LDF. This indicates that the inclusion of LDF contributes to an increase in lightness, with the effect becoming more pronounced at higher concentrations. The redness (a*) significantly decreases with the addition of LDF ($p < 0.05$). The yellowness (b*) was affected by the addition of LDF at a concentration of 6%. The observed reduction in redness and the increase in yellowness can be attributed to the natural soft yellow hue of the LDF powder. Consequently, reformulated *mortadellas* appear paler than the control, as illustrated in Figure 2. This change reflects the influence of LDF on the visual properties of the final product, driven by its inherent colouration. The analysis of chroma (C*) and hue (h°) values in the *mortadella* samples suggests that incorporating LDF results in a product with a slightly yellowish colour and moderate chromatic intensity. These attributes indicate that the product appears visually appealing while maintaining a natural aspect, avoiding an artificial appearance.

Table 8. Colour coordinates of *mortadellas* formulated with lemon dietary fibre (3% and 6%).

Formulation	*L**	*a**	*b**	*C**	*h*°
Control	70.93 ± 0.79 a	6.96 ± 0.11 a	10.79 ± 0.20 a	12.84 ± 0.16 a	57.17 ± 0.73 a
LDF (3%)	73.41 ± 0.93 b	5.79 ± 0.21 b	12.45 ± 1.23 a	13.59 ± 1.08 a	66.29 ± 2.54 b
LDF (6%)	73.96 ± 0.93 b	5.43 ± 0.24 b	14.33 ± 0.57 b	15.46 ± 0.56 b	68.00 ± 0.97 b

Control: *mortadella* without lemon by-products; LDF 3%: *mortadella* formulated with LDF at 3%; LDF 6%: *mortadella* formulated with LDF at 6%. Values followed by same letter within same column are not significantly different ($p > 0.05$), according to Tukey's HSD post hoc test ($p > 0.05$). Data are presented as mean values of replications ± SD, where n = 9.

Powell et al. (2019) [29], with the addition of 1% citrus fibre, reported colour coordinates similar to those observed in our study for alternatively cured all-pork bologna sausage under different lighting conditions, including retail display lights and darkness. This addition resulted in consistent L*, a*, and b* values across both dark and retail display lighting conditions. These findings suggest that the incorporation of citrus fibre contributes to stable colour attributes across various lighting environments [29]. Another study conducted by Gedikoğlu and Clarke (2019) aligns with our findings, reporting that the incorporation of citrus fibre in ground beef meatballs led to a decrease in L* and a* values, while the b* value increased with the addition of 10% lemon fibre [81].

3.6.5. Textural Properties

The textural properties (TPA) of *mortadellas* are presented in Table 9. The incorporation of LDF significantly influenced the hardness of the samples ($p < 0.05$). However, no significant differences were observed between the 3% and 6% LDF samples. Hardness increased with higher concentrations of lemon dietary fibre. These results could considerably affect the sensory attributes of *mortadellas*, particularly for the LDF (6%), as greater hardness may make the product more challenging to consume and less appealing to consumers. Springiness was significantly affected only with the incorporation of 6% LDF in comparison to the control. In contrast, cohesiveness and chewiness were not significantly influenced by the addition of lemon dietary enriched fibre ($p > 0.05$). A recent study conducted by Sarıçoban and Unal (2022) examined the effect of orange albedo on the textural properties of fermented sausages [82]. The results of hardness align with this study, showing that higher concentrations of orange albedo (especially 2.5% and 5%) led to increased hardness.

The control sample had the lowest hardness, while formulations with greater citrus-based powder incorporation exhibited a clear increasing trend in hardness.

Table 9. Textural properties (TPA) of *mortadellas* formulated with lemon dietary fibre (3% and 6%).

Formulation	Hardness (N)	Cohesiveness	Springiness (mm)	Chewiness (N mm)
Control	33.79 ± 0.85 [a]	0.27 ± 0.03 [a]	0.39 ± 0.04 [a]	3.48 ± 0.60 [a]
LDF (3%)	41.40 ± 2.57 [b]	0.23 ± 0.03 [a]	0.33 ± 0.03 [ab]	3.57 ± 0.21 [a]
LDF (6%)	45.10 ± 2.65 [b]	0.24 ± 0.02 [a]	0.30 ± 0.01 [b]	2.45 ± 0.34 [a]

Control: *mortadella* without lemon by-products; LDF 3%: *mortadella* formulated with LDF at 3%; LDF 6%: *mortadella* formulated with LDF at 6%. Values followed by same letter within same column are not significantly different ($p > 0.05$), according to Tukey's HSD post hoc test ($p > 0.05$). Data are presented as mean values of replications ± SD, where n = 6.

3.6.6. Residual Nitrite Level and Lipid Oxidation

Figure 9 illustrates the interrelation between residual nitrite levels and lipid oxidation in reformulated *mortadellas*. Residual nitrite levels were significantly higher in the control sample, followed by LDF (3%) and LDF (6%) ($p < 0.05$). This research demonstrates that lemon dietary fibre effectively reduces residual nitrite levels, being a promising approach to lowering dietary nitrite intake. This reduction is particularly relevant in mitigating the potential formation of harmful n-nitroso compounds, known to carry carcinogenic, teratogenic, and mutagenic risks [83]. Conversely, lipid oxidation values revealed an inverse trend, with higher oxidation concentrations corresponding to lower residual nitrite levels. Lipid oxidation is a chemical process that deteriorates fats in meat, resulting in rancid flavours and unpleasant odours. This oxidation process is closely linked to the effects of nitrite, which is commonly used as a preservative in processed meats. Sodium nitrite plays a critical role in reducing TBARs (thiobarbituric acid reactive substances) values, which are lipid oxidation indicators, helping maintain meat quality over time. Lipid oxidation was minimal in the control, increasing in LDF (3%) and reaching the highest value in LDF (6%). The control sample differed significantly from the LDF (3%) and LDF (6%) samples ($p < 0.05$), but no significant difference was observed between the LDF (3%) and LDF (6%). Additionally, it is important to note that lipid oxidation levels in all reformulated *mortadellas* with lemon dietary fibre remained below the rancidity threshold (≥1.0). A study conducted by Karwowska et al. (2020) supports this observation, demonstrating that reducing the amount of sodium nitrite in the samples increased lipid oxidation [84]. Specifically, on day 0, the TBAR value was lower (0.43 mg/kg) in samples containing 150 mg/kg of nitrite compared to those without nitrite (0.88 mg/kg). Over the 15-day storage period, this trend was consistent, further emphasising the protective role of sodium nitrite against lipid oxidation in meat products. In addition, the biologically active compounds are not as effective as nitrites in decreasing lipid oxidation; however, the polyphenols found in fruit wines were responsible for the reduction in MDA levels [85].

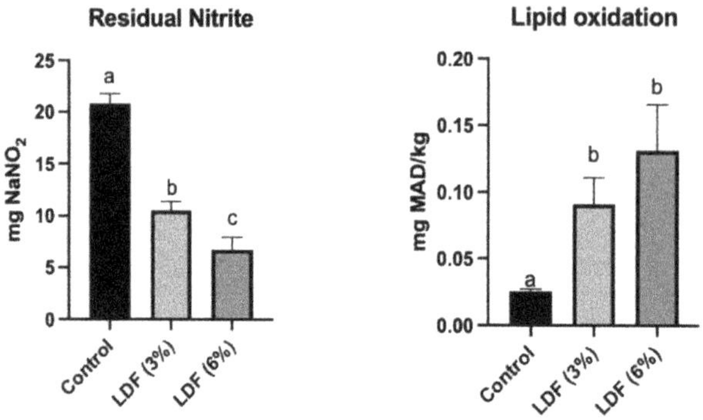

Figure 9. Residual nitrite (mg $NaNO_2$/sample) and lipid oxidation (mg MDA/kg sample) of mortadellas formulated with LDF (3% and 6%). For each parameter, results followed by same case letter are not significantly different according to Tukey's HSD post hoc test ($p > 0.05$). Data are presented as mean values of replications ± SD, where n = 3.

3.7. Sensory Evaluation

Figure 10 presents the sensory evaluation of reformulated *mortadellas* containing 3% and 6% fibre. The parameters evaluated included global and colour appearance, general quality, hardness, homogeneity, general flavour, acid, and bitter taste. All parameters showed significant differences between conditions ($p < 0.05$). Notably, the most affected parameters were flavour, homogeneity, hardness, and bitter taste. *Mortadella* labelled as a "source of fibre" (LDF (3%)) was more acceptable to the panellists compared to those labelled as having a "high content of fibre" (LDF (6%)). Panellists detected the bitter taste, normally associated with citrus products, in *mortadellas* with LDF (6%). Additionally, the increased fibre content impacted hardness, with the control being rated the most favourable (7.82), followed by LDF (3%) (7.23) and LDF (6%) (5.52). Homogeneity was another parameter significantly ($p < 0.05$) influenced by LDF concentration. Panellists preferred the control (7.61), followed by LDF (3%) (6.93), with LDF 6% being the least favoured (5.35). These results suggest that future studies should focus on developing *mortadellas* with 3% fibre, as this formulation was well accepted and maintained better sensory qualities compared to higher fibre concentrations.

Other studies have evaluated the impact of adding lemon co-products on the sensory properties of food products. For instance, Fernández-Ginés et al. (2004) studied two types of albedo (raw and cooked) at five concentrations (0%, 2.5%, 5%, 7.5%, and 10%) in bologna sausages [32]. The formulations that achieved sensory properties closest to conventional sausages were those containing 2.5% and 5% raw albedo, as well as 2.5%, 5%, and 7.5% cooked albedo. Similarly, Fu et al. (2015) evaluated the incorporation of lemon fibre (LF) recovered from lemon pomace (0%, 3%, 6%, and 9% LF) in dough and Mantou (steamed bread) to assess its effects on colour, flavour, texture, and overall acceptability [54]. The most pronounced sensory impacts were observed in products formulated with 6% and 9% LF, indicating that higher concentrations of LF may decrease sensory acceptability.

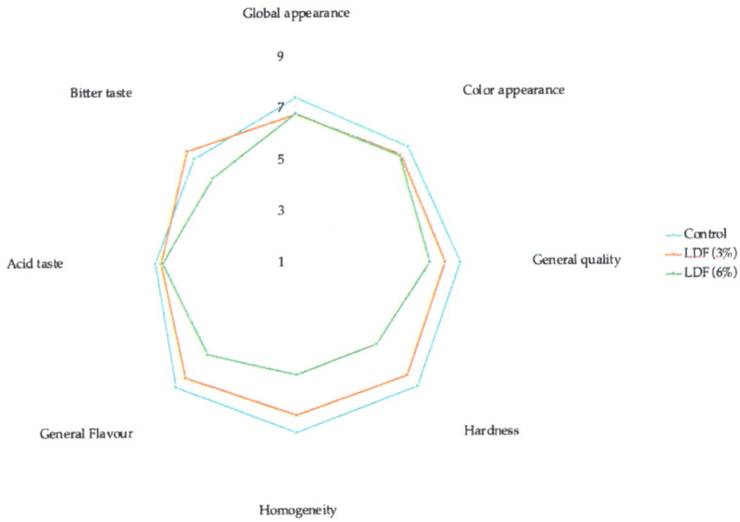

Figure 10. Sensory evaluation: A quantitative descriptive analysis was carried out on the different *mortadellas* reformulated with lemon dietary fibre (3% and 6%). Sensory score: 1—very unacceptable, 9—very acceptable.

4. Conclusions and Future Perspectives

Lemon co-products can be supplied as a potential functional and nutritional food ingredient containing abundant dietary fibre, as well as for their valuable functional properties, including technological and physicochemical benefits, along with preservation capabilities. The innovative LDF powder was obtained after the integrated extraction of (lemon hydrolat, essential oil, phenolic compounds extract, and pectin), which contains notable phenolic (free and bound) compounds that provide antioxidant properties and help reduce bacterial growth. These combined attributes make such powders highly suitable for selection as ingredients in the development of new food products. The incorporation of lemon dietary fibre (LDF) in *mortadellas* significantly influenced the colour, texture, and mineral composition of the products. Higher levels of LDF result in a paler colour and increased hardness in *mortadellas*. Additionally, LDF contributes to a reduction in sodium levels, which is highly beneficial for global population health. Furthermore, LDF reduces residual nitrite levels, though a slight increase in lipid oxidation accompanies this reduction; however, it remained below the rancidity threshold ($\geq$1.0) in all products, ensuring acceptable product quality. Sensory panellists evaluated the lemon dietary fibre-enriched *mortadellas* positively, demonstrating a distinct preference for the LDF (3%) formulation. In conclusion, this study highlights the potential of repurposing lemon co-products to develop value-added processed meat products with 3% and 6% LDF, which can be classified as "source of fibre" and "high in fibre", respectively. Additionally, further studies are recommended to assess the shelf-life of these products during storage, focusing on microbiological stability, texture, colour, lipid oxidation, and residual nitrite levels to ensure long-term quality and safety.

Author Contributions: Conceptualization, D.M., P.T. and M.P.; methodology, D.M., C.V.R., C.B.-M. and N.M.-T.; software, D.M.; validation, D.M., J.A.P.-Á., M.V.-M., P.T. and M.P.; formal analysis, D.M.; investigation, D.M., C.V.R., C.B.-M., N.M.-T., J.A.P.-Á., M.V.-M., P.T. and M.P.; resources, M.P.; data curation, D.M.; writing—original draft preparation, D.M.; writing—review and editing, D.M., C.V.R., C.B.-M., N.M.-T., J.A.P.-Á., M.V.-M., P.T. and M.P.; visualisation, D.M., C.V.R., C.B.-M., N.M.-T., J.A.P.-Á., M.V.-M., P.T. and M.P.; supervision, P.T. and M.P.; project administration, M.P.; funding acquisition, M.P. All authors have read and agreed to the published version of the manuscript.

Funding: This research was supported by the MEDISMART project—Mediterranean Citrus innovative soft processing solutions for S.M.A.R.T (Sustainable, Mediterranean, Agronomically evolved nutritionally enriched Traditional) products (reference number: 2019-SECTION2-20) and through Portuguese national funds from FCT—Fundação para a Ciência e Tecnologia (reference: PRIMA/0014/2019) and doctoral fellowship (reference 2022.12417.BD) to Daniela Magalhães.

Institutional Review Board Statement: The sensory analysis of this research article was performed ethically and according to the Guideline for Ethical and Professional Practices for the Sensory Analysis of Foods established by the Institute of Food Science and Technology (IFST), evaluated and supported by the Miguel Hernandez University Ethical Committee (reference number 2019.07.31 FPRL).

Informed Consent Statement: Informed consent was obtained from all subjects involved in the study.

Data Availability Statement: The data presented in this study are available on request from the corresponding author due to privacy.

Acknowledgments: The authors are grateful for the financial support of the MEDISMART project (reference number: 2019-SECTION2-20) and FCT—Fundação para a Ciência e Tecnologia (reference: PRIMA/0014/2019). In addition, the authors would like to thank the scientific collaboration of CBQF under the project UIDB/50016/2020. In addition, the author Daniela Magalhães would like to acknowledge FCT for the individual PhD grant (reference number: 2022.12417.BD).

Conflicts of Interest: The authors declare no conflicts of interest.

References

1. Rashwan, A.K.; Bai, H.; Osman, A.I.; Eltohamy, K.M.; Chen, Z.; Younis, H.A.; Al-Fatesh, A.; Rooney, D.W.; Yap, P.S. Recycling Food and Agriculture By-Products to Mitigate Climate Change: A Review. *Environ. Chem. Lett.* **2023**, *21*, 3351–3375. [CrossRef]
2. Magalhães, D.; Vilas-Boas, A.A.; Teixeira, P.; Pintado, M. Functional Ingredients and Additives from Lemon By-Products and Their Applications in Food Preservation: A Review. *Foods* **2023**, *12*, 1095. [CrossRef] [PubMed]
3. Vilas-boas, A.A.; Magalhães, D.; Campos, D.A.; Porretta, S.; Dellapina, G.; Poli, G.; Istanbullu, Y.; Demir, S.; San Martín, Á.M.; García-Gómez, P.; et al. Innovative Processing Technologies to Develop a New Segment of Functional Citrus-Based Beverages: Current and Future Trends. *Foods* **2022**, *11*, 3859. [CrossRef] [PubMed]
4. Rao, N.; Sharma, M.; Sharma, A. Development of Product Rich in Dietary Fiber and Antioxidant Prepared from Lemon Peel. *Glob. J. Biol. Agric. Health Sci.* **2016**, *5*, 120–123.
5. Yang, Y.Y.; Ma, S.; Wang, X.X.; Zheng, X.L. Modification and Application of Dietary Fiber in Foods. *J. Chem.* **2017**, *2017*, 9340427. [CrossRef]
6. Nie, Y.; Luo, F. Dietary Fiber: An Opportunity for a Global Control of Hyperlipidemia. *Oxidative Med. Cell. Longev.* **2021**, *2021*, 5542342. [CrossRef]
7. Terpstra, A.H.M.; Beynen, A.C.; Lapré, J.A.; De Vries, H.T.; De Vries, H.T. The Hypocholesterolemic effect of Lemon Peels, Lemon Pectin, and the Waste Stream material of Lemon Peels in Hybrid F_1B Hamsters. *Eur. J. Nutr.* **2002**, *41*, 19–26. [CrossRef]
8. Karaman, E.; Yılmaz, E.; Tuncel, N.B. Physicochemical, Microstructural and Functional Characterization of Dietary Fibers Extracted from Lemon, Orange and Grapefruit Seeds Press Meals. *Bioact. Carbohydr. Diet. Fibre* **2017**, *11*, 9–17. [CrossRef]
9. Borghi, S.; Pavanelli, W.R.; Kumar Saini, R.; Ranjit, A.; Sharma, K.; Prasad, P.; Shang, X.; Girinur Mallikarjuna Gowda, K.; Keum, Y.-S. Bioactive Compounds of Citrus Fruits: A Review of Composition and Health Benefits of Carotenoids, Flavonoids, Limonoids, and Terpenes. *Antioxidants* **2022**, *11*, 239. [CrossRef]
10. Rocchetti, G.; Gregorio, R.P.; Lorenzo, J.M.; Barba, F.J.; Oliveira, P.G.; Prieto, M.A.; Simal-Gandara, J.; Mosele, J.I.; Motilva, M.J.; Tomas, M.; et al. Functional Implications of Bound Phenolic Compounds and Phenolics–Food Interaction: A Review. *Compr. Rev. Food Sci. Food Saf.* **2022**, *21*, 811–842. [CrossRef]
11. Angulo-López, J.E.; Flores-Gallegos, A.C.; Ascacio-Valdes, J.A.; Contreras Esquivel, J.C.; Torres-León, C.; Rúelas-Chácon, X.; Aguilar, C.N. Antioxidant Dietary Fiber Sourced from Agroindustrial Byproducts and Its Applications. *Foods* **2023**, *12*, 159. [CrossRef] [PubMed]
12. Lario, Y.; Sendra, E.; García-Pérez, J.; Fuentes, C.; Sayas-Barberá, E.; Fernández-López, J.; Pérez-Alvarez, J.A. Preparation of High Dietary Fiber Powder from Lemon Juice By-Products. *Innov. Food Sci. Emerg. Technol.* **2004**, *5*, 113–117. [CrossRef]
13. Pathania, S.; Kaur, N. Utilization of Fruits and Vegetable By-Products for Isolation of Dietary Fibres and Its Potential Application as Functional Ingredients. *Bioact. Carbohydr. Diet. Fibre* **2022**, *27*, 100295. [CrossRef]
14. Barber, T.M.; Kabisch, S.; Pfei, A.F.H.; Weickert, M.O. The Health Benefits of Dietary Fibre. *Nutrients* **2020**, *12*, 3209. [CrossRef] [PubMed]

15. Makki, K.; Deehan, E.C.; Walter, J.; Bäckhed, F. The Impact of Dietary Fiber on Gut Microbiota in Host Health and Disease. *Cell Host Microbe* **2018**, *23*, 705–715. [CrossRef]
16. Thomson, C.; Garcia, A.L.; Edwards, C.A. Interactions between Dietary Fibre and the Gut Microbiota. *Proc. Nutr. Soc.* **2021**, *80*, 398–408. [CrossRef]
17. EFSA. Scientific Opinion on Dietary Reference Values for Carbohydrates and Dietary Fibre. *EFSA J.* **2016**, *8*, 1462. [CrossRef]
18. World Health Organization. *WHO Guideline Carbohydrate Intake for Adults and Children*; World Health Organization: New Delhi, India, 2023.
19. Venter, C.; Meyer, R.W.; Greenhawt, M.; Pali-Schöll, I.; Nwaru, B.; Roduit, C.; Untersmayr, E.; Adel-Patient, K.; Agache, I.; Agostoni, C.; et al. Role of Dietary Fiber in Promoting Immune Health—An EAACI Position Paper. *Eur. J. Allergy Clin. Immunol.* **2022**, *77*, 3185–3198. [CrossRef]
20. Kaczmarczyk, M.M.; Miller, M.J.; Freund, G.G. The Health Benefits of Dietary Fiber: Beyond the Usual Suspects of Type 2 Diabetes Mellitus, Cardiovascular Disease and Colon Cancer. *Metabolism* **2012**, *61*, 1058–1066. [CrossRef]
21. Fernández-López, J.; Cruz López-Marcos, M.; Bailina, C.; Viuda-Martos, M.; Angel Pérez-Alvarez, J. Properties of Dietary Fibers from Agroindustrial Coproducts as Source for Fiber-Enriched Foods. *Food Bioprocess Technol.* **2015**, *8*, 2400–2408. [CrossRef]
22. Viuda-Martos, M.; Ruiz-Navajas, Y.; Fernández-López, J.; Pérez-Álvarez, J.A. Effect of Added Citrus Fibre and Spice Essential Oils on Quality Characteristics and Shelf-Life of Mortadella. *Meat Sci.* **2010**, *85*, 568–576. [CrossRef] [PubMed]
23. Viuda-Martos, M.; Ruiz-Navajas, Y.; Fernández-López, J.; Pérez-Álvarez, J.A. Effect of Packaging Conditions on Shelf-Life of Mortadella Made with Citrus Fibre Washing Water and Thyme or Rosemary Essential Oil. *Food Nutr. Sci.* **2011**, *2*, 1–10. [CrossRef]
24. Das, A.K.; Nanda, P.K.; Madane, P.; Biswas, S.; Das, A.; Zhang, W.; Lorenzo, J.M. A Comprehensive Review on Antioxidant Dietary Fibre Enriched Meat-Based Functional Foods. *Trends Food Sci. Technol.* **2020**, *99*, 323–336. [CrossRef]
25. FAO. *Urbanization Is Transforming Agrifood Systems across the Rural–Urban Continuum Creating Challenges and Opportunities to Access Affordable Healthy Diets*; FAO: Rome, Italy, 2023; ISBN 978-92-5-138222-6.
26. Michel, M.; Eldridge, A.L.; Hartmann, C.; Klassen, P.; Ingram, J.; Meijer, G.W. Benefits and Challenges of Food Processing in the Context of Food Systems, Value Chains and Sustainable Development Goals. *Trends Food Sci. Technol.* **2024**, *153*, 104703. [CrossRef]
27. López-Vargas, J.H.; Fernández-López, J.; Pérez-Álvarez, J.Á.; Viuda-Martos, M. Quality Characteristics of Pork Burger Added with Albedo-Fiber Powder Obtained from Yellow Passion Fruit (*Passiflora edulis* Var. *Flavicarpa*) Co-Products. *Meat Sci.* **2014**, *97*, 270–276. [CrossRef] [PubMed]
28. Petridis, D.; Raizi, P.; Ritzoulis, C. Influence of Citrus Fiber, Rice Bran and Collagen on the Texture and Organoleptic Properties of Low-Fat Frankfurters. *J. Food Process Preserv.* **2014**, *38*, 1759–1771. [CrossRef]
29. Powell, M.J.; Sebranek, J.G.; Prusa, K.J.; Tarté, R. Evaluation of Citrus Fiber as a Natural Replacer of Sodium Phosphate in Alternatively-Cured All-Pork Bologna Sausage. *Meat Sci.* **2019**, *157*, 107883. [CrossRef]
30. Soncu, E.D.; Kolsarıcı, N.; Cicek, N.; Öztürk, G.S.; Arıcı, Y.K. The Comparative Effect of Carrot and Lemon Fiber as a Fat Replacer on Physico-Chemical, Textural, and Organoleptic Quality of Low-Fat Beef Hamburger. *Korean J. Food Sci. Anim. Resour.* **2015**, *35*, 370–381. [CrossRef]
31. Song, J.; Pan, T.; Wu, J.; Ren, F. The Improvement Effect and Mechanism of Citrus Fiber on the Water-Binding Ability of Low-Fat Frankfurters. *J. Food Sci. Technol.* **2016**, *53*, 4197–4204. [CrossRef]
32. Fernández-Ginés, J.M.; Fernández-López, J.; Sayas-Barberá, E.; Sendra, E.; Pérez-Álvarez, J.A. Lemon Albedo as a New Source of Dietary Fiber: Application to Bologna Sausages. *Meat Sci.* **2004**, *67*, 7–13. [CrossRef]
33. Haque, A.; Ahmad, S.; Azad, Z.R.A.A.; Adnan, M.; Ashraf, S.A. Incorporating Dietary Fiber from Fruit and Vegetable Waste in Meat Products: A Systematic Approach for Sustainable Meat Processing and Improving the Functional, Nutritional and Health Attributes. *PeerJ* **2023**, *11*, e14977. [CrossRef]
34. Koishybayeva, A.; Korzeniowska, M. Utilization and Effect of Apple Pomace Powder on Quality Characteristics of Turkey Sausages. *Foods* **2024**, *13*, 2807. [CrossRef] [PubMed]
35. Zhao, D.; Guo, C.; Liu, X.; Xiao, C. Effects of Insoluble Dietary Fiber from Kiwi Fruit Pomace on the Physicochemical Properties and Sensory Characteristics of Low-Fat Pork Meatballs. *J. Food Sci. Technol.* **2021**, *58*, 1524–1537. [CrossRef] [PubMed]
36. AOAC. *Official Methods of Analysis*, 15th ed.; AOAC: Rockville, MD, USA, 1990.
37. Ribeiro, T.B.; Oliveira, A.L.; Costa, C.; Nunes, J.; Vicente, A.A.; Pintado, M. Total and Sustainable Valorisation of Olive Pomace Using a Fractionation Approach. *Appl. Sci.* **2020**, *10*, 6785. [CrossRef]
38. Sluiter, A.; Hames, B.; Ruiz, R.; Scarlata, C.; Sluiter, J.; Templeton, D.; Crocker, D. *Determination of Structural Carbohydrates and Lignin in Biomass*; Laboratory Analytical Procedure (LAP) (Revised July 2011); NREL: Golden, CO, USA, 2008.
39. King, D.A. *AMSA Meat Color Measurement Guidelines*; American Meat Science Association: Kearney, MO, USA, 2012.
40. Chau, C.F.; Huang, Y.L. Comparison of the Chemical Composition and Physicochemical Properties of Different Fibers Prepared from the Peel of *Citrus sinensis* L. Cv. Liucheng. *J. Agric. Food Chem.* **2003**, *51*, 2615–2618. [CrossRef] [PubMed]

41. Tinh, N.T.T.; Sitolo, G.C.; Yamamoto, Y.; Suzuki, T. *Citrus limon* Peel Powder Reduces Intestinal Barrier Defects and Inflammation in a Colitic Murine Experimental Model. *Foods* **2021**, *10*, 240. [CrossRef]
42. Vilas-boas, A.A.; Campos, D.A.; Nunes, C.; Ribeiro, S. Polyphenol Extraction by Different Techniques for Valorisation of Non-Compliant Portuguese Sweet Cherries towards a Novel Antioxidant Extract. *Sustainability* **2020**, *12*, 5556. [CrossRef]
43. Oliveira, C.M.; Barros, A.S.; Silva Ferreira, A.C.; Silva, A.M.S. Influence of the Temperature and Oxygen Exposure in Red Port Wine: A Kinetic Approach. *Food Res. Int.* **2015**, *75*, 337–347. [CrossRef]
44. Bondet, V.; Brand-Williams, W.; Berset, C. Kinetics and Mechanisms of Antioxidant Activity Using the DPPH·Free Radical Method. *LWT—Food Sci. Technol.* **1997**, *30*, 609–615. [CrossRef]
45. Dávalos, A.; Gómez-Cordovés, C.; Bartolomé, B. Extending Applicability of the Oxygen Radical Absorbance Capacity (ORAC-Fluorescein) Assay. *J. Agric. Food Chem.* **2004**, *52*, 48–54. [CrossRef]
46. Ubeda, C.; Hidalgo, C.; Torija, M.J.; Mas, A.; Troncoso, A.M.; Morales, M.L. Evaluation of Antioxidant Activity and Total Phenols Index in Persimmon Vinegars Produced by Different Processes. *LWT* **2011**, *44*, 1591–1596. [CrossRef]
47. Botella-Martínez, C.; Viuda-Martos, M.; Angel Pérez-Alvarez, J.; Fernández-López, J. Total and Partial Fat Replacement by Gelled Emulsion (Hemp Oil and Buckwheat Flour) and Its Impact on the Chemical, Technological and Sensory Properties of Frankfurters. *Foods* **2021**, *10*, 1681. [CrossRef]
48. AOAC International. *Official Methods of Analysis of AOAC International*; AOAC International: Gaithersburg, MD, USA, 2012.
49. Bourne, M.C. Texture Profile Analysis. *Food Technol.* **1978**, *32*, 62–72.
50. *ISO 2918:1975*; ISO International Standard 2918. Meat and Meat Products: Determination of Nitrite Content. International Organization for Standardization: Genevè, Switzerland, 1975.
51. Buege, J.A.; Aust, S.D. Microsomal Lipid Peroxidation. *Methods Enzymol.* **1978**, *52*, 302–310.
52. Nempeque, L.V.J.; Cabrera, A.P.G.; Moncayo, J.Y.C. Evaluation of Tahiti Lemon Shell Flour (*Citrus latifolia* Tanaka) as a Fat Mimetic. *J. Food Sci. Technol.* **2021**, *58*, 720–730. [CrossRef]
53. Wang, L.; Xu, H.; Yuan, F.; Pan, Q.; Fan, R.; Gao, Y. Physicochemical Characterization of Five Types of Citrus Dietary Fibers. *Biocatal. Agric. Biotechnol.* **2015**, *4*, 250–258. [CrossRef]
54. Fu, J.T.; Chang, Y.H.; Shiau, S.Y. Rheological, Antioxidative and Sensory Properties of Dough and Mantou (Steamed Bread) Enriched with Lemon Fiber. *LWT* **2015**, *61*, 56–62. [CrossRef]
55. Xiao, R.; Qi, J.; Liao, J.; Wei, H.; Zhuo, T. Preparation of Citrus Fiber from Lemon Peel Residue: Effects of Structure and Endogenous Pectin Components on Emulsifying Properties of Citrus Fiber. *Int. J. Biol. Macromol.* **2024**, *283*, 137679. [CrossRef]
56. Parra, M.; Hortelano, D.; García-Sánchez, F.; Intrigliolo, D.S.; Rubio-Asensio, J.S. Effects of Drip Irrigation Design on a Lemon and a Young Persimmon Orchard in Semi-Arid Conditions. *Water* **2021**, *13*, 1795. [CrossRef]
57. Sánchez-Bravo, P.; Noguera-Artiaga, L.; Martínez-Tomé, J.; Hernández, F.; Sendra, E. Effect of Organic and Conventional Production on the Quality of Lemon "Fino 49". *Agronomy* **2022**, *12*, 980. [CrossRef]
58. Idamokoro, E.M.; Hosu, Y.S.; Oyedeji, O.O.; Miya, G.M.; Kuria, S.K.; Oyedeji, A.O. A Comparative Analysis of the Proximate and Mineral Composition of Whole *Citrus limon* and *Citrus clementina* as a Prospective Alternative Feed Resource for Livestock Farming in South Africa. *Front. Sustain. Food Syst.* **2022**, *6*, 1021175. [CrossRef]
59. González-Molina, E.; Domínguez-Perles, R.; Moreno, D.A.; García-Viguera, C. Natural Bioactive Compounds of *Citrus limon* for Food and Health. *J. Pharm. Biomed. Anal.* **2010**, *51*, 327–345. [CrossRef]
60. Shkembi, B.; Huppertz, T. Calcium Absorption from Food Products: Food Matrix Effects. *Nutrients* **2022**, *14*, 180. [CrossRef]
61. Panel, E.; Nda, A. Scientific Opinion on Dietary Reference Values for Phosphorus. *EFSA J.* **2015**, *13*, 4185. [CrossRef]
62. Czech, A.; Zarycka, E.; Yanovych, D.; Zasadna, Z.; Grzegorczyk, I.; Kłys, S. Mineral Content of the Pulp and Peel of Various Citrus Fruit Cultivars. *Biol. Trace Elem. Res.* **2020**, *193*, 555–563. [CrossRef] [PubMed]
63. Tapia, M.S.; Alzamora, S.M.; Chirife, J. Effects of Water Activity (Aw) on Microbial Stability as a Hurdle in Food Preservation. In *Water Activity in Foods: Fundamentals and Applications*, 2nd ed.; Wiley: Hoboken, NJ, USA, 2020; pp. 323–355.
64. Bakshi, G.; Ananthanarayan, L. Characterization of Lemon Peel Powder and Its Application as a Source of Pectin Degrading Enzyme in Clarification of Cloudy Apple Juice. *J. Food Sci. Technol.* **2022**, *59*, 2535–2544. [CrossRef]
65. Peng, F.; Ren, X.; Du, B.; Chen, L.; Yu, Z.; Yang, Y. Structure, Physicochemical Property, and Functional Activity of Dietary Fiber Obtained from Pear Fruit Pomace (Pyrus Ussuriensis Maxim) via Different Extraction Methods. *Foods* **2022**, *11*, 2161. [CrossRef]
66. Canteri, M.H.G.; Renard, C.M.G.C.; Le Bourvellec, C.; Bureau, S. ATR-FTIR Spectroscopy to Determine Cell Wall Composition: Application on a Large Diversity of Fruits and Vegetables. *Carbohydr. Polym.* **2019**, *212*, 186–196. [CrossRef]
67. Jiang, Z.; Zhang, M.; Huang, Y.; Ma, C.; Mu, S.; Li, H.; Liu, X.; Ma, Y.; Liu, Y.; Hou, J. Comparison and Characterization of the Structure and Physicochemical Properties of Three Citrus Fibers: Effect of Ball Milling Treatment. *Foods* **2022**, *11*, 2665. [CrossRef]
68. Núñez-Gómez, V.; San Mateo, M.; González-Barrio, R.; Periago, M.J. Chemical Composition, Functional and Antioxidant Properties of Dietary Fibre Extracted from Lemon Peel after Enzymatic Treatment. *Molecules* **2024**, *29*, 269. [CrossRef]
69. Huang, J.; Liao, J.; Qi, J.; Jiang, W.; Yang, X.Q. Structural and Physicochemical Properties of Pectin-Rich Dietary Fiber Prepared from Citrus Peel. *Food Hydrocoll.* **2021**, *110*, 106140. [CrossRef]

70. Al-Qassabi, J.S.A.; Weli, A.M.; Hossain, M.A. Comparison of Total Phenols Content and Antioxidant Potential of Peel Extracts of Local and Imported Lemons Samples. *Sustain. Chem. Pharm.* **2018**, *8*, 71–75. [CrossRef]
71. Carboni Martins, C.; Rodrigues, R.C.; Domeneghini Mercali, G.; Rodrigues, E. New Insights into Non-Extractable Phenolic Compounds Analysis. *Food Res. Int.* **2022**, *157*, 111487. [CrossRef] [PubMed]
72. Durmus, N.; Kilic-Akyilmaz, M. Bioactivity of Non-Extractable Phenolics from Lemon Peel Obtained by Enzyme and Ultrasound Assisted Extractions. *Food Biosci.* **2023**, *53*, 102571. [CrossRef]
73. Zappia, A.; Spanti, A.; Princi, R.; Imeneo, V.; Piscopo, A. Evaluation of the Efficacy of Antioxidant Extract from Lemon By-Products on Preservation of Quality Attributes of Minimally Processed Radish (*Raphanus sativus* L.). *Antioxidants* **2023**, *12*, 235. [CrossRef]
74. Presentato, A.; Scurria, A.; Albanese, L.; Lino, C.; Sciortino, M.; Pagliaro, M.; Zabini, F.; Meneguzzo, F.; Alduina, R.; Nuzzo, D.; et al. Superior Antibacterial Activity of Integral Lemon Pectin Extracted via Hydrodynamic Cavitation. *ChemistryOpen* **2020**, *9*, 628–630. [CrossRef]
75. Jeong, Y.; Han, Y. Effect on the Emulsification Stability and Quality of Emulsified Sausages Added with Wanggasi-Chunnyuncho (*Opuntia humifusa* f. *Jeollaensis*) Fruit Powders. *Food Sci. Anim. Resour.* **2019**, *39*, 953–965. [CrossRef]
76. Zhao, Y.; Hou, Q.; Zhuang, X.; Wang, Y.; Zhou, G.; Zhang, W. Effect of Regenerated Cellulose Fiber on the Physicochemical Properties and Sensory Characteristics of Fat-Reduced Emulsified Sausage. *LWT* **2018**, *97*, 157–163. [CrossRef]
77. Diao, X.; Zhu, J.; Huang, L.; Li, S.; Mao, X.; Li, C.; Ke, W. Effect of Citrus Fiber on Physicochemical Quality of Frankfurter Sausages: A Study on Lipid Oxidation and Protein Gel Characteristics. *LWT* **2024**, *209*, 116778. [CrossRef]
78. Wang, J.; Huang, X.H.; Zhang, Y.Y.; Li, S.; Dong, X.; Qin, L. Effect of Sodium Salt on Meat Products and Reduction Sodium Strategies—A Review. *Meat Sci.* **2023**, *205*, 109296. [CrossRef]
79. World Health Organization. *WHO Guideline Sodium Intake for Adults and Children*; World Health Organization: New Delhi, India, 2012.
80. Ibrahim, H.M.; Hassan, I.M.; Hamed, A.A. Application of Lemon and Orange Peels in Meat Products: Quality and Safety. *Int. J. Curr. Microbiol. Appl. Sci.* **2018**, *7*, 2703–2723. [CrossRef]
81. Gedikoğlu, A.; Clarke, A.D. Quality Attributes of Citrus Fiber Added Ground Beef and Consumer Acceptance of Citrus Fiber Added Turkish Meat-Balls. *Food Health* **2019**, *5*, 205–214. [CrossRef]
82. Sarıçoban, C.; Unal, K. Influence of Pre-Treated Bitter Orange Albedo on the Physicochemical, Textural and Sensory Properties of Fermented Sausages (Sucuk). *J. Food Sci. Technol.* **2022**, *59*, 1478–1486. [CrossRef] [PubMed]
83. Shakil, M.H.; Trisha, A.T.; Rahman, M.; Talukdar, S.; Kobun, R.; Huda, N.; Zzaman, W. Nitrites in Cured Meats, Health Risk Issues, Alternatives to Nitrites: A Review. *Foods* **2022**, *11*, 3355. [CrossRef] [PubMed]
84. Karwowska, M.; Kononiuk, A.; Wójciak, K.M. Impact of Sodium Nitrite Reduction on Lipid Oxidation and Antioxidant Properties of Cooked Meat Products. *Antioxidants* **2020**, *9*, 9. [CrossRef]
85. Čakar, U.; Čolović, M.; Milenković, D.; Pagnacco, M.; Maksimović, J.; Krstić, D.; Đorđević, B. Strawberry and Drupe Fruit Wines Antioxidant Activity and Protective Effect Against Induced Oxidative Stress in Rat Synaptosomes. *Antioxidants* **2025**, *14*, 155. [CrossRef]

foods

MDPI

Article

Implicit and Explicit Consumer Perceptions of Cashews: A Neuroscientific and Sensory Analysis Approach

Rocio Lopez-Navarro *, Luis Montero-Vicente, Carmen Escriba-Perez and Juan M. Buitrago-Vera

Department of Economics and Social Science, Universitat Politècnica de València, Camino de Vera w/o No., 46022 Valencia, Spain; luimonvi@upv.es (L.M.-V.); carespe@upv.es (C.E.-P.); jmbuitrago@upv.es (J.M.B.-V.)
* Correspondence: rolona2@upv.es

Abstract: This study investigated consumer perceptions of raw cashew nuts from two different private labels (private label A, PLA, and private label B, PLB), employing a combination of explicit (sensory analysis) and implicit (consumer neuroscience) methods. The objective was to analyse both conscious and unconscious responses to understand consumer preferences. Participants (n = 80) evaluated the samples, with electroencephalography (EEG) and electrodermal activity (EDA) as implicit methods, and hedonic scales, JAR scales, and the EsSense25 questionnaire used for explicit evaluations. The results revealed a clear preference for PLB, supported by higher global hedonic scores and a significant majority (65%) choosing PLB over PLA. EEG metrics calculated for participants' valence, frontal alpha asymmetry (FAA) for flavour indicated greater activity in the left frontal lobe for PLB, associated with positive emotions. Task engagement (TE) measurements revealed higher engagement with PLB during flavour evaluation. Penalty analysis identified that PLA was mainly penalised for a "too weak" aroma and flavour. The EsSense25 analysis showed that cashew consumption evoked predominantly positive emotions such as "pleasant", "satisfied", and "calm". In conclusion, the combination of implicit and explicit methods provided a comprehensive understanding of consumer preferences, highlighting the value of both approaches and the importance of sensory attributes in driving the overall liking of raw cashews. The findings have implications for product optimisation, market segmentation, and the development of marketing strategies in the cashew industry.

Keywords: cashews; consumer neuroscience; sensory analysis; consumer food preferences; implicit measures; explicit measures; EsSense Profile®

1. Introduction

Cashew nuts have gained increasing economic, nutritional, and consumer importance in recent years [1,2]. From an economic perspective, the global cashew nut market was valued at USD 5.95 billion in 2023, with a projected annual growth of 5.4% until 2030 [3,4]. This growth is largely due to increasing awareness of their nutritional benefits. They are also a rich source of monounsaturated fats, protein, vitamins and essential minerals such as magnesium, phosphorus, and zinc, which contribute to brain and cardiovascular health [5,6]. In addition, recent studies have shown that regular consumption of cashew nuts can help lower LDL cholesterol and triglycerides, thereby reducing the risk of cardiovascular disease [7]. However, as with many nuts, cashews have been found to contain mycotoxins, which may present a chemical hazard and compromise human health [8]. Studies have indicated that the concentrations of these toxins are generally lower in cashews when compared to other nuts, such as peanuts or walnuts, and that they do not pose a significant food safety risk [9,10].

Nevertheless, this parameter must be considered when evaluating the quality of this product in order to ensure that mycotoxin levels are safe for consumers. From a consumer perspective, the perception of cashews as a healthy and versatile snack has driven their demand, especially among health-conscious consumers seeking nutritious alternatives to traditional snacks [11,12]. This combination of economic, nutritional, and consumer preference factors has positioned cashews as a food of growing importance in the global health food market [13,14].

The development of novel foods depends, to a large extent, on knowing consumer preferences, which are influenced by a complex set of emotions, attitudes, or values that are difficult to measure. Some of the existing methods, such as questionnaires or interviews, generally measure conscious and rational responses to any stimulus by relying on different tools, such as the use of response scales, such as hedonic or just-about-right scales, when the objective is to assess consumer preferences [15,16]. But these methods have limitations, as they depend on people's ability and willingness to accurately report their attitudes and/or motivations, leading to biases in many cases, especially when the responses expected to be collected have a high emotional component [17]. Therefore, unconscious measures are increasingly used to assess consumer response, as they provide a more comprehensive view of people's reactions. The physiological mechanisms that consumers present in response to a stimulus can be so fast that they themselves are not aware of them; thus, through different neuroscience techniques, it is possible to measure emotions and spontaneous reactions of consumers and reflect them in a more unbiased and meaningful way [18].

Consumer neuroscience is a discipline that studies behaviour in response to a stimulus using biometric techniques [19], such as electrodermal activity (EDA), electroencephalography (EEG), eye tracking, and heart rate measurement. These tools can provide simple psychophysiological measures, such as EDA, which simply measures changes in skin conductance produced by the excitation of the sympathetic nervous system and directly associated with variations in the electricity transmitted by nerves and sweating [20]. An increase in skin electrical conductance signals is associated with increased arousal, generated by a state of alertness in the body, and is therefore used as an indication that the stimulus provokes an emotion in the consumer. Conversely, a decrease in this parameter reflects a lack of arousal [21].

EEG is a non-invasive technique that measures electrical changes in the cerebral cortex, providing information about emotional states [22]. This technique is distinguished by having a high temporal resolution in milliseconds [23] that accurately measures the response to changing stimuli. EEG measures the subject's brain activity to subsequently determine frontal alpha asymmetry (hereafter FAA) as a marker of affinity or rejection of stimuli (valence) [24]. FAA measures changes in the alpha brain rhythm (8–13 Hz) of the frontal lobe and is based on the fact that emotions related to rejection are detected in the right hemisphere while emotions associated with acceptance are detected in the left hemisphere [25]. These unconscious methods must be combined with conscious methodologies, such as questionnaires or other commercial research techniques [26].

On the other hand, to identify the motivations or emotions evoked by food, there are psychographic methods that measure consumers' emotional responses, such as the EsSense Profile®. The EsSense Profile® was specifically designed for the emotional evaluation of food. It consists of a questionnaire developed by [27], which includes 39 emotions classified as positive, negative, or neutral. After consuming the food, consumers are asked to identify which emotions they experienced by marking their responses on the questionnaire.

In recent years, the integration of explicit methodologies—such as questionnaires and the EsSense Profile—with implicit methodologies for studying food consumers has resulted in numerous studies across a wide variety of food products. For example, [28] compared three cider brands, two of which had a protected designation of origin and one that did not, using eye-tracking data analysis (EDA) and the EsSense Profile questionnaire. Similarly, [29] investigated consumer responses to five beer samples in different contexts, including a laboratory and a restaurant setting.

A study [30] focused on beverages, specifically various types of vegetable juices, and examined how purchase intention affects consumers' choices. Research on solid foods has examined chocolate pudding, focusing on consumer reactions to varying sucrose concentrations [31], as well as the effects of presentation formats for textured foods targeted at the senior population [32]. Additionally, some studies have measured implicit responses during different cooking phases across various meals [33,34] and explored consumer reactions to unfamiliar meals [35].

Despite the complexity and cost associated with neuroscientific techniques such as EEG (electroencephalography) and EDA (electrodermal activity), their application in consumer research is becoming increasingly justified. As technology advances, these tools are becoming more affordable and user-friendly, making them more accessible to both researchers and marketers [26].

Furthermore, the use of these techniques aligns with a growing trend in consumer food research that seeks to move beyond simple product liking and to explore a deeper understanding of consumer preferences [15]. By providing objective measures of unconscious responses, EEG and EDA offer valuable and reliable insights that complement traditional explicit measures [36]. This allows for a more comprehensive understanding of consumer behaviour.

This holistic approach is particularly relevant in the competitive food market, where recognising subtle differences in consumer perceptions can provide a significant advantage.

As mentioned previously, using both explicit and implicit methodologies in food consumer research is increasingly common; however, there is a lack of studies focusing specifically on nuts, especially cashews. This gap is significant given the increasing importance of nuts in the global health food market. Our study addresses this void by examining consumer perceptions of raw cashews from two supermarket private labels, using a combination of neuroscience techniques (EEG and EDA) and traditional explicit methods.

This approach contributes to the broader field of food consumer research and offers valuable insights that were previously unavailable, specifically for the nut industry. By applying these advanced techniques to cashew research, we create new opportunities to understand consumer preferences within the often-overlooked nut sector, which could lead to more effective marketing strategies.

Finally, the aims of this study were as follows:

- Examine the consumer's unconscious response with two neuroscience techniques, EEG and EDA;
- analyse the consumer's conscious response, both emotionally and in terms of acceptance;
- distinguish whether there are differences in consumer perception, both conscious and unconscious, between two private labels from to different supermarkets;
- analyse the possible food marketing implications of the combined use of implicit neuroscience-based methods in conjunction with explicit methods.

2. Materials and Methods

2.1. Participants

Following approval by the Institutional Review Board of Universitat Politècnica de València (P11_26-09-2023) in València, a total of 80 participants (18–64 years old and 54% female) including students and staff were recruited by means of convenience sampling. Because the study used techniques that collect neurophysiological data, participants should not be under treatment with any psychotropic drugs, should not suffer from neurological diseases [33], and should be right-handed to obtain a more homogeneous sample, because hemispheric specialisation in the cerebral cortex is different in left-handed individuals [37]. On the other hand, they also had to be non-smokers and not allergic to any type of nut.

All participants signed an informed consent and personal data protection before attending the tasting session, as required by the Declaration of Helsinki of the World Medical Association (WMA, 2013 [38]) and EU Regulation 2016/679 of the European Parliament and of the Council of 27 April 2016 on the protection of individuals with regard to the processing of personal data and the free movement of data and Organic Law 3/2018 of 5 December on the Protection of Personal Data and guarantee of digital rights. Finally, participants received a EUR 15 gift card for a Spanish department store as a reward for their collaboration in the study.

2.2. Sample Preparation

For the study, raw cashew nuts of two private label brands (PLA and PLB) with similar quality were purchased weekly from two different well-known local supermarkets. Once the cashew packs were opened, they were not used for more than 5 days, as recommended by the manufacturer on their packaging. As this was a blind test, all samples were served in white 60 mL cups coded with three random digits so that consumers could not identify them [39].

2.3. Implicit Methodologies: Applied Neuroscience Techniques

The techniques used in this study were electroencephalography (EEG) and electrodermal skin activity (EDA), both of which provided us with the physiological and unconscious responses of the participants. The responses of both were monitored and recorded using iMotions software version 9.3 (iMotions A/S, Copenhagen, Denmark).

EEG signals were measured using an Enobio 8 electroencephalography device (Neuroelectrics, Barcelona, Spain), which recorded brain electrical waves at a sampling frequency of 500 Hz for subsequent processing and calculation of the metrics detailed in the following section. The EEG device was used with a 7-channel frontal configuration at points determined by the international 10/20 system in the orbitofrontal (F7/F8), prefrontal (AF7/AF8), frontopolar (Fp1/Fp2), and frontal midline (Fz) areas, arranged in a neoprene band, so that the adjustment to the participant's head was adequate.

The electrodermal skin activity study, hereafter EDA, is a technique that measures changes in skin conductance (μS) produced by the involuntary response of the autonomic nervous system to a stimulus [40]. A SHIMMER™ device (SHIMMER Research Ltd., Dublin, Ireland) was employed at a sampling rate of 128 Hz, which consisted of a controller placed on the wrist of the non-dominant hand, from which two wires were connected to two 40 mm diameter AgCl pregelled electrodes and placed on the palm of the hand, specifically on the thenar and hypothenar eminences, areas where a greater number of sweat glands are found [41–43].

2.4. Explicit Methodologies: Affective Sensory Testing

In order to collect data from the explicit methodologies, a questionnaire was designed for participants to fill in parallel to the EEG and EDA data collection while evaluating the samples, including affective tests of acceptance and preference, as well as other emotional and socio-demographic aspects.

The structure of this consisted of five sections: the first three to assess visual appearance, aroma, and flavour, followed by overall liking and choice between the two private labels, and, finally, the EsSense25, a reduced version of EsSense Profile® [44].

The first section included questions related to visual attributes of both private labels. Acceptance of the appearance of each private label was assessed on a 9-point hedonic scale (1 "I don't like it at all", 9 "I like it very much") and, using 5-point verbal JAR scales, on size (1 "Too small" . . ., 5 "Too big").

The second section asked about the hedonic acceptability of the aroma of each of the samples and the intensity of the aroma using a JAR scale (1 "Too weak", 5 "Too intense"). The third section consisted of a question on overall flavour acceptability, with the same 9-point hedonic scale used for appearance and aroma and two further questions with a JAR scale for flavour intensity (1 "Too weak", "5 "Too intense") and texture (1 "Too soft", 5 "Too hard").

Finally, the overall liking of both samples was assessed using a 9-point hedonic scale as outlined above, followed by a bilateral paired comparison test, according to the standard ISO 5495 [45], wherein participants had to choose which of the two private labels they preferred after having tried them. Finally, the emotional responses were assessed using the EsSense25 framework with a check-all-that-apply (hereafter CATA) question included with the twenty-five EsSense25 emotions. The age range and gender of the participants were also recorded.

2.5. Consumer Session Procedure

The experiment was conducted within the facilities of the Universitat Politècnica de València, in a room set up for this purpose, with the appropriate temperature and lighting and free of external odours and noise [39]. This space was close to another sample preparation room but sufficiently far away to avoid interference with the study participants.

Once the consumers who met the requirements for an individual session lasting about 45 min had been summoned, they were sent an information sheet with the procedure and the neuroscience devices that were going to be used. Thus, once the participant arrived at the facility, after a brief explanation of the procedure and resolution of any doubts, the informed consent form was provided for signature. The participant was then seated in a chair in front of a screen that would show instructions to be followed at all times. Afterwards, the EEG and EDA devices were placed, and the signal from both was checked to ensure that it was correct to begin the study.

The experiment consisted of 5 distinct phases (Figure 1), which are described below.

The first three phases of the experiment corresponded to studying the perception of the two private labels described above (PLA and PLB) with neuroscience techniques. In these phases, the samples were presented in a sequential monadic manner, randomising the order of sample presentation per participant [39]. Finally, there was a fifth phase where participants' self-reported emotions were collected using the EsSense25 instrument.

First, the "Visual Phase" began with participants observing the external appearance of each sample for 20 s and then proceeding to evaluate them using the hedonic and JAR-type scales described in Section 2.4. The next phase consisted of assessing the aroma of the samples ("Aroma Phase"). In this phase, participants had to smell each sample for 20 s and then answer the aroma questions. The third phase was the evaluation of the flavour of the samples. Although it was similar to the previous ones, it had to be adapted to be able to capture the EEG signals adequately, as chewing interfered with the signal, and it was not possible to eliminate it during processing. This adaptation consisted of instructing the participant to taste the sample by chewing normally and then to remain at rest for 20 s to ensure that the EEG measurements were correct. As in the previous two phases, the flavour of both samples was assessed using hedonic scales and JAR.

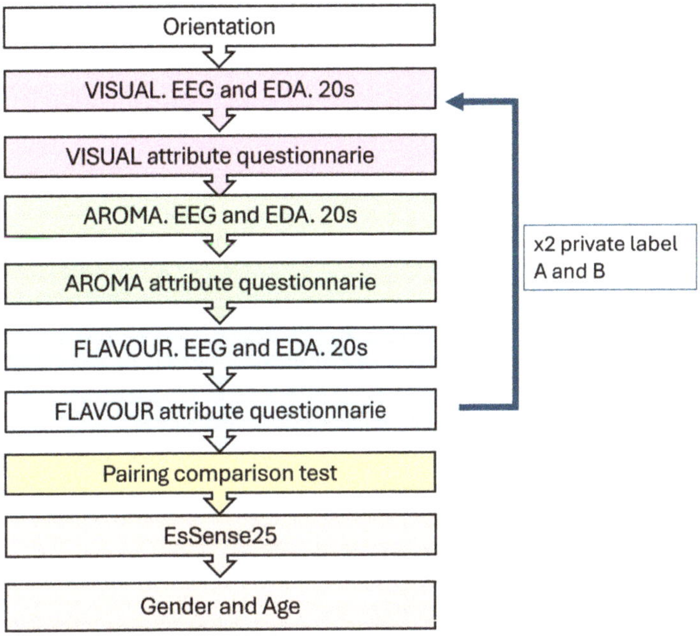

Figure 1. Graphical representation of the procedure followed in each session with the 80 consumers participating in the study.

The fourth phase consisted of the participants' overall liking of the two samples using a 9-point hedonic scale. Once they had rated the samples, they were asked to select their preferred sample, following the paired comparison test procedure described in Section 2.4.

Subsequent to the collection of the EEG and EDA signals, along with the sensory attributes for both samples obtained from the questionnaire, the fifth and final phase was initiated. This consisted of consumers answering the last question of the questionnaire, which corresponded to the EsSense25. For this, participants were asked to select the emotions they most identified with after consuming the raw cashew nuts through a check-all-that-apply (CATA) question. This question allowed consumers to choose from a list of twenty-five emotions, thereby providing insight into their emotional experiences associated with the consumption of the cashew nuts. Finally, participants were asked about their age and gender.

2.6. Data Processing

2.6.1. Electroencephalography (EEG)

The EEG data were processed according to the methods described by [46]. First, frequencies between 0.5 and 100 Hz were filtered using a Butterworth bandpass filter with zero phase delay, followed by a 50 Hz notch filter to mains noise. All values above 120 uV were then removed from the electrical waveform, as the typical EEG signal of an adult human being is between 10 and 100 uV. Once the signal was processed, the power spectral density (PSD) was calculated using fast Fourier transform (FFT) at theta (4–8 Hz) and alpha (8–12 Hz) frequencies for the calculation of the following metrics:

- Frontal alpha asymmetry (FAA): This was used to compare consumer valence towards the two private label of raw cashew nuts. It is the result of the difference of the Napierian logarithm of the power spectral density between the right and left frontal lobe [24], according to the following equation:

$$FAA = \ln \alpha PSD(F8, AF8, Fp1) - \ln \alpha PSD\ (F7,\ AF7, Fp1)$$

This metric is based on the theory that higher activity in the right frontal lobe is associated with negative emotions, while higher activity in the left frontal lobe is associated with positive emotions [47]. In the case of the alpha rhythm (8–13 Hz), an increase in alpha rhythm is inversely related to frontal lobe activity, i.e., a decrease in alpha frequency potential in the right frontal lobe implies greater activation in the left frontal lobe, and vice versa [48]. Higher FAA values indicate greater activation of the left frontal lobe, which is associated with approach motivation and a stronger affinity for the stimulus. Conversely, lower FAA values suggest greater activation of the right frontal lobe, indicating avoidance motivation and lower affinity for the stimulus. It is essential to interpret FAA scores in a relative context, comparing them across different stimuli to determine which one elicits stronger approach or weaker avoidance motivations [24,33,48]. FAA is a dimensionless metric.

- Task engagement (TE): Measures the participant's cognitive state in response to a stimulus, indicating their level of involvement or mental effort regarding this stimulus [49,50]. For this purpose, frontal midline activity (Fpz) will be monitored at the theta frequency (4–8 Hz), calculating its power spectral density in dB using the procedure described at the beginning of this section. An increase in theta power is directly related to an increase in cognitive workload, which results in greater interest, attention, or mental effort towards the task being performed at that moment [51].

2.6.2. Electrodermal Skin Activity, EDA

The conductance signals (μS) obtained for the study of the electrodermal activity were analysed using continuous decomposition analysis (hereinafter CDA). CDA consists of decomposing the signal of the skin conductance data into continuous signals of phasic (activation) and tonic (basal state) activity. This method takes advantage of the recovery of the signal characteristics of the underlying sudomotor nerve activity [40]. Prior to this analysis, the signal was pre-processed by applying a low-pass Butterworth filter (cutoff frequency: 5 Hz) to remove high-voltage line noise. Subsequently, the CDA considered all signals with an amplitude above 0.0005 μS and a duration above 500 ms as stimulus responses.

The metric calculated for each stimulus was the number of skin conductance response peaks per minute (nSCR), which is described as the number of detected peaks (phasic signal, dimensionless) that the consumer had during this stimulus [40] and has been widely used by different authors to study the intensity of the emotion elicited in the consumer [20,28,33].

2.6.3. Statistical Analysis

Statistical analyses of the obtained data are described below. All statistical analyses were conducted using the XLSTAT Premium, version 2024.4.0, statistical package (Addinsoft, Paris, France).

For the metrics obtained using EEG (FAA and TE) and EDA (nSCR) and the ratings obtained using hedonic scales, the Wilcoxon signed-rank test, a non-parametric test indicated for two related samples [52], was used, as the distribution of the data was non-normal. The Shapiro–Wilk test was used to check the normality of the data and showed that they were not normal. For both tests, the significance level was 5% ($p < 0.05$).

On the other hand, for the significance of the bilateral paired comparison test, where consumers had to choose which of the two samples they preferred overall, the tables in Appendix A of ISO 5495 [45] were used. In the case of this study, where the test was bilateral, when $\alpha = 0.05$ and $n = 80$, at least 50 consumers (62.5%) should select the same sample as their preferred sample.

The JAR data collected were analysed in two phases for each nut: a frequency analysis and then a penalty analysis.

For the frequency analysis, the frequencies marked as low (1 "Too much little" and 2 "Too little") were grouped for each attribute analysed and expressed as a percentage (%). The same was performed for the ideal point, JAR (3 "Just About Right"), and for the high level (4 "Too much" and 5 "Much Too much") to obtain a bar chart showing the percentages of the three levels obtained, which were named "Too little", "JAR", and "Too much".

Once the relative frequencies (in %) had been obtained, the next stage in dealing with the JAR data was the penalty analysis. This analysis relates the JAR data to overall product liking and aims to measure whether attributes that consumers have found to be high or below their ideal point produce a significant penalty in overall liking. To arrive at this, the overall liking scores for each level mentioned above ("Too little", "JAR", "Too much") were grouped and averaged. Penalties or mean drops were then obtained for the "Too much" and "Too little" levels by subtracting them from the average of the "JAR" level.

Once the mean drops were obtained, they were standardised, and a 3-level mean comparison test was performed at a 5% significance level to detect which penalties were significant and influenced the overall liking of the product.

Subsequently, to gain a better understanding of the data obtained in this analysis, a penalty graph was created. In this graph, the penalties calculated for each attribute are plotted against the percentage of consumers at each level.

Finally, the data collected using the EsSense25 tool were processed qualitatively, creating a ranking with the frequency (in %) with which each emotion was marked in order from highest to lowest. In order to make it less complex to see which emotions stood out the most, they were represented by a radial graph, thus generating the emotional profile of each one.

It is important to acknowledge the limitations of this study. The focus on only two private label brands within the same product category restricts the analysis of biometric measures to paired comparison tests for related samples [53]. To address this limitation and provide a more comprehensive understanding of consumer preferences, we have expanded our analysis to include additional approaches involving explicit measures, such as penalty analysis.

3. Results

3.1. Implicit Measures: Unconscious Consumer Response

Tables 1 and 2 show the average of the metrics calculated from the EEG and EDA signals recorded from the 80 consumers. Metrics were calculated to measure valence, arousal, and attention for the appearance, aroma, and flavour of each sample.

Table 1. Implicit measures mean to assess consumers' valence (FAA) and arousal level (nSCR) for the two private labels studied.

Attributes	FAA (EEG)			nSCR (EDA)		
	PLA	PLB	*p*-Value	PLA	PLB	*p*-Value
Appearance	−0.238	−0.251	0.869	6.038	6.188	0.519
Aroma	−0.204	−0.217	0.106	7.914	8.139	<0.0001 *
Flavour	−0.247	−0.241	<0.0001 ***	4.576	4.388	0.189

Wilcoxon test significant differences * for $p < 0.05$, *** $p < 0.0001$. FAA: Higher values (closer to zero for negatives) indicate stronger approach motivation. nSCR: Higher values indicate increased physiological arousal.

Regarding the measures related to valence (FAA) and arousal (nSCR) (Table 1), after performing the Wilcoxon test for related samples, significant differences were found, mainly in aroma and flavour. Appearance showed no differences in any of the calculated metrics ($p = 0.869$ and $p = 0.519$, respectively).

Table 2. Task engagement (dB) mean of consumer interest during the evaluation of the appearance, aroma, and flavour of both private labels.

Attributes	Task Engagement (dB)		
	PLA	PLB	p-Value
Appearance	5.366	5.174	0.729
Aroma	8.793	8.358	<0.0001 ***
Flavour	8.530	9.099	0.037 *

Wilcoxon test significant differences for * $p < 0.05$, *** $p < 0.0001$. TE: Higher values indicate greater cognitive workload and task engagement.

Specifically, higher arousal ($p < 0.0001$), by nSCR measurement, was observed for PLB, so it could be deduced that consumers showed greater emotional intensity for the aroma (nSCR = 8.139) of this private label. The valence data measured with FAA were not significant ($p = 0.106$).

Regarding flavour, consumers' preference was also for PLB, as it has a higher FAA value (−0.241), which implies a higher affinity for it. The metric calculated for skin conductance (nSCR) showed no significant differences between private labels, so the arousal level was similar among consumers.

In addition to valence and arousal, task engagement (TE) was also calculated using EEG by monitoring theta frequency activity (4–8 Hz) in each phase of the test and for each private label (Table 2). In this case, there were significant differences for aroma ($p < 0.001$) and flavour ($p = 0.037$). For aroma, a slight increase in involvement was observed in PLA. In contrast, for flavour, higher involvement is seen with PLB. These results reflect that in both the aroma and flavour of both private labels, there are attributes that have captured the attention of consumers, although the motivations leading to this are unknown.

To complement what has been obtained at the implicit level, as well as to specify which possible attributes are behind consumers' unconscious choices, the following section describes the results of the explicit methodologies based on sensory analysis that have been used in this research.

3.2. *Explicit Measures: Conscious Response*

3.2.1. Hedonic Scales and Paired Comparison Test

In addition to measuring their unconscious response, consumers were asked to rate the two samples to be studied (PLA and PLB) on a nine-point hedonic scale on the following attributes: appearance, aroma, flavour, texture, and overall liking; the results were then compared using the Wilcoxon non-parametric signed-rank test (Table 3).

Table 3. Results of the hedonic scores (1–9 points) mean of raw cashew nut attributes evaluated by consumers for each private label.

Attributes	PLA	PLB	p-Value
Appearance	6.088	6.488	0.120
Aroma	5.544	6.329	0.001 **
Flavour	6.325	6.838	0.060
Texture	6.638	6.988	0.065
Overall liking	6.363	6.950	0.017 *

Wilcoxon test significant differences for * $p < 0.05$, ** for $p < 0.01$.

Of these, only aroma (p = 0.001) and overall liking (p = 0.017) showed significant differences, and both showed higher values for PLB (Table 3). Appearance (p = 0.120), flavour (p = 0.060), and texture (p = 0.065) showed no significant differences.

To determine which private label of raw cashew nuts they preferred, consumers had to choose between the two samples presented following the procedure of a bilateral paired comparison test, as described in the ISO 5495 standard [45]. Table 4 shows the number of consumers who chose each sample. Of the 80 consumers who participated in the study, 52 (65%) chose PLB compared to 28 (35%) consumers who chose PLA. According to Table A2 of Appendix A of ISO 5495 [45] and for an alpha = 0.05, these results were statistically significant, which means that there is a clear preference for the PLB cashew sample.

Table 4. Percentage of consumers (n = 80) who overall chose each sample of raw cashew nuts.

	Paired Test	
	PLA	**PLB**
%	28 (35.00%)	52 (65.00%)

Minimum number of consumers = 50 for n = 80 and α = 0.05 according to ISO 5495:2005 Appendix A Table A2.

Considering these results, aroma and flavour were the attributes that determined the preference for PLB. In order to identify to what extent these attributes have influenced the overall liking, the following section describes the results obtained by means of the penalty analysis.

3.2.2. Penalty Analysis

JAR scales were used to evaluate five attributes of raw cashew nuts (size, colour, aroma, flavour, and texture). This was performed to see which attributes penalise the overall liking of the product the most, as well as to further investigate consumer opinion and find possible points of improvement for the product.

Figure 2a,b corresponds to the results grouped into three levels (too little, JAR, and too much) for each sample and the attributes analysed, using the procedure described in the methodology; the difference between the perception of aroma and flavour between the two private labels stands out. In the case of the PLA sample, 76% (Figure 2a) of consumers perceived the aroma of raw cashew nuts to be too weak for their liking, compared to 40% for PLB (Figure 2b). Regarding flavour, 54% of consumers found it less intense than they would like in the case of PLA, while for PLB, it was 36%.

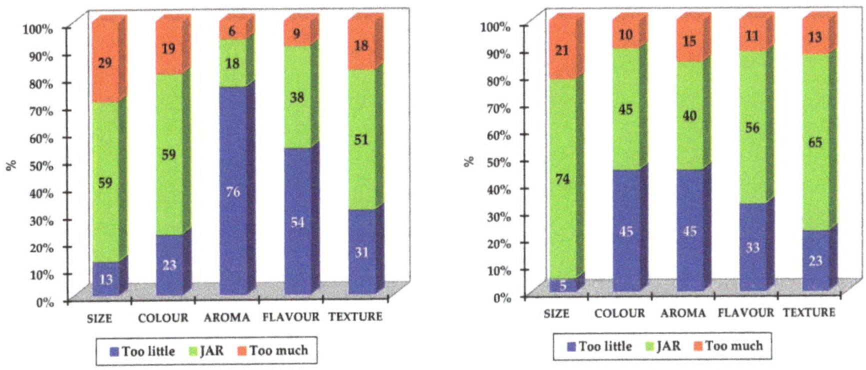

(**a**) PLA Sample (**b**) PLB Sample

Figure 2. Percentage of consumers aggregated according to three levels of intensity (too little, JAR, too much) for each attribute and for each private label of raw cashew nuts.

Once an initial frequency analysis was performed, the results of the penalty analysis were obtained, where the calculated mean drops were assessed for significance using multiple pairwise comparisons, as described in the methodology. The tables with the significance tests for each sample and attribute measured can be found in Appendix A. To better understand these results, a penalty graph was also created (Figure 3), which is the result of plotting the mean drops for the "Too little" and "Too much" levels against the percentage of consumers who had marked each one.

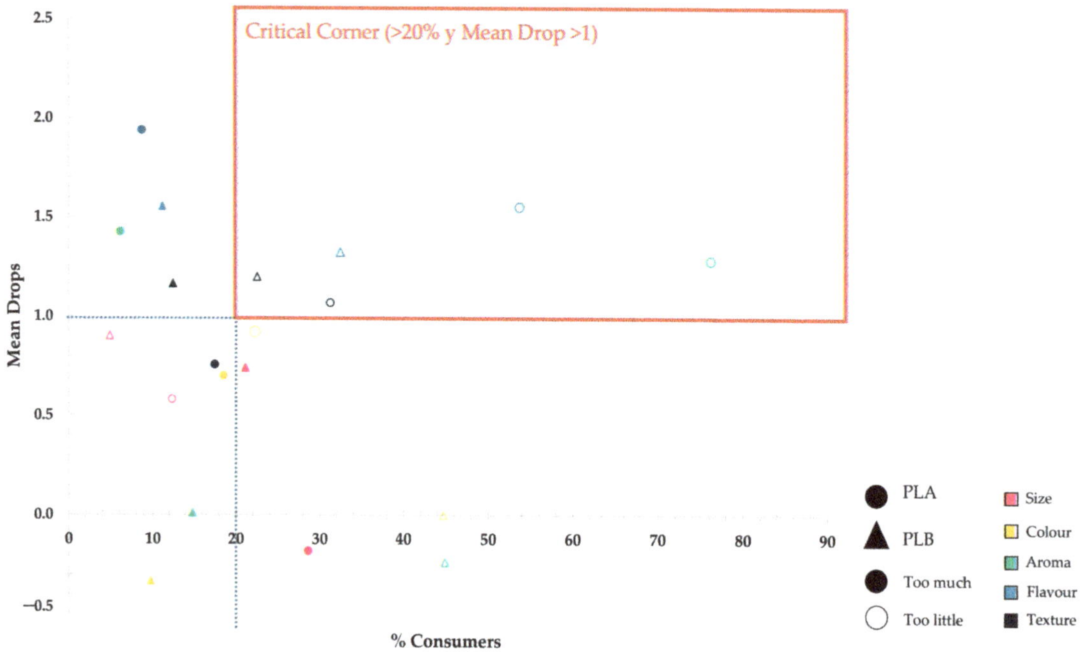

Figure 3. Graphical representation of the penalty analysis conducted for both private labels of raw cashew nuts. The dotted line delimits the mean drops greater than one on the x-axis and the percentage of consumers who have marked each level greater than 20% on the y-axis to obtain the critical corner (red lines).

To visually detect which attributes might require attention to improve both private labels of cashew nuts, the "critical corner" was delimited in the graph of penalties, which includes the attributes with mean drops higher than 1 and whose frequency exceeded 20% of consumers. Figure 3 shows that for PLA, the critical attributes are aroma (p = 0.002), flavour (p < 0.0001), and texture (p = 0.002), which are perceived as too weak in all three cases. For the PLB, the following penalised the overall liking: flavour (p < 0.0001) and texture (p = 0.002), perceived as too weak and too soft, respectively.

3.2.3. Emotional Profile of Raw Cashew Nuts. EsSense25

For the analysis of the EsSense25 lexicon, only emotions marked with a frequency of more than 20% were considered [24,30]. Figure 4 shows the frequency of emotions they identified with after consuming raw cashew nuts. It can be seen that positive emotions such as "Pleasant" (71.25%), "Satisfied" (50%), "Calm" (45%), or "Happy" (25%) predominate, which is common for emotions associated with food.

On the other hand, EsSense25 has also been useful when detecting participants with anomalous behaviour, such as those who had bad experiences with cashew nuts or other nuts, which could evoke negative feelings and influence the overall analysis. This was not the case in this study, as the number of negative emotions was negligible, less than 3%.

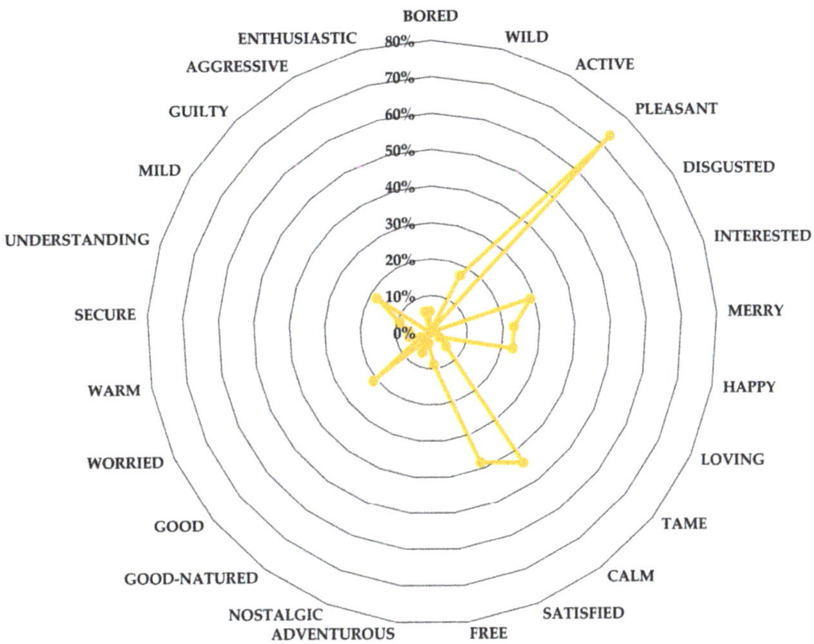

Figure 4. Radial representation of the emotional profile of raw cashew nuts using the EsSense25 lexicon.

4. Discussion

4.1. Comparison of Explicit vs. Implicit Methods

This research aimed to analyse consumers' conscious and unconscious responses to raw cashew nuts from two private labels (PLA and PLB). Implicit methods based on consumer neuroscience techniques and explicit methods such as sensory analysis were used.

The results reveal a clear preference for the PLB sample. The overall preference for PLB is evident in both implicit objective measurements using neuroscience techniques and subjective evaluations. Overall hedonic scores were significantly higher for PLB (Table 3), and the preference test revealed that a significant majority of consumers (65%) chose PLB over PLA (Table 4). These findings align with previous studies highlighting the importance of sensory as emotional experience in consumers' food acceptance and choice [30,54,55].

At the implicit level, the higher approach for PLB was reflected in the FAA values for flavour, indicating greater activity in the left frontal lobe, associated with positive emotions. This result suggests that the sensory experience with PLB generated a more favourable emotional response in participants, even at an unconscious level. Although no specific literature studying consumer perception of nuts using EEG was found, other studies have employed FAA as an indicator to examine valence for certain types of foods or related experiences. For instance, [37] found significant differences when comparing videos of different breakfast meals under varying hygienic conditions, specifically observing greater activity in the right frontal lobe for those presenting deterioration compared to the control, implying that consumers responded negatively to this stimulus as expected. Another study [33] reported similar results when comparing the FAA of consumers while cooking chicken and mealworms, with the latter resulting in lower FAA values, thus indicating greater rejection. The same authors repeated a study using two stimuli deemed pleasant [34], where consumers prepared the same Thai chicken meal with different quality ingredients—premium and basic. In this instance, none of the EEG implicit parameter measures employed were significant. This study is similar to ours in that the samples were comparable, and consumers were not shown the product brand. However, unlike their findings, we discovered significant differences in flavour (Table 1). This suggests that FAA could be useful for comparing two similar stimuli, a finding that warrants further verification through additional studies involving different types of nuts or other foods.

Task engagement (TE) is a metric based on measuring the brain activity of theta waves (4–8 Hz), frequencies whose increase in the frontal lobe is related to processes where cognitive effort is higher and our brain is allocating more resources [56]. In this study, significant differences in participants' attention and motivation during aroma and flavour evaluation were revealed (Table 2). A slight increase in engagement with PLA was observed during aroma evaluation, which might indicate that this attribute generated a higher cognitive processing need in the participants. However, the higher involvement with PLB during flavour evaluation suggests that this attribute was more motivational for consumers. Other authors such as [35] conducted research on food neophobia, where participants with higher food neophobia also showed higher levels of attention for an unfamiliar food product, aligning with the relationship between theta wave activity and mental workload. In our study, both stimuli were familiar to consumers. As noted earlier, the main differences lie in factors related to the aroma and flavour of both samples, which require greater attention, regardless of their valence.

Regarding the EDA metric, nSCR, significant differences were observed only in aroma, which provides limited insight into the arousal levels in both samples. Other studies, such as [33,34], have reported similar findings while studying familiar samples among consumers, where they also utilised EDA to examine arousal. In the first case study (mealworms vs. chicken), there was greater arousal in some phases of mealworm cooking, but in the second study where the samples were similar, no significant differences were found. Other authors found the same when studying responses to mixed vegetable juices [30], as did [57] with different types of beers.

On the other hand, when marketing variables such as brand or quality are included in experiments, more differences in consumer arousal exist. In a study conducted by [58] that examined private labels, specifically in the context of chocolate, researchers compared these private labels with national brands. They found that differences in arousal occurred only when consumers were aware of the brand. Under blind conditions, however, consumers demonstrated the same emotional response regardless of the brand. Also, brand familiarity played a significant role in the study conducted by [28], which examined how consumers perceive various types of cider with a protected designation of origin (PDO). The study found that consumers had a stronger emotional response when they were unfamiliar with the PDO logo compared to when they recognised it. These findings suggest that techniques like EDA may yield more insightful results when consumers have more information about a product. Therefore, it would be interesting for future studies on nuts to explore how knowing specific attributes—such as the brand or nuts origin—affects emotional responses in consumers.

Penalty analysis (Figure 3) revealed that the main deficiency of PLA was the perception of an aroma and flavour that was "too weak". In contrast, PLB was penalised for its flavour and texture being "too weak/soft", although to a lesser extent. This suggests that PLB achieved a more balanced and satisfying sensory profile for consumers, although there are still opportunities for improvement in flavour intensity and texture firmness.

The EsSense25 analysis revealed that raw cashew nut consumption predominantly evokes positive emotions, such as "Pleasant", "Satisfied", "Calm", and "Joyful" (Figure 4). This positive emotional profile aligns with the general perception of nuts as healthy and enjoyable snacks [11,59]. The absence of significant negative emotions suggests that the sensory experience with cashew nuts was generally well received by participants.

Consequently, aroma and flavour stood out as the key sensory attributes that differentiated the two private labels and guided the preference for PLB. Results from the hedonic scales showed significantly higher scores for aroma in PLB (Table 3), consistent with higher arousal as measured by skin conductance (nSCR) (Table 1). This finding underlines the importance of aroma as a determining factor in food perception and acceptance [60–62].

4.2. Marketing Implications

The findings of this research have important marketing implications for the raw cashew nut industry as well as for the development of novel foods:

1. Optimising the sensory experience: Raw cashew nut manufacturers can use the results of this study to optimise the sensory profile of their products, paying particular attention to aroma and flavour, to maximise consumer acceptance and preference. The information provided by the penalty analysis can guide the formulation of products with an intensity of aroma, flavour, and texture that matches consumer expectations and preferences.
2. Market segmentation: The information obtained in this study can be used to segment the raw cashew nut market based on consumers' sensory and emotional preferences. This would allow companies to tailor their products and marketing strategies to the specific needs and desires of each market segment.
3. Marketing strategies: Companies can use the findings of this research to design more effective marketing and communication strategies that highlight the sensory and emotional attributes of their raw cashew nut products. For example, advertising messages could be created that emphasise the aroma and flavour of cashew nuts or associate the consumption of cashew nuts with positive emotions.
4. Novel food development: Understanding the underlying emotions and preferences of consumers, gained through neuroscience techniques, can be used for the development of new food products that meet consumers' needs and desires more effectively. For example, products could be developed with specific flavour profiles that evoke positive emotions or offer a more intense and satisfying sensory experience.

4.3. Limitations and Future Research

The findings of this research offer valuable insights into the preferences of raw cashew nut consumers; however, there are several limitations to consider. Firstly, the sample size (n = 80) and the specific participant group (students and staff of the Universitat Politècnica de València) may restrict the generalisability of the results to broader populations. Secondly, the study's focus on only two private-label brands limits the complexity of the statistical analyses that can be conducted, such as multivariate comparisons.

Future research could benefit from exploring additional factors that influence consumer preferences and emotional responses, such as geographical origin, processing methods, packaging, and willingness to pay. Furthermore, increasing the sample size and diversifying the participant profiles could provide a more comprehensive understanding of consumer preferences.

5. Conclusions

The combination of implicit consumer neuroscience methodologies with explicit sensory analysis techniques has provided a holistic view of consumer preferences toward raw cashew nuts. The convergence of results between objective measurements of brain activity (FAA, TE, and nSCR) and participants' subjective evaluations reinforces the validity of the findings and underscores the importance of considering both conscious and unconscious responses in food perception research. Notably, this study represents one of the first applications of EEG and EDA to compare private-label nut brands, addressing a critical gap in the literature where nuts have been underrepresented in neuroscientific consumer studies.

From a practical perspective, these results demonstrate the potential of consumer neuroscience to inform targeted strategies for product optimisation and marketing. Additionally, the integration of penalty analysis with biometric metrics provides a framework for identifying sensory attributes that require prioritisation in product development.

The authors of this study believe that the results of this research demonstrate the potential of consumer neuroscience to provide valuable information for the optimisation and development of new food products and the improvement of marketing and communication strategies in the food industry.

Author Contributions: Conceptualisation, R.L.-N. and J.M.B.-V.; data curation, R.L.-N.; formal analysis, R.L.-N.; investigation, R.L.-N.; methodology, R.L.-N., L.M.-V., C.E.-P. and J.M.B.-V.; project administration, R.L.-N. and J.M.B.-V.; supervision, L.M.-V., C.E.-P. and J.M.B.-V.; validation, L.M.-V., C.E.-P. and J.M.B.-V.; writing—original draft, R.L.-N.; writing—review and editing, L.M.-V., C.E.-P. and J.M.B.-V. All authors have read and agreed to the published version of the manuscript.

Funding: This research received no external funding.

Institutional Review Board Statement: This study was conducted in accordance with the Declaration of Helsinki and approved by the Institutional Review Board of Universitat Politècnica de València (protocol code: P11_26-09-2023, approved prior to data collection).

Informed Consent Statement: Informed consent was obtained from all subjects involved in the study.

Data Availability Statement: The original contributions presented in the study are included in the article. Should further inquiries necessitate additional information, the corresponding author can be contacted directly.

Acknowledgments: This research was conducted as part of Rocío López-Navarro's PhD thesis at Universitat Politècnica de València. Rocio Lopez-Navarro would like to express their sincere gratitude to Juan M. Buitrago-Vera, Luis Montero-Vicente and Carmen Escriba-Perez, for their invaluable contributions, support, and guidance throughout this study. Their expertise, insights, and collaborative spirit have been instrumental in bringing this work to fruition.

Conflicts of Interest: The authors declare no conflicts of interest.

Abbreviations

The following abbreviations are used in this manuscript:

EDA	Electrodermal activity
EEG	Electroencephalography
FAA	Frontal alpha asymmetry
nSCR	Number of skin conductance response
JAR	Just about right
PLA	Private label A
PLB	Private label B
TE	Task engagement

Appendix A

Table A1. Results of multiple comparison of penalty analysis from PLA sample.

Attributes	Level	Frequencies	%	Sum (Value)	Average	Mean Drops	Standardised Difference	p-Value
SIZE	Too Small	10	12.50%	58.000	5.800	0.583		
	Just About Right	47	58.75%	300.000	6.383			
	Too Big	23	28.75%	151.000	6.565	−0.182	−0.509	0.613
COLOUR	Too clear	18	22.50%	104.000	5.778	0.924	2.296	0.025
	Just About Right	47	58.75%	315.000	6.702			
	Too dark	15	18.75%	90.000	6.000	0.702		
AROMA	Too weak	61	76.25%	375.000	6.148	1.281	3.154	0.002
	Just About Right	14	17.50%	104.000	7.429			
	Too intense	5	6.25%	30.000	6.000	1.429		
FLAVOUR	Low intensity	43	53.75%	250.000	5.814	1.553	5.163	<0.0001
	Just About Right	30	37.50%	221.000	7.367			
	Too intense	7	8.75%	38.000	5.429	1.938		
TEXTURE	Too soft	25	31.25%	144.000	5.760	1.069	3.152	0.002
	Just About Right	41	51.25%	280.000	6.829			
	Too hard	14	17.50%	85.000	6.071	0.758		

Table A2. Results of multiple comparison of penalty analysis from PLB sample.

Variable	Level	Frequencies	%	Sum (Value)	Average	Mean Drops	Standardised Difference	p-Value
SIZE	Too Small	4	5.00%	25.000	6.250	0.903		
	Just About Right	59	73.75%	422.000	7.153			
	Too Big	17	21.25%	109.000	6.412	0.741	2.098	0.039
COLOUR	Too clear	36	45.00%	249.000	6.917			
	Just About Right	36	45.00%	249.000	6.917			
	Too dark	8	10.00%	58.000	7.250	−0.333		
AROMA	Too weak	36	45.00%	255.000	7.083	−0.240	−0.656	0.514
	Just About Right	32	40.00%	219.000	6.844			
	Too intense	12	15.00%	82.000	6.833	0.010		
FLAVOUR	Low intensity	26	32.50%	162.000	6.231	1.325	4.153	<0.0001
	As I like it	45	56.25%	340.000	7.556			
	Too intense	9	11.25%	54.000	6.000	1.556		
TEXTURE	Too soft	18	22.50%	111.000	6.167	1.199	3.265	0.002
	Just About Right	52	65.00%	383.000	7.365			
	Too hard	10	12.50%	62.000	6.200	1.165		

References

1. Gonçalves, B.; Pinto, T.; Aires, A.; Morais, M.C.; Bacelar, E.; Anjos, R.; Ferreira-Cardoso, J.; Oliveira, I.; Vilela, A.; Cosme, F. Composition of Nuts and Their Potential Health Benefits—An Overview. *Foods* **2023**, *12*, 942. [CrossRef] [PubMed]
2. Wojdyło, A.; Turkiewicz, I.P.; Tkacz, K.; Nowicka, P.; Bobak, Ł. Nuts as Functional Foods: Variation of Nutritional and Phytochemical Profiles and Their in Vitro Bioactive Properties. *Food Chem. X* **2022**, *15*, 100418. [CrossRef]

3. Danso-Abbeam, G.; Fosu, S.; Ogundeji, A.A. Technical and Resource-Use Efficiencies of Cashew Production in Ghana: Implications on Achieving Sustainable Development Goals. *Sci. Afr.* **2021**, *14*, e01003. [CrossRef]
4. Monteiro, F.; Catarino, L.; Batista, D.; Indjai, B.; Duarte, M.C.; Romeiras, M.M. Cashew as a High Agricultural Commodity in West Africa: Insights towards Sustainable Production in Guinea-Bissau. *Sustainability* **2017**, *9*, 1666. [CrossRef]
5. Rico, R.; Bulló, M.; Salas-Salvadó, J. Nutritional Composition of Raw Fresh Cashew (*Anacardium occidentale* L.) Kernels from Different Origin. *Food Sci. Nutr.* **2016**, *4*, 329–338. [CrossRef]
6. Akyereko, Y.G.; Yeboah, G.B.; Wireku-Manu, F.; Alemawor, F.; Mills-Robertson, F.C.; Odoom, W. Nutritional Value and Health Benefits of Cashew Apple. *JSFA Rep.* **2023**, *3*, 110–118. [CrossRef]
7. Saini, R.K.; Meneguelli, T.S.; Wendling, A.L.; Claudia, A.; Kravchychyn, P.; Mayumi, D.; Prado Rocha, U.; Dionísio, A.P.; Bressan, J.; Stampini, H.; et al. Effects of Cashew Nuts (*Anacardium occidentale* L.) and Cashew Nut Oil on Intestinal Permeability and Inflammatory Markers during an Energy-Restricted 8-Week Intervention: A Randomized Controlled Trial (Brazilian Nuts Study). *Foods* **2024**, *13*, 2917. [CrossRef]
8. Wu, F.; Groopman, J.D.; Pestka, J.J. Public health impacts of foodborne mycotoxins. *Annu. Rev. Food Sci. Technol.* **2014**, *5*, 351–372. [CrossRef]
9. Cunha, S.C.; Sá, S.V.M.; Fernandes, J.O. Multiple mycotoxin analysis in nut products: Occurrence and risk characterization. *Food Chem. Toxicol.* **2018**, *114*, 260–269. [CrossRef]
10. Łozowicka, B.; Kaczyński, P.; Iwaniuk, P.; Rutkowska, E.; Socha, K.; Orywal, K.; Farhan, J.A.; Perkowski, M. Nutritional compounds and risk assessment of mycotoxins in ecological and conventional nuts. *Food Chem.* **2024**, *458*, 140222. [CrossRef]
11. Ghazzawi, H.A.; Muhanad Akash, K.A.I. Consumption Practices, Preferences and Barriers of Nuts Intake amongst University of Jordan Students. *Clin. Nutr. Exp.* **2019**, *24*, 45–53. [CrossRef]
12. Gutiérrez-Paz, C.; Rodríguez-Moreno, M.C.; Hernández-Gómez, M.S.; Fernández-Trujillo, J.P. The Cashew Pseudofruit (*Anacardium occidentale*): Composition, Processing Effects on Bioactive Compounds and Potential Benefits for Human Health. *Foods* **2024**, *13*, 2357. [CrossRef]
13. Lima, J.R.; Bruno, L.M.; Wurlitzer, N.J.; de SOUSA, P.H.M.; Holanda, S.A.d.M. Cashew Nut-Based Beverage: Development, Characteristics and Stability During Refrigerated Storage. *Food Sci. Technol.* **2020**, *41*, 60–64. [CrossRef]
14. Aslam, N.; Hassan, S.A.; Mehak, F.; Zia, S.; Bhat, Z.F.; Yıkmış, S.; Aadil, R.M. Exploring the Potential of Cashew Waste for Food and Health Applications- A Review. *Future Foods* **2024**, *9*, 100319. [CrossRef]
15. Giacalone, D.; Llobell, F.; Jaeger, S.R. "Beyond Liking" Measures in Food-Related Consumer Research Supplement Hedonic Responses and Improve Ability to Predict Consumption. *Food Qual. Prefer.* **2022**, *97*, 104459. [CrossRef]
16. Lawless, H.T.; Heymann, H. *Sensory Evaluation of Food: Principles and Practices*, 2nd ed.; Heymann, H., Ed.; Food Science Text Series; Springer: New York, NY, USA, 2010; ISBN 1-283-51066-9.
17. He, L.; Freudenreich, T.; Yu, W.; Pelowski, M.; Liu, T. Methodological Structure for Future Consumer Neuroscience Research. *Psychol. Mark.* **2021**, *38*, 1161–1181. [CrossRef]
18. Stasi, A.; Songa, G.; Mauri, M.; Ciceri, A.; Diotallevi, F.; Nardone, G.; Russo, V. Neuromarketing Empirical Approaches and Food Choice: A Systematic Review. *Food Res. Int.* **2018**, *108*, 650–664. [CrossRef]
19. Cherubino, P.; Martinez-Levy, A.C.; Caratù, M.; Cartocci, G.; Di Flumeri, G.; Modica, E.; Rossi, D.; Mancini, M.; Trettel, A. Consumer Behaviour through the Eyes of Neurophysiological Measures: State-of-the-Art and Future Trends. *Comput. Intell. Neurosci.* **2019**, *2019*, 1976847. [CrossRef]
20. Verastegui-Tena, L.; van Trijp, H.; Piqueras-Fiszman, B. Heart Rate, Skin Conductance, and Explicit Responses to Juice Samples with Varying Levels of Expectation (Dis)Confirmation. *Food Qual. Prefer.* **2019**, *71*, 320–331. [CrossRef]
21. Sharma, M.; Kacker, S.; Sharma, M. A Brief Introduction and Review on Galvanic Skin Response. *Int. J. Med. Res. Prof.* **2016**, *2*, 13–17. [CrossRef]
22. La Morena, A. *Neuromarketing Y Nuevas Estrategias de la Mercadotecnia: Análisis de la Eficiencia Publicitaria en la Diferenciación de Género Y la Influencia Del Marketing Sensorial Y Experiencial en la Decisión de Compra*; Universidad Complutense de Madrid: Madrid, Spain, 2016.
23. Huettel, S.A.; McKeown, M.J.; Song, A.W.; Hart, S.; Spencer, D.D.; Allison, T.; McCarthy, G. Linking Hemodynamic and Electrophysiological Measures of Brain Activity: Evidence from Functional MRI and Intracranial Field Potentials. *Cereb. Cortex* **2004**, *14*, 165–173. [CrossRef] [PubMed]
24. Briesemeister, B.B.; Tamm, S.; Heine, A.; Jacobs, A.M. Approach the Good, Withdraw from the Bad—A Review on Frontal Alpha Asymmetry Measures in Applied Psychological Research. *Psychology* **2013**, *4*, 261–267. [CrossRef]
25. Davidson, R.J.; Jackson, D.C.; Kalin, N.H. Emotion, Plasticity, Context, and Regulation: Perspectives from Affective Neuroscience. *Psychol. Bull.* **2000**, *126*, 890–909. [CrossRef] [PubMed]
26. Niedziela, M.M.; Ambroze, K. The Future of Consumer Neuroscience in Food Research. *Food Qual. Prefer.* **2021**, *92*, 104124. [CrossRef]

27. King, S.C.; Meiselman, H.L. Development of a Method to Measure Consumer Emotions Associated with Foods. *Food Qual. Prefer.* **2010**, *21*, 168–177. [CrossRef]
28. Mora, M.; Elzo-Aizarna, J.; Rozas-Fuertes, S.; Velilla-Echeita, L.; Vázquez-Araújo, L. Implicit Reaction vs. Explicit Emotional Response: Protected Designation of Origin in Apple Cider. *Food Qual. Prefer.* **2020**, *79*, 103773. [CrossRef]
29. Larrañaga-Ayastuy, E.; Mora, M.; Romeo-Arroyo, E.; Esteban, E.; Vázquez-Araújo, L. Electrodermal Response and Its Relationship with Explicit Response in Controlled and Real Contexts: A Case Study with Different Beer Styles. *J. Sens. Stud.* **2023**, *38*, e12799. [CrossRef]
30. Samant, S.S.; Seo, H.S. Influences of Sensory Attribute Intensity, Emotional Responses, and Non-Sensory Factors on Purchase Intent Toward Mixed-Vegetable Juice Products Under Informed Tasting Condition. *Food Res. Int.* **2020**, *132*, 109095. [CrossRef]
31. Martinez-Levy, A.C.; Moneta, E.; Rossi, D.; Trettel, A.; Peparaio, M.; Civitelli, E.S.; Di Flumeri, G.; Cherubino, P.; Babiloni, F.; Sinesio, F. Taste responses to chocolate pudding with different sucrose concentrations through physiological and explicit self-reported measures. *Foods* **2021**, *10*, 1527. [CrossRef]
32. Baranda, A.B.; Ríos, Y.; Llorente, R.; Naranjo, A.B.; da Quinta, N. Neuroscience tools to study the effect of the presentation form on food-evoked emotion for senior population. *Food Res. Int.* **2024**, *183*, 114158. [CrossRef]
33. Brouwer, A.M.; Hogervorst, M.A.; Grootjen, M.; van Erp, J.B.F.; Zandstra, E.H. Neurophysiological Responses during Cooking Food Associated with Different Emotions. *Food Qual. Prefer.* **2017**, *62*, 307–316. [CrossRef]
34. Brouwer, A.M.; Hogervorst, M.A.; van Erp, J.B.F.; Grootjen, M.; van Dam, E.; Zandstra, E.H. Measuring cook-ing experience implicitly and explicitly: Physiology, facial expression and subjective ratings. *Food Qual. Prefer.* **2019**, *78*, 103726. [CrossRef]
35. Stuldreher, I.V.; Kaneko, D.; Hiraguchi, H.; van Erp, J.B.; Brouwer, A.M. EEG measures of attention toward food-related stimuli vary with food neophobia. *Food Qual. Prefer.* **2023**, *106*, 104805. [CrossRef]
36. Bell, L.; Vogt, J.; Willemse, C.; Routledge, T.; Butler, L.T.; Sakaki, M. Beyond self-report: A review of physiological and neuroscientific methods to investigate consumer behavior. *Front. Psychol.* **2018**, *9*, 1655. [CrossRef]
37. Walsh, A.M.; Duncan, S.E.; Bell, M.A.; O'Keefe, S.F.; Gallagher, D.L. Integrating Implicit and Explicit Emotional Assessment of Food Quality and Safety Concerns. *Food Qual. Prefer.* **2017**, *56*, 212–224. [CrossRef]
38. AMM Declaración de Helsinki de La AMM- Principios Éticos Para Las Investigaciones Con Seres Humanos. Available online: https://www.wma.net/es/policies-post/declaracion-de-helsinki-de-la-amm-principios-eticos-para-las-investigaciones-medicas-en-seres-humanos/ (accessed on 27 May 2024).
39. *ISO 11136:2014*; Sensory Analysis-Methodology—General Guidance for Conducting Hedonic Tests with Consumers in a Controlled Area. International Organization for Standardization, ISO: Geneva, Switzerland, 2014.
40. Benedek, M.; Kaernbach, C. A Continuous Measure of Phasic Electrodermal Activity. *J. Neurosci. Methods* **2010**, *190*, 80–91. [CrossRef] [PubMed]
41. Fowles, D.C.; Christie, M.J.; Edelberg, R.; Grings, W.W.; Lykken, D.T.; Venables, P.H. Publication Recommendations for Electrodermal Measurements. *Psychophysiology* **1981**, *18*, 232–239. [CrossRef]
42. Saga, K. Structure and Function of Human Sweat Glands Studied with Histochemistry and Cytochemistry. *Prog. Histochem. Cytochem.* **2002**, *37*, 323–386. [CrossRef]
43. van Dooren, M.; de Vries, J.J.G.G.J.; Janssen, J.H. Emotional Sweating across the Body: Comparing 16 Different Skin Conductance Measurement Locations. *Physiol. Behav.* **2012**, *106*, 298–304. [CrossRef]
44. Nestrud, M.A.; Meiselman, H.L.; King, S.C.; Lesher, L.L.; Cardello, A.V. Development of EsSense25, a shorter version of the EsSense Profile®. *Food Qual. Prefer.* **2016**, *48*, 107–117. [CrossRef]
45. *ISO 5495:2005*; Sensory Analysis-Methodology—Paired Comparison Test. International Organization for Standardization, ISO: Geneva, Switzerland, 2005.
46. Coan, J.A.; Allen, J.J.B. Frontal EEG Asymmetry and the Behavioral Activation and Inhibition Systems. *Psychophysiology* **2003**, *40*, 106–114. [CrossRef] [PubMed]
47. Davidson, R.J. Anterior Cerebral Asymmetry and the Nature of Emotion. *Brain Cogn.* **1992**, *20*, 125–151. [CrossRef] [PubMed]
48. Allen, J.J.B.; Coan, J.A.; Nazarian, M. Issues and Assumptions on the Road from Raw Signals to Metrics of Frontal EEG Asymmetry in Emotion. *Biol. Psychol.* **2004**, *67*, 183–218. [CrossRef]
49. Biercewicz, K.; Wiscicka-Fernando, M. Customers' Emotions and Openness to Product Co-Creation: An Empirical Analysis Based on EEG Data. *Eur. Res. Stud. J.* **2023**, *XXVI*, 49–69.
50. Berka, C.; Levendowski, D.J.; Lumicao, M.N.; Yau, A.; Davis, G.; Zivkovic, V.T.; Olmstead, R.E.; Tremoulet, P.D.; Craven, P.L. EEG Correlates of Task Engagement and Mental Workload in Vigilance, Learning, and Memory Tasks. *Aviat. Space. Environ. Med.* **2007**, *78*, B231–B244. [PubMed]
51. McEvoy, L.K.; Smith, M.E.; Gevins, A. Test–Retest Reliability of Cognitive EEG. *Clin. Neurophysiol.* **2000**, *111*, 457–463. [CrossRef]
52. Wilcoxon, F. Individual Comparisons by Ranking Methods. In *Breakthroughs in Statistics: Methodology and Distribution*; Kotz, S., Johnson, N.L., Eds.; Springer: New York, NY, USA, 1992; pp. 196–202, ISBN 978-1-4612-4380-9.
53. Corder, G.W.; Foreman, D.I. *Nonparametric Statistics: A Step-by-Step Approach*, 2nd ed.; Wiley: Hoboken, NJ, USA, 2014.

54. Gutjar, S.; Dalenberg, J.R.; de Graaf, C.; de Wijk, R.A.; Palascha, A.; Renken, R.J.; Jager, G. What Reported Food-Evoked Emotions May Add: A Model to Predict Consumer Food Choice. *Food Qual. Prefer.* **2015**, *45*, 140–148. [CrossRef]
55. Carvalho, F.M.; Spence, C. Methods for Studying Multisensory Perception in Consumer Research. In *Consumer Research Methods in Food Science*; Gómez-Corona, C., Rodrigues, H., Eds.; Springer: New York, NY, USA, 2023; pp. 195–224, ISBN 978-1-0716-3000-6.
56. Klimesch, W. EEG Alpha and Theta Oscillations Reflect Cognitive and Memory Performance: A Review and Analysis. *Brain Res. Rev.* **1999**, *29*, 169–195. [CrossRef]
57. Gonzalez Viejo, C.; Fuentes, S.; Howell, K.; Torrico, D.D.; Dunshea, F.R. Integration of non-invasive biometrics with sensory analysis techniques to assess acceptability of beer by consumers. *Physiol. Behav.* **2019**, *200*, 139–147. [CrossRef]
58. Rita, P.; Guerreiro, J.; Omarji, M. Autonomic emotional responses to food: Private label brands versus National Brands. *J. Consum. Behav.* **2021**, *20*, 440–448. [CrossRef]
59. Neale, E.P.; Tran, G.; Brown, R.C. Barriers and Facilitators to Nut Consumption: A Narrative Review. *Int. J. Environ. Res. Public Health* **2020**, *17*, 9127. [CrossRef] [PubMed]
60. Turhan, K.N.; Gurkaynak, N.; Sadikzade, R. Neuromarketing Concepts in Food Studies. In Proceedings of the TIPTEKNO 2023—Medical Technologies Congress, Famagusta, Cyprus, 10–12 November 2023. [CrossRef]
61. Berčík, J.; Paluchová, J.; Neomániová, K. Neurogastronomy as a Tool for Evaluating Emotions and Visual Preferences of Selected Food Served in Different Ways. *Foods* **2021**, *10*, 354. [CrossRef] [PubMed]
62. Berčík, J.; Gálová, J.; Vietoris, V.; Paluchová, J. The Application of Consumer Neuroscience in Evaluating the Effect of Aroma Marketing on Consumer Preferences in the Food Market. *J. Int. Food Agribus. Mark.* **2023**, *35*, 261–282. [CrossRef]

foods

MDPI

Article

The Effect of Incorporating Fermented Elderberries (*Sambucus nigra*) into Bread: Quality, Shelf Life, and Biological Enhancement

Natália L. Seixas [1,2,3], Vanessa B. Paula [2,4], Teresa Dias [2,3], Luís G. Dias [2,3,*] and Letícia M. Estevinho [2,3]

[1] Doctoral School, University of León (ULE), Campus de Vegazana, 24007 León, Spain; natalia.seixas@ipb.pt
[2] Centro de Investigação de Montanha (CIMO), Instituto Politécnico de Bragança, Campus de Santa Apolónia, 5300-252 Bragança, Portugal; vanessapaula@ipb.pt (V.B.P.); tdias@ipb.pt (T.D.); leticia@ipb.pt (L.M.E.)
[3] Laboratório para a Sustentabilidade e Tecnologia em Regiões de Montanha, Instituto Politécnico de Bragança, Campus de Santa Apolónia, 5300-253 Bragança, Portugal
[4] Instituto Politécnico de Bragança, Campus de Santa Apolónia, 5300-253 Bragança, Portugal
* Correspondence: ldias@ipb.pt

Abstract: Elderberries, known for their antioxidant, anti-inflammatory, and antiviral properties, have traditionally been used to prevent and treat infections and boost the immune system. By increasing the quantity and quality of certain compounds, fermentation can potentially make them more effective as food additives. The aim of this study was to evaluate the effect of incorporating fermented elderberries on the bioactivity and shelf life of a traditional bread. The elderberry fermentation process was optimised using *Saccharomyces cerevisiae*, guided by a Plackett–Burman experimental design. The aim was to assess the impact of incorporating fermented elderberries into bread on its bioactive properties and shelf life. The fermentation of the elderberries was found to enhance their bioactive compounds and antioxidant activity. The total phenolic content ranged from 8.63 to 20.56 mg GAE/g (in samples without and with 2% extract, respectively). The antioxidant capacity, measured using the FRAP method, also showed a significant increase with the addition of the extract (from 9.16 to 26.66 mg Fe (II) E/g of the sample). Furthermore, bread enriched with fermented elderberry extracts demonstrated an extended shelf life during the study period.

Keywords: elderberries (*Sambucus nigra*); fermentation; bread; shelf life; biological activity

1. Introduction

Sambucus nigra, known as elderberry, is abundant in the forests and uncultivated areas of Portugal [1,2]. Its berries have been used in traditional medicine since ancient times, and in recent decades, growing scientific evidence has highlighted their biological activities, especially their antimicrobial, antioxidant, and anti-inflammatory properties [3–5]. These properties are closely linked to the presence of phenolic compounds (flavonoids and phenolic acids), organic acids, anthocyanin pigments, reducing sugars, pectin, tannins, vitamins A and C, and mineral elements [6,7]. Several studies have demonstrated that adding elderberries or their extracts to foods such as wine, juice, tea, dairy, and meat products significantly improves their nutritional value and increases consumer acceptance [8]. This dual benefit makes elderberries an attractive addition to functional foods, offering health benefits alongside appealing sensory qualities.

Although some bioactive compounds are reduced during the production and processing of elderberry-based foods, they remain present in significant quantities. Key factors such as temperature, time, and pH play a crucial role in the preservation of these bioactive compounds.

Fermentation, a well-established process for creating new food products such as wine and vinegar, is gaining attention for its potential to not only extend shelf life and produce natural flavours, but also to improve the nutritional quality of foods [9]. Microorganisms involved in fermentation, whether acting individually or in combination, utilise food constituents—specifically proteins, carbohydrates, and lipids—to produce flavour and aroma compounds such as aldehydes, esters, and ketones [10]. Each microbial strain exhibits distinct aroma-forming capabilities and metabolises these substrates uniquely. As a result, fermentation can yield products with novel organoleptic properties and improved nutritional value.

Feng et al. [11] and Hang et al. [12] reported that the fermentation of elderberries with lactic acid bacteria (LAB) led to an increase in total phenolic compounds and total amino acids, suggesting that fermentation with LAB enhances both the nutritional value and bioactive properties of elderberries.

Bread is one of the most widely consumed foods worldwide and plays a vital role in a balanced diet due to its high nutritional value, being rich in complex carbohydrates (primarily starch), fibre, B-complex vitamins, and minerals [13]. According to Portuguese legislation, bread is defined as the product obtained through the kneading, fermentation, and baking of wheat, rye, triticale, or maize flour, either individually or in combination. It may also include salt and other ingredients, such as additives [14,15]. Additionally, bread serves as an excellent medium for delivering functional ingredients to consumers in a familiar and widely accepted form. Due to their beneficial properties, the inclusion of natural additives improves the nutritional profile of bread and can also have a positive impact on human health [16,17].

The main objective of this study was to optimise the fermentation process of elderberries by *Saccharomyces cerevisiae* using a Plackett–Burman experimental design. The aim was also to assess the effect of incorporating fermented elderberries on the bioactivity and shelf life of a traditional bread loaf.

2. Materials and Methods

2.1. Sampling

In this study, the dehydrated elderberries were supplied by the InovTerra Association, located in Vila Pouca de Salzedas, Portugal, in 2021. The dehydrated sample contained some impurities from the tree and handling processes, which were manually removed prior to use. Before homogenisation, the elderberries were frozen at −24 °C and were subsequently milled into a fine powder using an IKA Tube Mill (Staufen, Germany).

2.2. Reagents

The reagents magnesium sulphate ($MgSO_4$), monopotassium phosphate (KH_2PO_4), urea, and sodium hydroxide (NaOH) were purchased from Merck KGaA (Darmstadt, Germany). The Rose Bengal agar culture medium was supplied by Liofilchem (Roseto degli Abruzzi, Italy). The modified iron sulphite agar culture medium, Baird-Parker (BP) culture medium, and *Bacillus cereus* Selective Agar (MYP) culture medium were purchased from HiMedia Laboratories (Modautal, Germany). The SimPlate kit for counting total coliforms/*Escherichia coli* and the 1–2 Test Kit for *Salmonella* were supplied by Biocontrol® (Bellevue, WA, USA). Plate Count Agar (PCA) medium was provided by Panreac (AppliChem, Barcelona, Spain). Ethanol, 99.8% pure, was obtained from Carlo Erba Reagents (Chaussee du Vexin, France). The reagents Folin–Ciocalteau, 2,2-diphenyl-1-picrylhydrazyl (DPPH•), aluminium chloride, sodium carbonate, quercetin, gallic acid, Trolox, Ferric-

Reducing Antioxidant Power (FRAP), acetic acid, 2,4,6-tri(2-pyridyl)-s-triazine (TPTZ), iron(III) chloride ($FeCl_3$), iron(II) sulphate heptahydrate ($FeSO_4.7H_2O$), hydrochloric acid (HCl), and other reagents used in our study were purchased from Sigma Chemical Co. (St. Louis, MO, USA). Petroleum ether was purchased from Chem-LAB (Zedelgem, Belgium). Potassium dihydrogen phosphate (KH_2SO_4) and selenium (Se) were supplied by Scharlau (Barcelona, Spain).

2.3. Optimisation of Fermentation Conditions

The optimisation of fermentation conditions for elderberry was performed using the Plackett–Burman design with 15 experiments (Table 1), involving 6 parameters at 3 levels each, as follows: urea (1.00, 1.50, and 2.00 g/L), $MgSO_4$ (0.50, 1.00, and 1.50 g/L), KH_2PO_4 (0.50, 1.00, and 1.50 g/L), and pH of the culture medium (4.50, 4.75, and 5.00); inoculum concentration of *Saccharomyces cerevisiae* was tested at 10^5, 10^6, and 10^7 colony forming units (CFU/mL); amounts of elderberries (50.0, 150.0, and 250.0 g), with a final volume of 250 mL.

Table 1. Experimental design to optimise fermentation conditions for elderberries.

Assay	Inoculum (CFU/mL)	Urea (g/L)	$MgSO_4$ (g/L)	KH_2PO_4 (g/L)	pH	Sample (g)
1	10^6	1.50	1.00	1.00	4.75	150
2	10^7	2.00	1.50	0.50	4.50	50
3	10^5	2.00	0.50	1.50	5.00	50
4	10^7	1.00	0.50	0.50	5.00	50
5	10^7	2.00	0.50	1.50	5.00	250
6	10^5	2.00	1.50	0.50	5.00	250
7	10^5	1.00	1.50	0.50	5.00	250
8	10^6	1.50	1.00	1.00	4.75	150
9	10^7	1.00	1.50	1.50	4.50	250
10	10^7	2.00	0.50	0.50	4.50	250
11	10^5	1.00	0.50	1.50	4.50	250
12	10^7	1.00	1.50	1.50	5.00	50
13	10^5	1.00	0.50	0.50	4.50	50
14	10^5	2.00	1.50	1.50	4.50	50
15	10^6	1.50	1.00	1.00	4.75	150

The fermentation tests were conducted in 500 and 1000 mL Erlenmeyer flasks. The salts required for the different experimental conditions were accurately weighed and added to each flask. Then, 150 mL of distilled water was added to dissolve the salts, and the initial pH of the medium was adjusted. The solution's volumes were adjusted to 250 mL and were sterilised by autoclaving at 121 °C for 15 min. After sterilisation, the flasks were allowed to cool and then the appropriate amount of sample was added to each flask, according to the experimental design. The flasks containing the samples were then pasteurised in a water bath at 65 °C for 15 min. After cooling on ice, the inoculum (*Saccharomyces cerevisae*) was added. The Erlenmeyer flasks were incubated at 25 °C in an orbital incubator (Incubated Shaker SIF6000R). Fermentations was monitored daily for 144 h by evaluating the following parameters: flask weight, pH, and colony forming units (CFU).

2.4. Preparation of Fermented Elderberry Extract

For the preparation of the fermented elderberry extract (FEE), the optimal fermentation process was selected based on the experimental design described above. The selected conditions were 1.0 g/L urea; 0.5 g/L $MgSO_4$; 0.5 g/L KH_2PO_4; 50 g sample; and pH 4.5. The ideal inoculum concentration was determined to be 10^5 CFU/mL, based on the experimental design. At the end of the fermentation, the product was frozen at −18 °C and lyophilized (−60 °C under vacuum) for later hydroethanolic extraction. The extraction was performed using a hydroethanolic solution (40:60, v/v) at pH 0.9, which had been optimised for the extraction of total phenolic compounds from elderberries. A total of 3.16 g of the lyophilised sample was weighed, and 50 mL of hydroethanolic solution was added. The mixture was stirred overnight and then filtered. After filtration, the extract was concentrated on a rotary evaporator (using a water bath at 40 °C and rotating the flask at 200 rpm) (model IKA V8).

2.5. Bread Production

Bread was prepared using type 65 flour (40 g)—this type of flour is a commercial, moderately refined flour that is ideal for baking, with a medium ash content, a good elasticity, and a balanced flavour—baker's fresh yeast (probiotic microorganism) (0.3 g), salt (Portuguese "Vatel" salt) (0.10 g), water (30 mL), and different concentrations of FEE (0.0%, 0.5%, and 2.0%) to produce loaves of approximately 50 g each. Bread production started with the mixing of the following ingredients: flour, water, yeast, and salt (the yeast was hydrated in warm water). The dough was worked (kneaded) for 10 to 15 min until it reached a smooth and elastic texture. Then, it was left for the first fermentation, resting for 1 h until it doubled in size.

After the initial fermentation, the dough was shaped into the desired form and underwent a second fermentation (30 min). The bread was then baked in an oven at 250 °C for approximately one hour. Once baked, the bread was left to cool and stored at room temperature in a dry environment. A total of six loaves were prepared: two without FEE (control), two with 0.5%, and two with 2.0% FEE. The bread was stored in paper bags purchased from a supermarket to simulate typical consumer storage conditions. The samples were subjected to analysis immediately after production (T0) and two days later (T2) for microbiological, chemical, and physicochemical analyses.

2.6. Microbiological Analysis

The purpose of the microbiological analyses was to evaluate the quality of the product obtained from the fermented elderberries (at 0 h and after 144 h) and the bread samples (at time 0 and after 2 days).

The quantification of moulds and yeasts was performed using DRBC (Rose Bengal) agar medium. Growth was observed from the second day of incubation at 25 °C for yeasts and up to the fifth day of incubation for moulds [18].

Sulphite-reducing clostridial spores were analysed and quantified by incorporation after the inactivation of the sample. The tubes were placed in anaerobic conditions and incubated at 37 °C for 5 days [19].

The SimPlate kit (Biocontrol) was used to test and quantify total coliforms and *Escherichia coli*, following the manufacturer's recommendations [20].

Total coliforms were assessed by observing the colour change in the wells of the KIT plate from blue to pink, while the detection and quantification of *E. coli* were performed by counting fluorescing wells under ultraviolet (UV) light at 365 nm. The results were interpreted using the conversion table provided by the manufacturer.

Staphylococcus aureus enumeration was performed with Baird-Parker medium (BP-HIMEDIA), supplemented with egg yolk, and the spread technique was used. Incubation was performed for 48 h at 37 °C [21]. After this period, characteristic colonies (grey colonies with a transparent halo) and non-characteristic colonies were selected to test for the presence of positive coagulase. These were inoculated into Brain Heart Infusion (BHI) broth and incubated at 37 °C for 24 h. Rabbit plasma was then added and incubated for

12 h. *S. aureus* colonies were identified if clot formation was observed at the bottom of the tube.

Salmonella spp. were tested using the 1–2 TEST kit from BioControl and Ambifood (Porto, Portugal) [22]. After the pre-enrichment and enrichment of the samples at 37 °C, the kits were prepared according to the method recommended by the manufacturer. The kits were incubated at 37 °C for 24 h, after which the presence or absence of an immunoprecipitation band was observed, indicating positivity for *Salmonella*.

Bacillus cereus was quantified by plating 0.1 mL of each dilution onto Petri dishes containing selective agar for *B. cereus* (MYP) [23]. Plates were incubated at 30 °C for 24 to 48 h. Positive colonies were identified by their pink colour.

The culture medium used to count total mesophilic aerobic bacteria was Plate Count Agar (PCA), and the incorporation technique was used. Plates were incubated at 30 °C for 48 to 72 h [24].

In general, microbial counts were expressed in colony forming units (CFU) per gram or millilitre of sample.

2.7. Bioactive Compounds and Antioxidant Activity

Bioactive compounds and antioxidant activity were analysed for both the fermented berries and the bread enriched with the extract. The elderberry extract was also analysed for these compounds.

To carry out these analyses, hydroethanolic extracts were prepared from the elderberry, the fermented product, and the bread by weighing 1 g of each freeze-dried product, adding 25 mL of the hydroethanolic solution, stirring for one hour, filtering, and evaporating. The extract obtained was then dissolved in 10 mL of the ethanolic solution.

The method reported by Singleton et al. [25], with modifications, was used for the determination of total phenolic compounds (TPCs). For this analysis, 0.5 mL of sample, 2.5 mL of Folin–Ciocalteu reagent (10%), and 2 mL of 75 g/L sodium carbonate were pipetted. The solution was left in the dark for 2 h and its absorbance was read at 760 nm (Spectrophotometer UV-3100PC, VWR International LLC, Avantor, Radnor, PA, USA). The results were calculated from a calibration curve using gallic acid (GA) as a standard and with concentrations ranging from 4 to 60 mg/L, expressed in mg GAE per g of sample.

The flavonoid content (TFC) was determined using the method described by Woisky and Salatino [26] with modifications. For this analysis, 2.5 mL of sample and 2.5 mL of 2% (*w*/*v*) $AlCl_3$ were pipetted. The solution obtained was allowed to rest in the dark for 1 h. After this period, the absorbance was measured at 420 nm. Quercetin (Q) was used as a standard to construct the calibration curve using concentrations between 0.48 and 3 mg/L. The results were expressed in mg QE per g of sample.

Antioxidant capacity was assessed using the DPPH• and FRAP methods.

- DPPH• (2,2-diphenyl-1-picrylhydrazyl)

This method was performed as described by Hatano et al. [27] with slight modifications. For this test, 0.3 mL of sample and 2.7 mL of DPPH• reagent (2.0×10^{-4} M) were pipetted. The solutions were kept in the dark for 60 min and the absorbance was measured at 517 nm. The results were then expressed in mg TroloxE/g of sample, using a calibration curve with concentration ranges between 12.6 and 100.8 mg/L, established using Trolox.

- FRAP (Ferric-Reducing Antioxidant Power)

Antioxidant activity was also determined using the FRAP method according to Berker et al. [28]. FRAP reagent was prepared by mixing 0.3 M acetate buffer (pH 3.6), 10 mM TPTZ, and 20 mM $FeCl_3$ in a 10:1:1 ratio. For the analysis, 0.1 mL of sample was pipetted, followed by the addition of 3 mL of FRAP reagent and 0.3 mL of deionized water.

The solutions were incubated in the dark for 6 min, and absorbance was measured at 595 nm. The results were expressed in mg of Fe(II)E per gram of sample, using a calibration curve constructed with ferrous sulphate heptahydrate at concentrations between 25 and 350 mg/L.

2.8. Physicochemical Properties

The physicochemical composition of the bread samples was assessed by determining the following parameters: moisture, ash content, protein content, fat content, and pH. Energy and carbohydrate content were also calculated.

The moisture content was determined by AOAC 925.10 [29], measuring weight loss in an oven at 105 °C until a constant weight was achieved. The moisture percentage was calculated as the mass difference between the initial weight and the weight after drying.

The ash content was determined according to AOAC 923.03 [30] via the incineration of the samples in a muffle furnace at 600 °C (BOX Furnace—model 51894). The quantification of the mineral residue was calculated as the difference between the sample´s mass and the ash´s mass. The pH of the bread samples was evaluated according to [31], using a potentiometer (Mettler Toledo model, Mumbai, India) with a combined pH electrode. The method recommended by AOAC 920.85 [32] was used to quantify the total fat content of the bread samples. This method recommends Soxhlet extraction (Behrotest, Labor-Technik, Düsseldorf, Germany) using petroleum ether as solvent. Approximately 2 g of the sample was placed in a cartridge, which was placed in a previously weighed flat-bottomed flask containing 170 mL of light petroleum, and the system was heated for 6 h. After extraction, the solvent was distilled off and the residue was transferred to an oven at 105 °C for one hour. After cooling, the flask was weighed. To determine the total protein content of bread, we used AOAC 950.36 [33]. Two grams of the homogenised sample was weighed and placed in a Kjedahl tube. A volume of 12 mL of concentrated sulfuric acid, along with two catalyst tablets (composed of K_2SO_4 and Se), was added. The tube was placed in the mineralisation unit (DK 8-Heating Digester, Italy) at 420 °C for 60 min. After cooling the samples to 50 °C, distillation was started in the UDK Semi-Automatic Distillation Unit, followed by titration with 0.2 mol/L HCl. This method makes it possible to quantify the protein content of the samples analysed. To determine the percentage of protein in the bread samples, the percentage of free nitrogen present in the samples is first calculated and then the percentage of protein is calculated using the following formula: % protein = % nitrogen × 5.7.

2.9. Statistical Analysis

All data treatments were performed using the statistics software R 4.4.0 GUI 1.80 Big Sur Intel build (The R Foundation for Statistical Computing, Vienna, Austria), a free software environment for statistical computing and graphics. The statistical analysis included mean comparisons using Student's *t*-test, one-way ANOVA, two-way ANOVA, and the Generalised Linear Model (GLM). The data were tested for normality and homogeneity of variance assumptions using the Shapiro–Wilk test on residuals and Levene's Test, respectively. When the assumption of variance homogeneity was violated, the validity of the ANOVA results was compromised. Therefore, the GLM approach was applied to confirm whether there were significant differences in the means of the response variable. Taking into account the interactive terms and the factors under study, this offers a robust alternative due to its flexibility and ability to handle heteroscedasticity. GLMs extend traditional linear models by allowing the specification of a link function that relates the mean of the response variable to the linear predictors (an identity link function was used), as well as a variance structure that accommodates non-constant variance, enabling the model to adapt to the characteristics of the data. To determine which group means are statistically different, pairwise comparisons were performed using Tukey's Honest Significant Difference (HSD) test.

3. Results and Discussions

3.1. *Optimising Fermentation Conditions*

For optimising the fermentation conditions for elderberry, 15 different experiments were carried out, varying six parameters (X1: inoculum concentration; X2: urea; X3: $MgSO_4$; X4: KH_2PO_4; X5; pH; and X6: sample quantity). Three concentration levels were used. It was found that variables X4 and X6 were the most significant contributors to yeast growth, suggesting that they should be selected for further tests. An analysis of Figure 1 shows that growth was optimal in the presence of 0.5 g/L KH_2PO_4 and 50 g of sample.

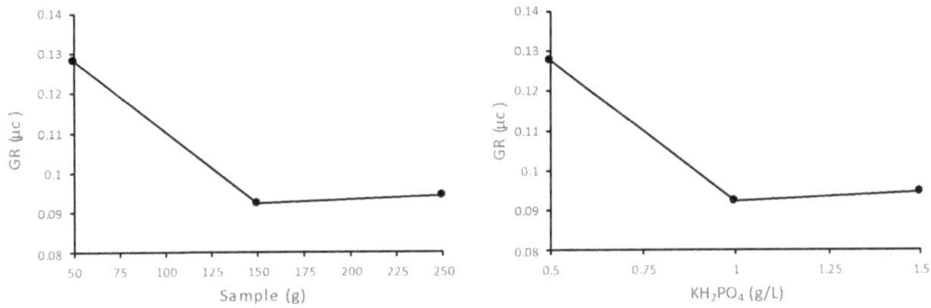

Figure 1. Influence of the variables that contributed to the growth rate (GR).

In other words, the maximum specific growth rate (0.128 µC) was obtained at the lowest concentration of KH_2PO_4 tested for a sample quantity of 50 g. In this context, all the fermentations in our study were carried out under the following conditions: 1.0 g/L urea; 0.5 g/L $MgSO_4$; 0.5 g/L KH_2PO_4; 50 g sample; and pH 4.50. The optimal inoculum concentration, determined based on experimental design analysis, was found to be 10^5 CFU/mL.

3.2. *Microbial and Chemical Profile of the Fermented Elderberries (FEs)*

3.2.1. Microbiological Analysis of the Fermented Elderberries

From the microbiological analyses (yeast and moulds, total coliforms/*E. coli*, *S. aureus*, *Clostridium*, *Salmonella*, and *B. cereus*) of the fermented elderberries, only yeast growth was detected (5.53 and 4.44 CFU/mL for T0 and T144, respectively). Analysis of the data showed statistically significant differences ($p < 0.001$) in the growth of moulds and yeasts between the two time points. The decrease in yeast population after 144 h of elderberry fermentation may be due to several factors, such as nutrient depletion, ethanol accumulation, pH changes, or competition with other microorganisms that contribute to the decrease in viable yeast cells.

3.2.2. Chemical Analysis of the Fermented Elderberries

The average results obtained for the elderberry extract (EE) and the fermentations are summarised in Table 2. All tests were performed in triplicate. Analyses were performed at time zero (T0) and after 144 h (T144).

Table 2. Results of the chemical analysis of the fermented product and application of ANOVA.

	EE	T0	T144
TPC (mg GAE/g S)	40.80 ± 0.21 [c]	79.63 ± 4.39 [b]	178.30 ± 1.78 [a]
TFC (mg QE/g S)	1.27 ± 0.05 [c]	13.66 ± 0.69 [b]	27.87 ± 0.34 [a]
DPPH• (mg TroloxE/g S)	21.69 ± 0.08 [c]	62.90 ± 1.09 [b]	116.05 ± 1.89 [a]
FRAP (mg. Fe (II)E/g S)	139.81 ± 1.29 [c]	204.14 ± 7.60 [b]	425.44 ± 4.88 [a]

Results expressed in mean of analysis ± standard deviation. EE: elderberry extract; TPC: total phenolic content; TFC: total flavonoids content; DPPH•: 2,2-Diphenyl-1-picrylhydrazyl; FRAP: Ferric-Reducing Antioxidant Power; S: sample; GAE: gallic acid equivalent; QE: quercetin equivalent; TroloxE: Trolox equivalent; Fe (II)E: ferrous ion equivalent. In each variable, different letters indicate significant mean differences (a–c).

Table 2 shows that all the evaluated parameters exhibited a twofold increase, both with the addition of the fermentation medium components and from T0 to T144. Fermentation enhanced the antioxidant properties with DPPH• and FRAP, by 2.64 and 2.08 times. This increase is likely attributed to the production or biotransformation of phenolic compounds and flavonoids by the yeast. Furthermore, the bioactive compounds, as indicated by total phenolic content (TPC) and total flavonoid content (TFC), showed a twofold increase, with values rising by 2.24 and 2.04 times, respectively, during the fermentation process.

In fermented products, Kwaw et al. [34] also observed an increase in the content of anthocyanins, phenolic compounds, flavonoids, and antimicrobial activity, which corroborates with the results obtained in this work. The increase in bioactive compounds and biological properties in the fermentation medium (T0) compared to the elderberry extract may be related to the reagents incorporated in the medium, i.e., urea, KH_2PO_4, and $MgSO_4$. KH_2PO_4 is a buffer that influences the action of the enzymes that carry out the reactions, such as hydroxylation or methylation, responsible for the degradation of the various compounds present in the fermentation medium. The antioxidant capacity of anthocyanins is also directly influenced by pH [35].

The ANOVA results regarding the TPC, TFC, and antioxidant capacity (DPPH• and FRAP) values showed significant differences (p-values < 0.001) between the values obtained in the three matrices analysed (letters a, b, and c).

The increase in total phenolic compounds can be attributed to the action of the yeast, suggesting that not only does it release these compounds from the plant material, but it also modifies them during its cellular metabolism.

According to YEO et al. [36], the yeast *S. cerevisiae* produces secondary metabolites during fermentation (aldehydes, esters, acids, phenols, biomass, and methanol, among others), which, in many situations, are responsible for different biological activities such as of an antimicrobial, antifungal, or antiviral nature. Also, according to the study conducted by Sabel et al. [37], yeasts, especially those of the genus *Saccharomyces*, were more resistant to the phenolic compounds tested compared to lactic and acetic bacteria, suggesting that yeasts are more resistant to the action of phenolic compounds.

3.3. Microbial and Chemical Profile of Bread

Figure 2 shows the three breads prepared with different concentrations of fermented elderberry extract (FEE: 0.0%, 0.5%, and 2.0%), highlighting their distinct colours.

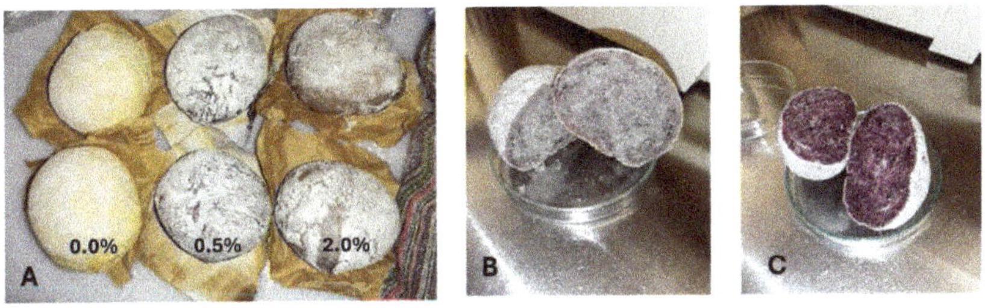

Figure 2. Photos of bread made with (**A**) 0.0%, 0.5%, and 2.0% of fermented elderberry extract; (**B**) 0.5% of fermented elderberry extract; and (**C**) 2.0% of fermented elderberry extract.

3.3.1. Microbiological Analysis of Bread

All the microorganisms studied (mould and yeast, total coliforms/*E. coli*, *S. aureus*, *Clostridium*, and total mesophilic) were absent in all the samples analysed, as expected due to the high baking temperature. After 2 days of storage, only the total mesophilic microorganisms were quantified; there was a decrease as a result of the increase in the concentration of FEE added to the bread (0.0%—5.18 ± 0.14 log CFU/mL; 0.5%—4.93 ± 0.03 log CFU/mL; 2.0%—3.87 ± 0.04 log CFU/mL). The number of aerobic mesophiles in the samples without added extract was significantly different from that in the bread with 2% extract ($p = 0.001$). According to Silva et al. [38], the number of aerobic mesophilic microorganisms is an indicator of the quality of products, raw materials, and handling conditions, i.e., low levels indicate the good general hygiene of a product; in this context, the samples studied can be considered satisfactory and do not represent increased risks to consumer health [39–42]. These results suggest that elderberries could be used to extend the shelf life of bread, as their ability to ensure microbiological safety was superior to that of bread containing only conventional ingredients.

3.3.2. Chemical Analysis of Bread

Table 3 shows the average values obtained for total phenolics and flavonoids, as well as antioxidant activity (DPPH and FRAP methods), for the three types of bread produced, in addition to the results of the two-way ANOVA study. Figure 3 presents the boxplots of the two-way ANOVA data of the TPC, TFC, DPPH, and FRAP results.

Table 3. Results of the chemical analysis of the bread with and without fermented elderberry extract, measured at 0 and 2 days of storage.

Bread	Time (Days)	TPC (mg eq. GAE/g S)	TFC (mg eq. QE/g S)	FRAP (mg Fe (II)E/g S)
0.0% FEE	0	8.63 ± 1.23 [c]	1.02 ± 0.03 [d]	9.16 ± 0.16 [c]
	2	10.25 ± 0.33 [c]	2.90 ± 0.03 [a]	10.45 ± 0.28 [c]
0.5% FEE	0	13.08 ± 0.44 [b]	0.65 ± 0.02 [e]	14.30 ± 0.44 [b]
	2	13.22 ± 0.62 [b]	1.72 ± 0.05 [c]	14.16 ± 0.22 [b]
2.0% FEE	0	20.56 ± 0.72 [a]	2.36 ± 0.12 [b]	25.61 ± 0.45 [a]
	2	20.40 ± 0.12 [a]	1.77 ± 0.05 [c]	25.66 ± 0.51 [a]

Results expressed in mean of analysis ± standard deviation. FEE: fermented elderberry extract; TPC: total phenolic content; TFC: total flavonoids content; DPPH: 2,2-Diphenyl-1-picrylhydrazyl; FRAP: Ferric-Reducing Antioxidant Power; S: sample; GAE: gallic acid equivalent; QE: quercetin equivalent; TroloxE: Trolox equivalent; Fe (II)E: ferrous ion equivalent. In each variable, different letters indicate significant mean differences (a–e).

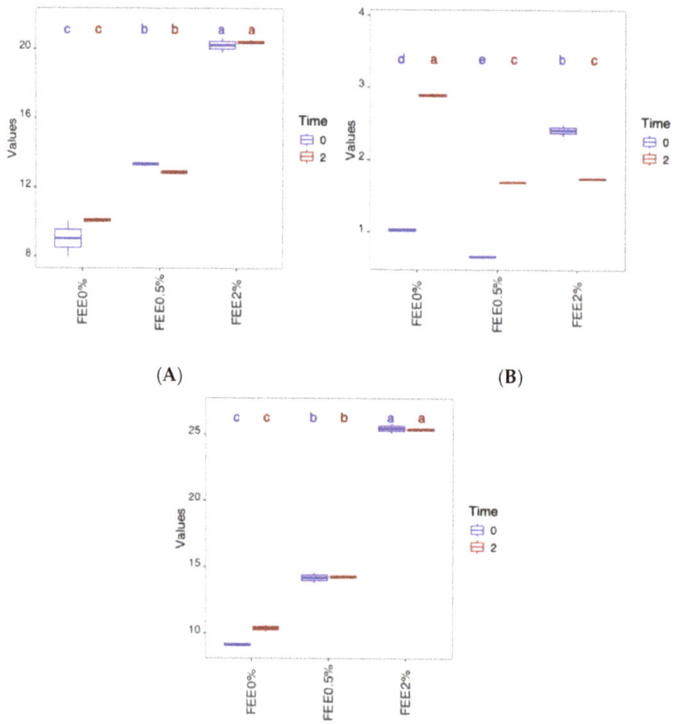

(C)

Figure 3. Two-way ANOVA with box plots for bread with and without fermented elderberry extract, measured at 0 and 2 days of storage. (**A**) TPC, (**B**) TFC. and (**C**) FRAP results. TPC: total phenolic content; TFC: total flavonoids content; FRAP: Ferric-Reducing Antioxidant Power. In each variable, different letters indicate significant mean differences (a–e).

From analysing Table 3, it was observed that there were no statistically significant differences in the quantification of TPC from T0 to T2 in all the samples studied, which means that they remained stable during the shelf life of the bread. The same can be observed for the antioxidant capacity determined using the FRAP method. It can also be seen that the extract factor (0.0%, 0.5%, and 2.0% FEE) had a significant effect on all the parameters studied ($p < 0.001$). There were significant differences, both between T0 and T2 and with the amount of extract added ($p < 0.001$), for the antioxidant capacity determined using TFC.

Karakaya et al. [43] studied the behaviour of anthocyanins (pigments belonging to the flavonoid group) over time in bakery products (bread, baguette bread, and biscuits) enriched with this compound and found that baking and storage (room temperature) did not affect the stability of anthocyanins, while maintaining their bioactive properties. This confirms the results obtained in this study relating to the behaviour of total phenolics.

Figure 3A shows that the amount of phenols increased as the FEE content added to the bread increased. However, in the samples where FEE was incorporated, the content of these compounds remained practically constant throughout the storage time. The phenol content of the control was the lowest, increasing from time 0 to the final time. Analysing the two times simultaneously for the same situation, it was found that in the bread incorporated with extracts, the phenolic content increased with the amount of extract added and remained practically constant during storage.

Analysing Figure 3B, for the two times analysed, the sample with 2.0% FEE had the highest total flavonoid content (2.36 ± 0.12 and 2.90 ± 0.03 mg eq Q/g sample, respectively), while the bread sample with 0.5% FEE had the lowest value (0.65 ± 0.02 mg eq Q/g sample). There were no significant differences between the samples with added FEE on the second day of storage. These results show that increasing the concentration of FEE added to the bread did not increase the total flavonoid content at T2.

The results shown in Figure 3C indicate that on the day the bread was made (T0), the highest value of antioxidant capacity was obtained in the control sample (247.23 ± 0.94 mg TroloxE/g sample), with this formulation also having the highest value at the end of the storage process. In the bread samples with FEE added, an increase in antioxidant activity was observed as a function of the incorporated extract content. For all the bread samples analysed, there was an increase in antioxidant activity from T0 to T2, as assessed using the DPPH method, with the sample containing 0.5% FEE showing the greatest increase.

The results obtained in the different bread samples at T0 and T2 for the antioxidant capacity assessed using the FRAP method (Figure 3C showed an increase due to the addition of FEE to the bread samples. Comparing the two times studied and the three samples, in the control sample (0.0% FEE) there was a slight increase in antioxidant activity from T0 to T2, while this variation was practically zero in the other two formulations.

The increase in antioxidant capacity assessed using the FRAP method may be related to the TPC of the bread samples, since the behaviour of these two parameters (Figure 3A,C) was similar [17].

3.3.3. Physicochemical Analysis

Table 4 summarises the results obtained for the parameters of protein, fat, moisture, ash, pH, carbohydrates, and energy for the bread samples at the two times studied (T0 and T2 days).

Table 4. Results of the physicochemical analysis of the bread with and without fermented elderberry extract, measured at 0 and 2 days of storage.

Bread	Time (Days)	Protein (%)	Moisture (%)	Ash (%)	pH	CH (g/100 g)	EV (Kcal/100 g)
0.0% FEE	0	8.44 ± 0.06 ^{a}	33.24 ± 1.13 ^{a}	1.01 ± 0.17 ^{b}	5.57 ± 0.05 ^{a}	86.97 ^{a}	381.93 ^{a}
	2	8.33 ± 0.03 ^{ab}	33.74 ± 0.26 ^{a}	1.50 ± 0.11 ^{b}	5.39 ± 0.01 ^{ab}	86.33 ^{a}	379.73 ^{a}
0.5% FEE	0	8.27 ± 0.05 ^{ab}	32.28 ± 0.62 ^{a}	1.32 ± 0.38 ^{b}	5.30 ± 0.06 ^{bc}	86.76 ^{a}	380.68 ^{a}
	2	7.93 ± 0.29 ^{b}	34.16 ± 0.14 ^{a}	1.24 ± 0.06 ^{b}	5.11 ± 0.04 ^{c}	87.21 ^{a}	380.85 ^{a}
2.0% FEE	0	8.17 ± 0.05 ^{ab}	3.22 ± 0.59 ^{a}	2.07 ± 0.74 ^{a}	4.21 ± 0.08 ^{d}	86.02 ^{a}	377.84 ^{a}
	2	8.31 ± 0.00 ^{ab}	31.75 ± 0.21 ^{a}	2.32 ± 0.15 ^{a}	4.12 ± 0.07 ^{d}	85.99 ^{a}	377.47 ^{a}

Results expressed in means of analysis ± standard deviation. FEE: fermented elderberry extract; CH: carbohydrates; EV: energy value. In each variable, different letters indicate significant mean differences (a–e).

Table 4 shows that, overall, the results obtained in the three samples for the different parameters studied decreased from T0 to T2. In the case of the control sample (0.0% FEE), protein and pH stand out, with significant differences between the two times studied, whereas in the sample with 0.5% FEE, all the parameters decreased except moisture, which remained constant over time. The 2.0% FEE sample only showed significant differences in the protein and fat parameters. The values obtained for protein content (between 7.93 ± 0.29 and 8.44 ± 0.06%) are within the range recommended by the General Directorate of Health [44] for white bread (8.40%).

The fat content of the samples studied was low (values ranging from 0.122 ± 0.003 to 0.030 ± 0.002 %) compared to that described by Dr. Ricardo Jorge at the National Health Institute [45], who indicates that the value for fat in white bread is 2.20%. These results can be explained by the fact that no type of fat (olive oil, oil, or butter) was added to the bread during its preparation.

The obtained moisture values ranged from 31.75 ± 0.21 to 35.22 ± 0.59 %, and there was no trend in relation to the addition of FEE and storage time (T0 and T2) for the different samples evaluated. Martini et al. [46] also obtained average moisture values ranging from 26.28 to 33.87 % for hot dog buns.

The ash content ranged from 1.01 ± 0.17 to 2.32 ± 0.15%. The values of this parameter in the sample with 2.0% FEE at T2 were almost double those observed in the control at T0, showing statistically significant differences between these two samples. The values obtained for ash in this study, ranging from 1.16 to 2.04 %, were identical to those reported by Martini et al. [46] in French bread, hot dog bread, and biscuits. According to Coutinho et al. [47], the ash content of samples is used to assess the purity and authenticity of foods, as it reflects the amount of inorganic minerals present in their composition.

Harbers and Nielsen [48] also reported that ash content is made up of inorganic substances of nutritional importance, i.e., the higher its content in food, the greater its contribution to health. In this context, the bread sample with the highest percentage of elderberry extract (2.0% FEE) was the one with the highest percentage of ash, which means that the consumption of bread with the highest amount of extract will have a greater contribution to human health.

The pH values obtained for the three samples at the two times studied ranged from 4.12 ± 0.07 to 5.57 ± 0.05. These values decreased from T0 to T2, as well as with the increase in the percentage of FEE incorporated into the bread. These results may be related to the increase in the amount of anthocyanins (the most abundant phenolic compound in elderberries) in the bread samples, to which a higher concentration of FEE was added. Bratu et al. [49] showed that in dried berry infusions, the greater the presence of anthocyanins in the samples, the lower the pH.

The total carbohydrate content of the different bread samples studied ranged from 85.99 to 87.21 g/100 g of bread. These values are slightly higher than those stipulated by the Directorate General for Health for bread known as 'white bread' or 'wheat bread' (57.3 g/100 g). These quantities may be related to the lower fat content obtained in this work.

The energy values obtained (377.47 and 381.93 Kcal/100 g) were also higher than those established by the Directorate General for Health (289 Kcal/100 g bread). This study also showed that the energy value of bread decreased slightly, although the differences were not significant (p = 0.0638), as the percentage of FEE added to the samples increased, suggesting that bread with a higher amount of extract can be considered more beneficial for human health.

Regarding fat, since the values obtained were very small and some of the errors were too high, it was considered more appropriate not to carry out a statistical analysis for this parameter.

Considering the overall results of the project, it can be concluded that elderberry holds significant potential as a functional ingredient in fermented products. However, further studies are essential to gain a better understanding of its variability and to expand upon the knowledge of its impact on various food applications. Assessing the variability of elderberries across Portugal is crucial for evaluating the feasibility of standardising the extract, including its chemical profile and biological properties. Additionally, selecting suitable probiotic microorganisms for the fermentation process and optimising the extract's characteristics are critical steps to enhance its application in the food industry.

We are not aware of any studies relating to the incorporation of elderberries in traditional bread. There are also no data in the literature on the optimisation of the fermentation process with a probiotic yeast, or on the effect of using a fermented product on the characteristics of the bread, particularly its preservation and bioactivity.

4. Conclusions

Elderberries are rich in phenolic compounds and flavonoids, with potential for use as a functional food. Fermentation with *S. cerevisiae* showed that lower amounts of sample and KH_2PO_4 improved the process, increasing total phenolic compounds and antioxidant activity. After pasteurisation, the fermented product showed a low microbial load, making it bioactively interesting for incorporation into bread.

When fermented berries (FEE) were added to the bread, there was an increase in total phenolic compounds, antioxidant activity (especially via the FRAP method), and an improvement in conservation, with a reduction in the count of mesophilic aerobic microorganisms. At the same time, there were physicochemical changes, such as an increase in ash and a decrease in pH.

The results suggest that fermented elderberries show great potential for enhancing bread by improving its bioactive properties, enriching its nutritional value, and extending its shelf life.

However, future research should evaluate the sensory aspect and toxicity of this product.

Author Contributions: Conceptualization: L.M.E. and L.G.D.; methodology: N.L.S. and L.M.E.; validation: L.M.E. and T.D.; formal analysis: N.L.S. and V.B.P.; investigation: N.L.S. and V.B.P.; resources: L.M.E.; data curation: N.L.S., V.B.P., T.D., L.G.D. and L.M.E.; writing—original draft preparation: N.L.S.; writing—review and editing: L.M.E. and T.D.; supervision: L.G.D. and L.M.E. All authors have read and agreed to the published version of the manuscript.

Funding: This work was supported by national funds through FCT/MCTES (PIDDAC): CIMO, UIDB/00690/2020 (DOI:10.54499/UIDB/00690/2020), and UIDP/00690/2020 (DOI:10.54499/UIDP/00690/2020); and SusTEC, LA/P/0007/2020 (DOI: 10.54499/LA/P/0007/2020).

Data Availability Statement: The original contributions presented in this study are included in the article. Further inquiries can be directed to the corresponding author.

Conflicts of Interest: The authors declare no conflicts of interest.

References

1. Cunha, A.P. *Plantas e Produtos Vegetais em Cosmética e Dermatologia*; Fundação Calouste Gulbenkian: Lisboa, Portugal, 2011.
2. Viapiana, A.; Wesolowski, M. The phenolic contents and antioxidant activities of infusions of *Sambucus nigra* L. *Plants Foods Hum. Nutr.* **2017**, *72*, 82–87. [CrossRef] [PubMed]
3. Senica, M.; Stampar, F.; Veberic, R.; Mikulic-Petkovsek, M. Processed elderberry (*Sambucus nigra* L.) products: A beneficial or harmful food alternative? *LWT Food Sci. Technol.* **2016**, *72*, 182–188. [CrossRef]
4. Wieland, L.S.; Piechotta, V.; Feinberg, T.; Ludeman, E.; Hutton, B.; Kanji, S.; Seely, D.; Garritty, C. Elderberry for prevention and treatment of viral respiratory illnesses: A systematic review. *BMC Complement. Med. Ther.* **2021**, *21*, 112. [CrossRef]
5. Liu, D.; He, X.Q.; Wu, D.T.; Li, H.B.; Feng, Y.B.; Zou, L.; Gan, R.Y. Elderberry (*Sambucus nigra* L.): Bioactive compounds, health functions, and applications. *J. Agric. Food Chem.* **2022**, *70*, 4202–4220. [CrossRef] [PubMed]
6. Duymus, H.G.; Göger, F.; Başer, K.; Hüsnü, C. In vitro antioxidant properties and anthocyanin compositions of elderberry extracts. *Food Chem.* **2014**, *155*, 112–119. [CrossRef] [PubMed]
7. Cao, H.; Bai, M.; Lou, Y.; Yang, X.; Zhao, C.; Lu, K.; Zhang, P. Optimization of the Brewing Process and Analysis of Antioxidant Activity and Flavor of Elderberry Wine. *Fermentation* **2023**, *9*, 276. [CrossRef]

8. Domínguez-Valencia, R.; Cittadini, A.; Pateiro, M.; Munekata, P.E.; Lorenzo, J.M. Elderberry Lipophilic and Hydrophilic Bioactive Compounds: Characterization and Extract Encapsulation. *Foods* **2023**, *12*, 4233. [CrossRef] [PubMed]
9. Ricci, A.; Cirlini, M.; Levante, A.; Dall'Asta, C.; Galaverna, G.; Lazzi, C. Volatile profile of elderberry juice: Effect of lactic acid fermentation using *L. plantarum*, *L. rhamnosus* and *L. casei strains*. *Food Res. Int.* **2018**, *105*, 412–422. [CrossRef] [PubMed]
10. Smid, E.J.; Kleerebezem, M. Production of aroma compounds in lactic fermentations. *Annu. Rev. Food Sci. Technol.* **2014**, *5*, 313–326. [CrossRef]
11. Feng, X.; Wu, Z.; Weng, P. Characterization of metabolites of elderberry juice fermented by *Lactobacillus bulgaricus* BNCC336436 and *Streptococcus thermophilus* ABT-T using LC–MS/MS. *J. Food Meas. Charact.* **2022**, *16*, 4486–4496. [CrossRef]
12. Hang, J.; Wu, Z.; Weng, P. Changes of Quality and Volatile Flavor Compounds of Elderberry Juice Fermented by Lactic Acid Bacteria. *J. Chin. Inst. Food Sci. Technol.* **2022**, *22*, 291–299.
13. Gębski, J.; Jezewska-Zychowicz, M.; Szlachciuk, J.; Kosicka-Gębska, M. Impact of nutritional claims on consumer preferences for bread with varied fiber and salt content. *Food Qual. Prefer.* **2019**, *76*, 91–99. [CrossRef]
14. ASAE. O Pão, Abril de 2017. Available online: https://www.asae.gov.pt/perguntas-frequentes1/area-alimentar/pao.aspx (accessed on 12 September 2024).
15. Portaria nº52/2015, 2015, Características dos Diferentes Tipos de Pão. Available online: https://diariodarepublica.pt/dr/detalhe/portaria/52-2015-66603017 (accessed on 12 September 2024).
16. Collar, C. Bread: Types of bread. In *Encyclopedia of Food and Health*; Elsevier: Amsterdam, The Netherlands, 2016; pp. 500–507.
17. Zammel, N.; Saeed, M.; Bouali, N.; Elkahoui, S.; Alam, J.M.; Rebai, T.; Kausar, M.A.; Adnan, M.; Siddiqui, A.J.; Badraoui, R. Antioxidant and anti-Inflammatory effects of *Zingiber officinale roscoe* and *Allium subhirsutum*: In silico, biochemical and histological Study. *Foods* **2021**, *10*, 1383. [CrossRef] [PubMed]
18. *ISO 21527-2:2008*; Microbiology of Food and Animal Feeding Stuffs—Horizontal Method for the Enumeration of Yeasts and Moulds—Part 2: Colony Count Technique in Products with Water Activity Less Than or Equal to 0.95. International Organization for Standardization: Geneva, Switzerland, 2008.
19. *ISO 15213:2003*; Microbiology of Food and Animal Feeding Stuffs—Horizontal Method for the Enumeration of Sulfite-Reducing Bacteria Growing Under Anaerobic Conditions. International Organization for Standardization: Geneva, Switzerland, 2003.
20. *AOAC Official Method 2005.03*; Detection and Confirmed Quantitation of Coliforms and E. coli in Foods. SimPlate Coliform and E. coli Color Indicator. In Official Methods of Analysis of AOAC International. AOAC Publications: New York, NY, USA, 2003.
21. *ISO 6888-1:1999/Amd 1:2003*; Microbiology of Food and Animal Feeding Stuffs—Horizontal Method for the Enumeration of Coagulase-Positive Staphylococci (Staphylococcus aureus and Other Species)—Part 1: Technique Using Baird-Parker Agar Medium. International Organization for Standardization: Geneva, Switzerland, 2003.
22. *AOAC 989.13*; Motile Salmonella in All Foods. Immunodiffusion (1-2 TEST) Method. In Official Methods of Analysis of AOAC International. AOAC Publications: New York, NY, USA, 2013.
23. *ISO 7932:2004*; Microbiology of Food and Animal Feeding Stuffs—Horizontal Method for the Enumeration Of Presumptive Bacillus cereus—Colony-Count Technique at 30 Degrees. International Organization for Standardization: Geneva, Switzerland, 2004.
24. *ISO 4833-1:2013(en)*; Microbiology of the Food Chain—Horizontal Method for the Enumeration of Microorganisms—Part 1: Colony Count at 30 °C by the Pour Plate Technique. International Organization for Standardization: Geneva, Switzerland, 2013.
25. Singleton, V.L.; Orthofer, R.; Lamuela-Raventos, R.M. Analysis of total phenols and other oxidation substrates and antioxidants by means of Folin–Ciocalteu reagent. *Method Enzymol.* **1999**, *299*, 152–178.
26. Woisky, R.G.; Salatino, A. Analysis of propolis: Some parameters and procedures for chemical quality control. *J. Apic. Res.* **1998**, *37*, 99–105. [CrossRef]
27. Hatano, T.; Kagawa, H.; Yasuhara, T.; Okuda, T. Two new flavonoids and other constituents in licorice root: Their relative astringency and radical scavenging effects. *Chem. Pharm. Bull.* **1988**, *36*, 2090–2097. [CrossRef]
28. Berker, K.I.; Güçlü, K.; Tor, İ.; Apak, R. Comparative evaluation of Fe (III) reducing power-based antioxidant capacity assays in the presence of phenanthroline, batho-phenanthroline, tripyridyltriazine (FRAP), and ferricyanide reagents. *Talanta* **2007**, *72*, 1157–1165. [CrossRef] [PubMed]
29. *AOAC 925.10*; Solids (Total) and Moisture in Flour. AOAC International: Rockville, MD, USA, 1925.
30. *AOAC 923.03*; Ash of Flour. Direct Method. AOAC International: Rockville, MD, USA, 1923.
31. *AOAC 943.02*; pH of Flour. Potenciometric Method. AOAC International: Rockville, MD, USA, 1943.
32. *AOAC 920.85*; Fat (Crude) or Ether Extract in Flour. In Methods of the Association of Official Analytical Chemists. AOAC International: Rockville, MD, USA, 1920.
33. *AOAC 950.36*; Protein in Bread. AOAC International: Rockville, MD, USA, 1950.
34. Kwaw, E.; Ma, Y.; Tchabo, W.; Apaliya, M.T.; Wu, M.; Sackey, A.S.; Xiao, L.; Tahir, H.E. Effect of Lactobacillus strains on phenolic profile, color attributes and antioxidant activities of lactic-acid-fermented mulberry juice. *Food Chem.* **2018**, *250*, 148–154. [CrossRef]

35. Nicolescu, A.; Babotă, M.; Barros, L.; Rocchetti, G.; Lucini, L.; Tanase, C.; Mocan, A.; Bunea, C.I.; Crişan, G. Bioaccessibility and bioactive potential of different phytochemical classes from nutraceuticals and functional foods. *Front. Nutr.* **2023**, *10*, 1184535. [CrossRef]
36. Yeo, S.K.; Ewe, J.A. Effect of fermentation on the phytochemical contents and antioxidant properties of plant foods. In *Advances in Fermented Foods and Beverages*; Woodhead Publishing: Sawston, UK, 2015; pp. 107–122.
37. Sabel, A.; Bredefeld, S.; Schlander, M.; Claus, H. Wine phenolic compounds: Antimicrobial properties against yeasts, lactic acid and acetic acid bacteria. *Beverages* **2017**, *3*, 29. [CrossRef]
38. Silva, N.; Junqueira, V.C.A.; de Arruda Silveira, N.F.; Taniwaki, M.H.; Gomes, R.A.R.; Okazaki, M.M. *Manual de Métodos de Análise Microbiológica de Alimentos e Água*; Editora Blucher: São Paulo, Brazil, 2017.
39. Dias, I.; Laranjo, M.; Potes, M.E.; Agulheiro-Santos, A.C.; Ricardo-Rodrigues, S.; Fialho, A.R.; Véstia, J.; Fraqueza, M.J.; Oliveira, M.; Elias, M. Autochthonous starter cultures are able to reduce biogenic amines in a traditional portuguese smoked fermented sausage. *Microorganisms* **2020**, *8*, 686. [CrossRef] [PubMed]
40. Fraqueza, M.J.; Laranjo, M.; Alves, S.; Fernandes, M.H.; Agulheiro-Santos, A.C.; Fernandes, M.J.; Potes, M.E.; Elias, M. Dry-cured meat products according to the smoking regime: Process optimization to control polycyclic aromatic hydrocarbons. *Foods* **2020**, *9*, 91. [CrossRef]
41. Rodrigues, J.C.F. Avaliação da Qualidade Higiênico-Sanitária em Abatedouro Frigorífico de Bovinos. Ph.D. Thesis, Instituto Federal de Educação, Ciência e Tecnologia Goiano, Campus Rio Verde, Rio Verde, Brazil, 2019.
42. Prado, J.C.S.; dos Santos Fontenelle, R.O.; Coutinho, M.G.S.; da Silva, L.C.; Neves, A.M. Avaliação microbiológica e parasitológica da água proveniente de chuvas armazenadas em cisternas localizadas em uma zona rural do município de Santana do Acaraú, CE. *Rev. Uningá* **2019**, *56*, 141–150. [CrossRef]
43. Karakaya, S.; Simsek, S.; Eker, A.T.; Pineda-Vallido, C.; Dupont, D.; Perez, B.; Viadel, B.; Sanz-Buenhombre, M.; Rodriguez, A.G.; Kertész, Z.; et al. Stability and bioaccessibility of anthocyanins in bakery products enriched with anthocyanins. *Food Funct.* **2016**, *7*, 3488–3496. [CrossRef] [PubMed]
44. DGS—Direção Geral da Sáude. 2021. Available online: https://alimentacaosaudavel.dgs.pt/alimento/pao/ (accessed on 30 September 2024).
45. Instituto Nacional de Saúde Dr. Ricardo Jorge, Tabela de Composição de Alimentos. 2019. Available online: https://jspapp.test.insa.foodcase-services.com/# (accessed on 30 September 2024).
46. Martini, N.O.; Escobar, T.D.; Kaminski, T.A. Caracterização físico-química de pães do tipo francês, bolacha e de cachorro-quente. *Rev. Inst. Adolfo Lutz* **2016**, *75*, 1–8. [CrossRef]
47. Coutinho, A.J.M.; Santos, M.M.R.; dos Santos, E.F.; Novello, D. Cupcake adicionado de farinha de resíduos de brócolis: Análise físico-química e sensorial entre crianças. *Multitemas* **2021**, *26*, 5–20. [CrossRef]
48. Harbers, L.; Nielsen, S. *El análisis de cenizas. Em S. Suzanne Nielsen (coord.) Análisis Aliment*; Editorial ACRIBIA, S.A: Zaragoza, Spain, 2003; pp. 121–132.
49. Bratu, M.M.; Birghila, S.; Popescu, A.; Negreanu-Pirjol, B.S.; Negreanu-Pirjol, T. Correlation of antioxidant activity of dried berry infusions with the polyphenols and selected microelements contents. *Bull. Chem. Soc. Ethiop.* **2018**, *32*, 1–12. [CrossRef]

foods

MDPI

Article

From a Coriander Mayonnaise to a Vegan Analogue: Assessing pH and Salt Influence in a *Saccharomyces cerevisiae* Yeast Protein Extract and *Chlorella vulgaris* Mixed System

Pedro Coelho [1], Carmo Serrano [1], Norton Komora [1,2] and Anabela Raymundo [1,*]

[1] Linking Landscape, Environment, Agriculture and Food (LEAF), School of Agronomy, Lisbon University (ISA), Tapada da Ajuda, 1349-017 Lisboa, Portugal; pmcoelho@isa.ulisboa.pt (P.C.)
[2] Mendes Gonçalves SA, Zona Industrial, Lote 6, 2154-909 Golegã, Portugal
* Correspondence: anabraymundo@isa.ulisboa.pt

Abstract: History aside, traditional mayonnaise faces a rising animal welfare concern dietary approach and remains dependent on cold environments throughout the supply chain due to food safety. Nowadays, consumers are able to find alternative formulas from vegetable sources with relevant emulsifying capacity. However, sensory characteristics may differ from the traditional expected product. A mixed system composed of the innovative ingredient heterotrophic white *Chlorella vulgaris* and a disruptive emulsifier, yeast protein extract (YPE), was assessed to transform traditional coriander mayonnaise into an analog product. The effect of pH and salt (NaCl) content was also evaluated. The mixed system depicts a promising stability since the average Sauter diameter of both is similar (7.94 µm and 7.49 µm), also observed in the unimodal droplet size distribution. Viscoelastic behavior has slightly different responses for the plateau model (278.951 Pa and 252.053 Pa), while increasing the salt content reflects an approximation regarding firmness (0.059 N and 0.057 N) and adhesiveness (0.372 N.s and 0.361 N.s). Introduction of microalgae increases bioactivity, mainly TPC (+118.84 ugGAeq/g) and antioxidant activity—RSA (+31.29 ugTEAC/g) and FRAP (+35.26 ugTEAC/g). Despite the color deviation, the sensorial analysis of both products enlightened the absence of major perception.

Keywords: mayonnaise; yeast protein extract (YPE); microalgae *C. vulgaris*; pH; salt

1. Introduction

The food sector raises a contemporary concern driven by sustainability issues, mainly combining nutritious food offers and limited natural resource management. Growing trends align with the preference and consumption of foods containing plant-based proteins [1]. This shift generates opportunities for food researchers and companies to lead disruptive innovation. Flexitarian lifestyle consumers may lead the promotion of non-conventional alternative foods consumption if the circle of research-industry-consumers respect a clear pact of transparent and helpful information associated with a competitive price [2]. New millennium research approaches mayonnaise production in order to develop similar products from alternative nonanimal proteins, facing a rising need to find valid options for egg proteins [3]. Current challenges remain around sensory characteristics as aroma, color, flavor, and texture. Alternative protein incorporation may very well improve the physicochemical, nutritional, and functional properties of mayonnaises [4]. Recent studies have found potential in plant-origin protein mixed systems incorporated in emulsions, such as grass pea sweet miso with lupin and faba bean [5] and white fruit flour from apple, nectarine, pear, and peach with lupin and faba bean [6]. Another study compared emulsions produced with solo proteins of aquafaba, faba bean, green pea, and lupin [7]. Other disruptive approaches studied analogues of proteins like single-cell proteins from the yeast *Saccharomyces cerevisiae* [8], the green-colored cyanobacteria *Spirulina* [9], and the insect *Tenebrio molitor* [10]. These range of ingredients depict potential as emulsifiers based on their high protein content, alongside the possible nutritional boost in terms of bioactive compounds and minerals.

In traditional mayonnaise, egg fractions have an important emulsifier role. Egg yolk is a standard ingredient, mainly due to the presence of lecithin, which is a complex mixture of phospholipids acting as a natural emulsifier [11]. Egg white is also commonly added because of its rich nutritional value and functional properties, specifically emulsifying, foaming, gelling, and thickening [12]. Whole or single fraction addition can be in liquid or powder form. An emerging eco-friendly consciousness among the new generation of consumers raises questions towards the food industry about food safety and security, sustainability, and animal welfare. Representing a historical and widely consumed sauce, traditional mayonnaise composition complies with these emerging concerns. At first, the addition of unpasteurized eggs concerned consumers' health in terms of food safety and quality, before a gap of unmonitored contamination risk by *Salmonella* [4]. Secondly, this emulsion-based sauce's components are strictly dependent on refrigerated supply chains, requiring impeccable quality control parameters to avoid microbial contamination [13]. Thirdly, containing a high proportion of unsaturated fats, mayonnaise becomes susceptible to oxidation acceleration in the presence of small quantities of transition metals, further developing off-flavors [14]. A mitigation strategy resides in the addition of antioxidants such as rosemary extract and Ethylenediaminetetraacetic acid (EDTA). Although these are not naturally present in mayonnaise components, they work as a chelating agent that sequesters ions vulnerable to oxidation [15].

The present work focuses on the substitution of animal protein in readily available coriander mayonnaise with a mixed system of *Saccharomyces cerevisiae* yeast protein extract (YPE) and white *Chlorella vulgaris* (Microalgae). In one hand, YPE stands recently as an emerging novelty in both food and feed, representing a disruptive source of both macro and micronutrients, a versatile flavoring agent controlled by Maillard reactions, and a production line processes enhancer such as reducing yogurt fermentation time, fining wines, and extending processed meat products physical properties [16,17]. Regarding its many applications and technological advantages, yeast proteins are considered safe and economically accessible for human consumption [18]. Preliminary studies state that *S. cerevisiae* contains polymers capable of emulsifying and stabilizing emulsions [19]. Yeast extract emulsions' kinetic stability is guaranteed by the droplets' electrostatic repulsion, viscoelastic behavior, and resemblance to pickering-style stabilization. These mechanisms increase the emulsion resistance against physical instabilization phenomena [20]. On the other hand, microalga are not only suitable for inclusion in daily human diets to enhance nutritive intake [21], but they also contain functional properties capable of stabilizing oil-in-water emulsions [22]. Microalga's natural composition favors new sensory experiences, offers functional properties with technological interest, prevents oxidation and color change processes in foods, and aligns with sustainable production and harvesting sources. Several studies underline the potential of *Chlorella vulgaris* as a dietary supplement with a positive impact on human health maintenance. Nutritive and vitaminic richness in *C. vulgaris* composition relates to antidiabetic, antihyperlipidemic, antihypertensive, antioxidant, and immunomodulatory properties [23]. Chlorella *vulgaris* is a heterotrophic microalga, requiring certain levels of sunlight exposure in a CO_2 environment to attain necessary sources of energy and carbon [24].

Therefore, microalga can be considered a healthy substitute for the lipoproteins in egg yolk and can also work as a natural substitute for antioxidants and colorants, making it possible to obtain greener Clean Label products. The recent decade rose among the consumers a food philosophy guided by healthy eating habits, motivating the consumption of safe foods with a valid nutritive value to support the immune system [25]. Developed countries' societies lean towards Clean Label products, under the values of convenience, health concerns, and sustainability. Convenience refers to a rising consumption of pre-cooked or ready-to-eat meals. Health concerns focus on fighting unbalanced consumption habits related to diseases and allergies/intolerances towards certain types of food components. Sustainability corroborates with a greater acceptance of the climatic changes due to constant pollution from conventional agricultural and nonorganic food production practices [26]. Clean Label products are responsible for reducing the number of ingredients in their formulation, substituting most of their synthetic components with natural sources, and sharing transparency with the consumer. As far as consumers' perception is concerned, coloring agents and stabilizer agents are less natural ingredients [27]. A mixed system of YPE and microalgae biomass (heterotrophic white *Chlorella vulgaris*) may change the perception of the consumer and answer the contemporary gap between consumption intent and market availability. This emulsifying system was evaluated, as well as the impact of pH and salt on the stability of the emulsified system.

The main goals of the following research are the development of a vegan clean label coriander mayonnaise analog with overall improvements associated with the mixed system application and the study of possible physical changes in the final product due to unforeseen pH and salt concentration deviations during the production chain.

2. Materials and Methods

2.1. Materials

Main ingredients were supplied by Casa Mendes Gonçalves (Golegã, Portugal), in order to mimic their coriander mayonnaise formula: colza oil, alcohol vinegar, modified starch, egg yolk powder, salt, sugar, coriander, colorant (shade yellow; E-104), potassium sorbate (E-200), citric acid (E-330), xanthan gum (E-415) and ethylenediaminetetraacetic acid (EDTA; E-385). Usually, the company produces the sauces with sunflower oil; however, due to the recent Russia–Ukraine war, global markets were affected, and the price of vegetable oils rose [28], directly impacting the food industry. Yeast protein extract (M2344-1) retrieved from *Saccharomyces cerevisiae* was kindly supplied by Proenol SA (Canelas, Portugal), a biotech Portuguese company leading the supply of this ingredient to continuous research focused on its application in food products. Heterotrophic white *Chlorella vulgaris* results from controlled environment mutagenesis [29], produced by Allmicroalgae (Pataias, Portugal).

2.2. Methods

2.2.1. Mayonnaise Production

The control formula's main components were disclosed by the company Casa Mendes Gonçalves in order to study alternative sustainable proteins to produce an analog to the market-available coriander mayonnaise sauce. To achieve a free animal food product, egg yolk powder was substituted by a mixed system (CvY) composed of 0.8% of *Saccharomyces cerevisiae* yeast protein extract and 1% heterotrophic white *Chlorella vulgaris*. After a battery of attempts with different percentages in mixed systems for each protein (Figure A1), the present analysis focused on a concentration that better fitted the requirements from the company in comparison with the current target (Control). Also, changes in pH and salt concentration were assessed to study the full extent of its influence in terms of rheological response. Batches were produced in a Thermomix kitchen robot (Vorwerk, Wuppertal, Germany), under the following settings: temperature = 20 °C, medium velocity = 4 position, and emulsification time = 40 s. The final formula in Table 1 was applied in 1 kg batches for an initial scale-up viability assessment.

Table 1. Coriander mayonnaise formulas (control; 0.8% YPE + 1% *C. vulgaris* (CvY) with different concentrations of citric acid (CvY_ca) and added salt—NaCl (CvY_salt)).

Ingredients	Formulation—g/100 g of Sample					
	Control	CvY	CvY_ca0.1	CvY_ca0.4	CvY_salt0.1	CvY_salt0.5
Colza oil	55.76	55.76	55.76	55.76	55.76	55.76
Distilled water	31.42	31.42	**31.54**	**31.24**	**31.62**	**31.22**
Alcohol vinegar	4	4	4	4	4	4
Modified starch	0.3	0.3	0.3	0.3	0.3	0.3
Egg yolk powder	**1.8**	-	-	-	-	-
YPE	-	**0.8**	0.8	0.8	0.8	0.8
Salt (NaCl)	0.3	0.3	0.3	0.3	**0.1**	**0.5**
Granulated sugar	3.5	3.5	3.5	3.5	3.5	3.5
Coriander	2	2	2	2	2	2
Shade yellow	0.3	0.3	0.3	0.3	0.3	0.3
Potassium sorbate	0.26	0.26	0.26	0.26	0.26	0.26
Citric acid	0.22	0.22	**0.1**	**0.4**	0.22	0.22
Xanthan gum	0.125	0.125	0.125	0.125	0.125	0.125
EDTA	**0.0072**	-	-	-	-	-
White *C. vulgaris*	-	**1.0072**	1.0072	1.0072	1.0072	1.0072
Total (g)	100	100	100	100	100	100

Powdered sources of proteins, white *C. vulgaris* biomass and *S. cerevisiae* (YPE), were hydrated [8] and homogenized in distilled water upon a AREX heating magnetic stirrer (VELP Scientifica Srl, Usmate, Italy) for 30 min at room temperature. Before emulsification, the remaining dried ingredients were added to the previously hydrated mixed system. Emulsification occurred with the colza oil pouring, followed by vinegar addition. Each mayonnaise was analyzed on the following day in terms of pH, rheology behavior, texture, color, and droplet size distribution. Antioxidant potential and total phenolic compounds quantification were also assessed.

2.2.2. Physicochemical Characterization

Rheological characterization was carried out using a rotational rheometer, the Haake Mars III Modular Advanced Rheometer System (Thermo Fisher Scientific Inc., Karlsruhe, Germany), connected with the Peltier cooling system (Thermo Fisher Scientific Inc., Karlsruhe, Germany) and the air compression system Eheim professional (Eheim, Deizisau, Germany).

Frequency sweep tests were performed with an ensemble of a conic probe and plate system. Steady-state flow measurements were performed with a serrated parallel plate system with a 1.5 mm gap. Both ensemble systems had 35 mm, and tests ran at 20 °C.

Viscoelasticity

Frequency sweep tests, or mechanical spectra, were assessed by means of small amplitude oscillatory shear (SAOS) measurements within the predefined linear viscoelastic region—LVR (accessed by a stress test at 1 Hz) for a frequency range of 0.01–100 Hz. The output refers to the mechanical spectrum (G′ and G″ as a function of frequency) and the loss tangent (tan δ = G″/G′). The plateau modulus ($G^0{}_N$) was also estimated as the value of G′ observed for the minimum value of the loss tangent [30].

Viscosity

Viscosity tests were assessed by means of steady-state flow curves in a shear rate range of 10^{-4} to 500 s^{-1}. The Williamson model (1) was applied to adjust the test outputs using the Origin 2019 software (OriginLab, Northampton, MA, USA) for a better understanding of the viscosity *versus* shear rate curves [31].

$$\eta = \frac{\eta_0}{1 + (k\dot{\gamma})^m} \tag{1}$$

In the Williamson model (1), η_0 is the zero shear rate limiting viscosity at low shear rates (Pa.s); k is the consistency coefficient (s) and a dimensionless shear-thinning index; and m is the slope of the power law shear-thinning region. Both tests were carried out in triplicate.

Texture Evaluation

To compare the final results, a texture profile analysis (TPA) was performed in a TA.XT plus texturometer (Stable Micro-Systems, Godalming, UK). TPA was performed with a 15 mm double penetration at 1 mm/s of a 19 mm diameter cylindrical acrylic probe in cylindrical glass jars (100 mL with 66 mm diameter, 56 mm height) filled with each sample up to 32 mm in height. Each sample was tested 5 times at room temperature. The main test outputs are firmness (N) and adhesiveness (N.s).

Evaluation of the Droplet Size Distribution

Droplet size distribution (DSD) was assessed to achieve an initial correlation with the final product structure and physical stability. Tests were carried through a Partica LA-960V2 laser scattering particle size distribution analyzer (Horiba, Kyoto, Japan). The minimal necessary portion of emulsion was introduced as per the equipment software's running test indication. Test outputs are the $d_{3,2}$ (2) and $d_{4,3}$ (3) diameters, also span (4), after a triplicate measurement.

Sauter diameter ($d_{3,2}$):

$$d_{3,2} = \frac{\sum n_i d_i^3}{\sum n_i d_i^2} \tag{2}$$

De Brouckere diameter ($d_{4,3}$):

$$d_{4,3} = \frac{\sum n_i d_i^4}{\sum n_i d_i^3} \tag{3}$$

Span:

$$Span = \frac{d(90) - d(10)}{d(50)} \tag{4}$$

In terms of diameters, n_i represents the number of droplets with a diameter equivalent to d_i [5]. While $d_{3,2}$ indicates the overall droplets' average diameter, $d_{4,3}$ indicates particle size changes caused by an increased sensibility to destabilize processes, such as droplet aggregation [32]. In terms of span, d(x) is the x volume percentile of droplets with diameters smaller or equal to the predefined values, meaning that a higher value of span indicates a diverse droplet polydispersity [33].

Color Comparison and pH Measurement

Color differences and comparisons were assessed based on the CIELAB color coordinate system (L*, a*, b*). Coordinate parameter (luminosity and chromatic range) measurements were analyzed through a Chroma Meter CR-400 colorimeter (Konica Minolta Business Technologies, Inc., Tokyo, Japan) calibrated according to the Konica Minolta white standard: Y = 86.7; x = 0.3160; y = 0.3233. The total color change (ΔE) was calculated between microalgae samples and the target using the following Formula (5):

$$\Delta E = \sqrt{\left(L_i^* - L_0^*\right)^2 + \left(a_i^* - a_0^*\right)^2 + \left(b_i^* - b_0^*\right)^2} \tag{5}$$

For each sample, 10 replicates were considered at different points of the sample spread in a petri dish.

To determine the final pH, samples were measured in triplicate with a pH electrode connected to a phM92 Lab Meter (Radiometer, Copenhagen, Denmark). Equipment calibration was performed with standard buffer solutions at pH 4, 7, and 9.

2.2.3. Antioxidant Activity

Extraction Methodology

Antioxidant activity was assessed through 1,1-diphenyl-2-picrylhydrazyl (DPPH) Radical Scavenging Activity (RSA) and Ferric Reducing Antioxidant Power (FRAP) methods. Total Phenolic Compounds (TPC) were measured by the Folin–Ciocaltêu (FC) methodology. Sample preparation began with an extraction previously adapted and optimized to assess microalga food products antioxidant activity [34]. To achieve both aqueous and oil phase, 2 g of sample + 2 mL of diluted Methanol (Honeywell, Morris Plains, NJ, USA) in deionized water ($MeOH:H_2O$/70:30 ratio) + 2 mL of Hexane (Honeywell, Morris Plains, NJ, USA) were mixed for 1 h in a Reax 2 shaker (Heidolph Instruments, Schwabach, Germany) and duly centrifuged in a Z383-K high performance centrifuge (Labnet International, Edison, NJ, USA) at 5000 rpm/4 °C/10 min. After centrifugation, a hydroalcoholic fraction (supernate) was recovered, and the process was repeated in triplicate. The recovered fraction was evaporated in volumetric glass flasks submerged in a 40 °C bath through a R-300 Rotavapor® (BUCHI, Barcelona, Spain) connected to a V-700 vacuum pump (BUCHI, Barcelona, Spain) reaching 58 bars. The evaporation process resulted in dry matter for each fraction. Each dry matter was weighted and dissolved in 2 mL of Dimethyl sulfoxide (DMSO) (Fisher Scientific International, Inc., Pittsburgh, PA, USA) in an LBX V05 vortex stirrer (Labbox Labware, Barcelona, Spain) and further filtered through a syringe coupled to a 0.20 µm nylon filter with a 25 mm diameter.

DPPH

The Radical Scavenging Activity assay was performed as previously mentioned [35] and adjusted for microalga food products research [34]. A combination of 20 µL of the previously extracted sample with 180 µL of a methanolic DPPH solution in a concentration of 60 µM was placed in NUNCTM 96-well microplates (Nalge Nunc International, Rochester, NY, USA) and underwent an incubation period for 30 min. Afterward, absorbance was measured at 517 nm in a Libra S22 UV/Visible Spectrometer (Biochrom, Cambridge, UK). Analysis outputs were expressed as Trolox equivalents per gram (µgTroloxEq/g).

FRAP

The Ferric Reducing Antioxidant Power assay consists of the reduction of Fe^{3+}-TPTZ (ferric-tripyridyl triazine complex) to Fe^{2+} (ferrous complex), which is noticeable due to a color change under an intense blue spectrum [36]. FRAP solution was prepared with 10 mL of acetate buffer (300 mM), pH adjusted to 3.6 using acetic acid, blended with 1 mL of ferric chloride hexahydrate (20 mM) dissolved in distilled water, and 1 mL of 2,4,6-Tris(2-pyridyl)-s-triazine (TPTZ—10 mM) dissolved in chloridric acid (HCl—40 mM). The analysis was adapted as previously described [37] with minor changes to enable execution in 96-well microplates. Inserted in the NUNCTM 96-well microplates (Nalge Nunc International, Rochester, NY, USA), 25 µL of the previously extracted sample combined with 175 µL of the FRAP solution (prewarmed to 37 °C) was left in triplicate to incubate for 30 min away from light at room temperature. Afterward, absorbance was measured at 595 nm in a Libra S22 UV/Visible Spectrometer (Biochrom, Cambridge, UK). Analysis outputs were expressed as Trolox equivalents per gram (µgTroloxEq/g).

Total Phenolic Compounds (TPC)

The Total Phenolic Compounds (TPC) assay followed the 96-well microplate Folin–Ciocalteu (FC) (PanReac AppliChem, Barcelona, Spain) technique [38], previously adjusted and optimized to microalga extracts [34]. Test preparation started with the addition of 100 µL of FC reagent (1:4) to 20 µL of the extract and a settle time of 5 min at room temperature. Each triplicate was added 80 µL of 7.5% sodium carbonate solution, and the microplate was kept for 2 h in the dark at room temperature. The absorbance was measured at 760 nm using a Thermo Scientific™ Multiskan™ GO Microplate Spectrophotometer (Fisher Scientific International, Inc., Pittsburgh, PA, USA). Analysis outputs were expressed as gallic acid equivalents per gram (µgGAEeq/g).

2.2.4. Sensorial Analysis

Acceptability and purchase intent towards the final product have a major influence on scale-up viability. For this purpose, a ranking test based on ISO 8587:2006 was firstly performed with 20 voluntary participants [39], qualitatively targeting the main organoleptic parameters (aspect, color, flavor, odor, texture, global evaluation) in a hedonic scale between 1 (extremely dislike) to 9 (extremely like). Secondly, the purchasing intent for each product was asked in a quantitative range from 1 (Certainly would not buy) to 5 (Certainly would buy). Both egg mayonnaise (Control—378) and mixed system microalgae + YPE mixed system mayonnaise (CvY—259) were taken into comparison with random codes. The tasting panel was run by the common standards used within the research center LEAF (Linking Landscape, Environment, Agriculture and Food) from Instituto Superior de Agronomia, Portugal. Volunteer participants were provided with an informed consent in accordance with the ethical standards of the local committee responsible for human experimentation and with The Code of Ethics of the World Medical Association (Declaration of Helsinki of 1975, as revised in 2013).

2.2.5. Statistical Analysis

Results obtained in the conducted analysis were submitted to statistical treatment in order to ascertain the significant differences between the output averages. Statistical analysis was performed in GraphPad Prism software (version 9.0) to carry out the analysis of variance (ANOVA), using Fisher's test to compare two samples and the Tukey's test to compare three or more samples. Both tests used a 95% confidence level ($\alpha = 0.05$). Different letters indicate significantly different results ($p < 0.05$) in the data presented.

3. Results and Discussion

3.1. pH Measurement

The pH value of each sample was measured to achieve a clearer comprehension of its impact in the following analysis. The stable pH for each sample is shown in Table 2.

Table 2. pH measurement and comparison between Control and mixed system (CvY) samples.

Sample	pH
Control	3.92 ± 0.01 [a]
CvY	3.89 ± 0.02 [a]
CvY_ca0.1	4.10 ± 0.03 [b]
CvY_ca0.4	3.65 ± 0.02 [c]
CvY_salt0.1	3.91 ± 0.03 [a]
CvY_salt0.5	3.87 ± 0.02 [a]

Different letters indicate significantly different results ($p < 0.05$) according to Tukey's test.

The pH of the Control does not present significant differences compared to the microalgae + YPE systems with the usual citric acid concentration (0.22 g/100 g) or when NaCl is added at concentrations of 0.1 or 0.5 g/100 g. However, in cases where citric acid addition was increased or decreased, there is a significant alteration in the stable pH values of the emulsions, proportional to the concentration of acid added (0.1 or 0.4%).

3.2. *Rheological Behaviour*

3.2.1. Frequency Sweeps

From the mechanical spectra (variation of G′ and G″ in the LVR). The storage modulus corresponds to the elastic capacity associated with a solid-like behavior of viscoelastic fluids, whereas the loss modulus corresponds to the thick capacity associated with a liquid-like behavior [40]. Overall, G′ values present higher values than the G″ values, indicating a common elastic behavior. Both moduli depict a gradual increase along the frequency, as observed in Figure 1, leading to an expectation of a stable emulsion with a gel-like behavior [7].

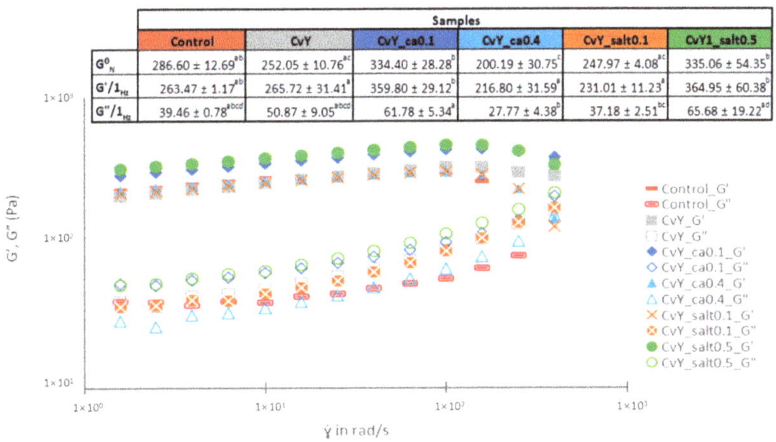

	Samples					
	Control	CvY	CvY_ca0.1	CvY_ca0.4	CvY_salt0.1	CvY1_salt0.5
G^0_N	286.60 ± 12.69 [ab]	252.05 ± 10.76 [ac]	334.40 ± 28.28 [b]	200.19 ± 30.75 [c]	247.97 ± 4.08 [ac]	335.06 ± 54.35 [b]
$G'/1_{Hz}$	263.47 ± 1.17 [ab]	265.72 ± 31.41 [a]	359.80 ± 29.12 [b]	216.80 ± 31.59 [a]	231.01 ± 11.23 [a]	364.95 ± 60.38 [b]
$G''/1_{Hz}$	39.46 ± 0.78 [abcd]	50.87 ± 9.05 [abcd]	61.78 ± 5.34 [a]	27.77 ± 4.38 [b]	37.18 ± 2.51 [bc]	65.68 ± 19.22 [ad]

Figure 1. Mechanical spectra and plateau model (G^0_N), storage (G′), and loss (G″) moduli at 1 Hz (6.28 rad/s) comparison between Control and samples with the mixed system (CvY). Different letters indicate significantly different results ($p < 0.05$) for each row according to Tukey's test.

By substituting egg yolk (Control) with microalgae + YPE (CvY), it is possible to achieve a similar viscoelastic behavior (similar shape of the spectra). However, the YPE promotes an increase in the complexity of the entanglement network established between the protein films of neighboring droplets, enhancing their structural interactions and overall system stability, which is reflected in higher plateau modulus (G^0_N) values [6].

The control emulsion and the mixed systems CvY, CvY_ca0.4, and CvY_salt0.1 exhibit G' and G'' (1 Hz) values, as well as G^0_N values, without significant differences ($p < 0.05$). These results suggest that the physical stability of these emulsions is similar, as evidenced by their comparable viscoelastic properties. Both viscoelastic functions are critical parameters in predicting the stability and performance of emulsions under stress [41]. The plateau modulus (G^0_N) further supports the network structure's resilience, a key factor in determining the long-term stability of such systems. These findings align with the established principles of soft matter rheology, particularly in systems where the balance between elastic and viscous components plays a pivotal role in maintaining structural integrity under various conditions [42].

It is important to highlight that the mixed systems exhibit a synergistic effect [43], making it possible to obtain emulsions that are structurally similar to the control emulsion prepared with egg. This highlights the potential of these systems as an important approach for the development of vegan emulsions, aligning with the growing consumer trend toward plant-based and sustainable food products. The increasing demand for vegan alternatives reflects a shift in dietary preferences driven by environmental concerns, health benefits, and ethical considerations. Similar studies have demonstrated the successful use of other plant-based proteins in vegan emulsions, such as soy protein [44], pea protein [45], lentil protein [46], and lupinus and fava proteins [7], showcasing their ability to stabilize emulsions while offering functional and nutritional benefits.

Although, from an industrial perspective, the pH values and salt concentration are typically well-defined according to the specifications of a given product, it is crucial to understand the sensitivity of emulsified systems to variations in these two parameters. Such knowledge is fundamental for optimizing formulation stability, especially under processing and storage conditions where fluctuations in pH or ionic strength can impact the emulsions' rheological properties, microstructure, and shelf life [47,48]. Insights into these sensitivities contribute to the development of more robust and versatile formulations suitable for a wide range of applications.

With respect to pH variation, two conditions were studied: CvY_ca0.1 (pH—4.10 ± 0.03) and CvY_ca0.4 (pH—3.65 ± 0.02), with citric acid concentrations of 0.1 and 0.4 g/100 g, respectively. The pH values obtained were significantly different, and the system with the lower pH exhibited a significant reduction ($p < 0.05$) in the viscoelastic properties compared to the other conditions studied. This behavior may be attributed to the pH approaching the isoelectric point of the mixed system, where interactions between protein particles become less favorable due to charge neutralization. Similar behavior has been observed by other authors; for instance, in soy protein emulsions, lower pH values resulted in decreased viscoelastic properties [49]. Likewise, Vélez-Erazo et al. [50] reported that a reduction in pH in pea protein emulsions led to a significant decrease in gelation and stability properties, which was associated with the proximity to the isoelectric point. This behavior reflects changes in protein interactions, which directly affect the formation and stability of emulsions. In addition, emulsions' rheological response may also be adjusted by the pH, probably due to changes in the proteins' electrostatic interactions [51].

To evaluate the impact of salt (NaCl) addition on the rheological behavior of emulsions, two concentrations were studied: 0.1 (CaY_salt0.1) and 0.5 (CaY_salt0.5) g/100 g. It was observed that the emulsion with higher salt content showed an enhancement in its structure, corresponding to higher values of G', G'', and G^0_N. This indicates a strengthening of the emulsion network, likely due to the increased ionic strength. Salt content's main impact in both moduli evidences a reinforcement in the viscoelasticity capacities, ultimately improving final structure and stability [52].

The presence of salt influences the electrostatic interactions between protein molecules, which can lead to changes in the protein–protein network, enhancing the stability and rheological properties of the emulsions. Similar results have been reported in other studies; for example, Sriprablom et al. [40] found that increasing the ionic strength in whey protein-stabilized emulsions enhanced their viscoelastic properties, attributed to the screening of electrostatic repulsions between protein molecules. Additionally, Mańko-Jurkowska and Domian [53] observed that higher salt concentrations in chickpea protein emulsions resulted in stronger gel formation and improved stability due to the ionic crosslinking effect, which promoted better structuring of the protein network. Such findings highlight the critical role of ionic strength in emulsion formation and stabilization.

3.2.2. Viscosity Measurements

Both the control and mixed systems of CvY mayonnaises' steady-state flow curves depict a shear-thinning behavior, with an expected tendency towards a zero-shear rate-limiting viscosity. This behavior is closely aligned with the oil droplet deflocculation mechanism [54]. In Figure 2, it is possible to compare the behavior of the Control and CvY, as the tests were performed with different concentrations of citric acid and salt (NaCl).

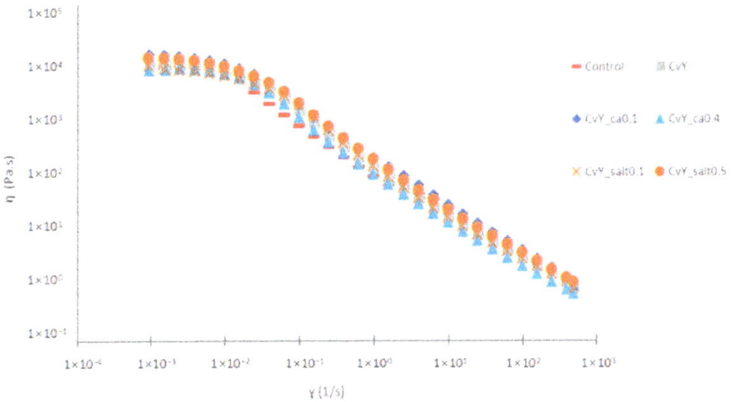

Figure 2. Flow curves and Williamson model adjustment parameters comparison between samples: Control; CvY; CvY_ca0.1; CvY_ca0.4; CvY_salt0.1; and CvY_salt0.5.

Adjusted Williamson curves parameters are presented in Table 3.

Table 3. Williamson model fitting curves parameters comparison between Control and mixed system (CvY) samples.

Samples	η_0 (Pa.s)	k (s)	m	R^2
Control	$9.20 \times 10^3 \pm 6.23 \times 10^2$ ab	$7.57 \times 10^1 \pm 1.65 \times 10^1$ A	0.89 ± 0.01 ^{i}	0.980
CvY	$8.96 \times 10^3 \pm 1.23 \times 10^3$ a	$5.57 \times 10^1 \pm 1.55 \times 10^1$ A	0.92 ± 0.03 ^{iI}	0.991
CvY_ca0.1	$1.59 \times 10^4 \pm 5.96 \times 10^3$ b	$7.23 \times 10^1 \pm 3.74 \times 10^1$ A	0.91 ± 0.06 ^{iI}	0.989
CvY_ca0.4	$7.57 \times 10^3 \pm 3.01 \times 10^2$ a	$4.80 \times 10^1 \pm 4.76 \times 10^0$ A	0.96 ± 0.02 ^{I}	0.989
CvY_salt0.1	$6.94 \times 10^3 \pm 5.71 \times 10^2$ a	$4.25 \times 10^1 \pm 1.39 \times 10^1$ A	0.94 ± 0.01 ^{iI}	0.992
CvY_salt0.5	$1.14 \times 10^4 \pm 2.35 \times 10^2$ ab	$4.74 \times 10^1 \pm 2.23 \times 10^0$ A	0.94 ± 0.01 ^{iI}	0.996

Different letters indicate significantly different results ($p < 0.05$) for each column according to Tukey's test.

The control and the emulsion with the binary system YPE + microalgae (CvY) do not show significant differences ($p < 0.05$) in the values of zero-shear rate limiting viscosity

(η_0). This suggests that both systems exhibit similar flow behavior under conditions of very low shear rates, where the material is dominated by its intrinsic viscosity and does not undergo shear thinning. Such behavior is often observed in complex systems, where viscosity at low shear rates is an important indicator of the network structure and molecular interactions [55].

When testing the impact of pH on the viscosity (CvY_ca0.1 and CvY_ca0.4), it was found that the results obtained are consistent with those observed in terms of linear viscoelastic behavior. That is, the more acidic sample—CvY_ca0.4 with a pH value of 3.65 ± 0.02—shows a significantly ($p < 0.05$) lower value of η_0. This result aligns with the previous observation; the pH value reached is close to the isoelectric point of the mixed system. In such situations, the limiting viscosity of emulsions should be lower, as the protein–protein interactions are weakened due to charge neutralization at the isoelectric point, resulting in a less structured network [56]. These changes in viscosity can be attributed to the reduced ability of proteins to form stable networks at pH values close to their isoelectric points.

Regarding salt concentration, no significant differences ($p > 0.05$) were observed between the two NaCl concentrations studied (0.1 and 0.5 g/100 g). This result does not align with what was observed in terms of the mechanical spectra, where a higher NaCl concentration induced an increase in the degree of structuring of the emulsion. However, this increased structuring did not impact the flow behavior. A similar situation was observed by other authors. Quintana et al. [57] reported that emulsions stabilized by whey protein isolate exhibited enhanced viscoelastic properties at higher ionic strengths, but the zero-shear viscosity remained unchanged. Similarly, Sriprablom et al. [40] demonstrated that whey protein stabilized emulsions showed increased gel-like behavior under higher salt concentrations, although the flow properties, such as shear-thinning behavior, remained consistent across different salt levels. These findings suggest that while ionic strength can significantly influence the network structure, it may not always translate to changes in bulk flow behavior.

3.3. *Texture Profile Analysis (TPA)*

Texture parameters obtained from the TPA profile are relevant aspects that increase the probability of alternative products' success among consumers. This relies on the perceived texture during the tasting experience. Perceived texture splits into three moments. The first moment refers to the extra oral, mainly dependent on appearance. The second moment refers to the intra oral, characterized by first mandibular compression, chewing action, and the swallowing process. The third moment relates to the physical behavior of the product during ingestion, mainly in parameters as hardness, fracturability, viscosity, and adhesiveness [27]. TPA parameters were compared based on Table 4.

With the exception of the CvY_ca0.1, all the mixed systems depicted a significant decrease in firmness and adhesiveness when compared with the Control. Adding microalga, especially *C. vulgaris*, seems to generate softer textures in the food products [58]. This behavior does not reflect what was observed in terms of linear viscoelastic behavior, as in this case, the emulsions CvY_ca0.1 and CvY_salt0.5 exhibited a higher degree of structuring compared to the control, which should correspond to higher firmness values. However, firmness results do not always reflect the behavior evaluated under oscillatory conditions for a low range of applied stresses. This discrepancy arises because the types of stresses applied are fundamentally different. Similar discrepancies have been reported by other authors. Chen and Dickinson [59] observed that emulsions with enhanced viscoelastic moduli in oscillatory tests did not exhibit corresponding increases in firmness during large deformation tests due to differences in the nature of applied stress. Additionally, protein-stabilized emulsions exhibited stronger network structures under oscillatory shear, but this structural enhancement did not translate to firmness measurements under large deformation conditions, highlighting the importance of the testing methodology.

Table 4. TPA parameters comparison between samples.

Samples	Firmness (N)	Elasticity	Cohesiveness	Adhesiveness (N.s)
Control	0.059 ± 0.001 ^{a}	0.969 ± 0.001 ^{AC}	0.857 ± 0.004 ^{i}	0.372 ± 0.005 ^{I}
CvY	0.047 ± 0.001 ^{b}	0.965 ± 0.006 ^{AB}	0.781 ± 0.006 $^{ii, iii}$	0.256 ± 0.002 ^{II}
CvY_ca0.1	0.078 ± 0.002 ^{c}	0.956 ± 0.001 ^{B}	0.838 ± 0.013 ^{i}	0.595 ± 0.007 ^{III}
CvY_ca0.4	0.043 ± 0.001 ^{d}	0.977 ± 0.003 ^{C}	0.783 ± 0.014 ^{ii}	0.184 ± 0.003 ^{IIII}
CvY_salt0.1	0.039 ± 0.001 ^{e}	0.968 ± 0.007 ^{ACD}	0.748 ± 0.013 ^{iii}	0.181 ± 0.011 ^{IIII}
CvY_salt0.5	0.057 ± 0.001 ^{d}	0.959 ± 0.005 ^{BD}	0.784 ± 0.027 ^{ii}	0.361 ± 0.011 ^{I}

Different letters indicate significantly different results ($p < 0.05$) for each column according to Tukey's test.

The elasticity values of the emulsions do not deviate from the Control. This texture parameter is not appropriate for discriminating the behavior of emulsions, as noted by other authors. For example, Graça et al. [60] demonstrated that firmness and adhesiveness were sufficient for characterizing emulsions, emphasizing that these textural parameters provide a more direct assessment of the structural properties and stability of such systems.

Cohesiveness is also not usually a relevant parameter for discriminating emulsions. It is observed that there is a general decrease of cohesiveness in the mixed systems, compared with the control.

3.4. Droplet Size Distribution (DSD)

Assessing the DSD of the emulsions requires an intrinsic comparison of both the Sauter diameter ($d_{3,2}$) and the De Brouckere diameter ($d_{4,3}$). The Sauter diameter, or surface-area-weighted mean diameter, provides insights into the ratio of the total droplet volume to the total surface area, which is critical for understanding the efficiency of emulsifier coverage. In contrast, the De Brouckere diameter, or volume-weighted mean diameter, reflects the contribution of larger droplets to the overall size distribution, which significantly affects stability and coalescence behavior. The combined analysis of these two parameters allows for a comprehensive understanding of emulsion characteristics and has been widely used in emulsion science [61].

From the $d_{3,2}$ measurement, it is possible to obtain the average diameter for most droplets, where smaller values mean smaller and condensed droplets, mirroring a more stable system [62]. From the $d_{4,3}$, or volumetric diameter, it is possible to evaluate the susceptibility to droplet size changes linked to destabilizing processes or emulsion stability physical degradation [32]. The droplet size distribution of all samples was analyzed (Figure 3), and from the curves, the $d_{3,2}$ and $d_{4,3}$ values were calculated.

It was observed that all emulsions prepared with the mixed system CvY exhibited a unimodal distribution, similar to the control. This type of distribution is generally associated with more stable emulsions. A unimodal distribution indicates a more uniform droplet size, which minimizes differences in droplet behavior under stress conditions and reduces the likelihood of phenomena such as creaming, coalescence, or Ostwald ripening. Uniform droplet sizes also promote efficient emulsifier coverage, enhancing the kinetic stability of the emulsion. This relationship has been reported by McClements [63], who noted that emulsions with a narrower and unimodal droplet size distribution tend to exhibit better physical stability due to the uniformity in their interfacial properties.

On the other hand, it is observed that the smaller the average droplet diameter, the greater the stability of the emulsion. This is because smaller droplets have a higher surface-to-volume ratio, which enhances the effectiveness of emulsifier molecules in stabilizing the interface and reduces the driving force for coalescence. Additionally, smaller droplets are less prone to creaming due to their slower rise velocity, as described by Stokes's law. Xue et al. [64] demonstrated that emulsions with smaller droplet sizes exhibit prolonged stability, attributed to their ability to resist destabilization mechanisms like coalescence and Ostwald ripening more effectively.

Samples	$d_{3,2}$	$d_{4,3}$	Span
Control	7.94 ± 0.35	6.11 ± 0.05^{A}	1.39 ± 0.02^{ac}
CvY	7.49 ± 0.01	22.02 ± 2.18^{B}	1.64 ± 0.07^{b}
CvY_ca0.1	8.15 ± 0.05	19.41 ± 0.37^{B}	1.41 ± 0.02^{a}
CvY_ca0.4	13.85 ± 0.01	38.85 ± 5.02^{C}	1.45 ± 0.09^{a}
CvY_salt0.1	9.66 ± 0.01	14.96 ± 4.83^{AB}	1.24 ± 0.05^{c}
CvY_salt0.5	10.08 ± 0.05	14.59 ± 3.33^{AB}	1.23 ± 0.05^{c}

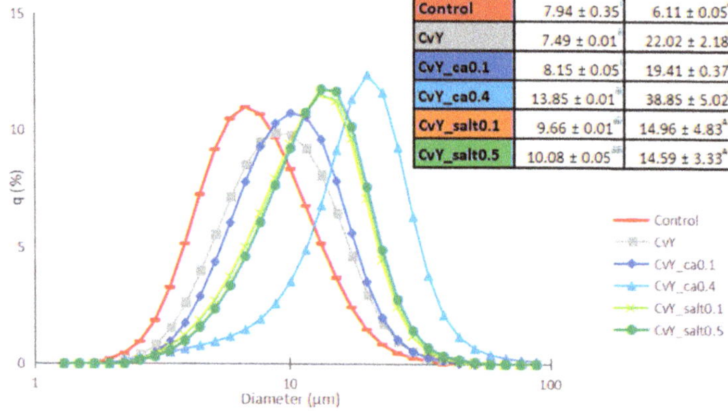

Figure 3. Droplet size distribution, Sauter diameter ($d_{3,2}$), De Brouckere diameter ($d_{4,3}$), and span comparison between samples: Control; CvY; CvY_ca0.1; CvY_ca0.4; CvY_salt0.1 and CvY_salt0.5. Different letters indicate significantly different results ($p < 0.05$) for each column according to Tukey's test.

The auter diameter of the CvY is slightly smaller than the Control. CvY_ca0.1 is smaller than CvY_ca0.4. This decrease may be linked with the slight increase in viscosity. Higher viscosity values correlate with a lower probability of coalescence [65].

Changes in the droplet size between emulsions with different pH levels reflect a decrease in the Sauter diameter ($d_{3,2}$) as pH increases. Decreasing the pH results in larger Sauter diameters, a fact closely related to the protein's lower diffusion rate around the O/W interface, explained by a raise in the denaturation degree of the protein. In contrast, increasing the pH leads to lower droplet sizes [54].

Moreover, decreasing the pH reduces the electrostatic repulsion, possibly influencing the molecular conformation functional properties since the molecules bind to negatively charged groups, further decreasing the number of similarly charged groups. This interchange explains the narrow distribution observed [66]. Based on this, stronger emulsion stability can be expected with higher values of pH when the droplet size is smaller. The good stability of the emulsions at higher pH values can be attributed to their strong electrostatic repulsion [67].

Different salt concentrations deviate from the Control since it is possible to observe larger droplets. This tendency may be explained by the interaction of salt and the particle-interface mixed system attraction during mixing, resulting in a lack of particle capacity to adsorb to the interface when compared with the presence of egg yolk or due to the particle trapping energy decrease [68]. Despite the differences between the Control and CvY with different salt concentrations, these concentrations did not show different values, suggesting that salt does not have major effects on the droplet size distribution [57,69]; however, they may differ in terms of emulsion stability. Nonetheless, salt does not influence the ionic strength on the physicochemical stability of O/W emulsion, meaning that the main dependency remains in the ionic species and ionic strength of the continuous phase [70].

3.5. Antioxidants and Bioactive Compounds

Mayonnaise lipid portion is susceptible to oxidative deterioration derived from the auto-oxidation of unsaturated and polyunsaturated fats oil fraction. This oxidation extent may result in negative organoleptic impacts on flavor, aroma, color, and nutritional value [71]. The food industry plays a crucial role in the application of effective strategies against the lipid oxidation of mayonnaise, mainly through the addition of antioxidants or the use of other lipid sources that are naturally rich in compounds capable of antioxidant activity [72,73]. Commercial mayonnaise usually resorts to EDTA as an antioxidant and preservative component [74]. Antioxidant activity potential was compared in Table 5. While the Control contains EDTA, the mixed system CvY relies on the properties of the heterotrophic microalgae white *C. vulgaris*'s natural antioxidants.

Table 5. Total phenolic compounds (TPC), Radical Scavenging Activity (DPPH), and Ferric Reducing Antioxidant Power (FRAP) comparison between mayonnaise with egg plus EDTA (Control) and mayonnaise with mixed system of YPE plus microalgae (CvY).

Samples	TPC ugG Aeq/g	DPPH ugTEAC/g	FRAP ugTEAC/g
Control	84.21 ± 10.56 [a]	85.97 ± 12.18 [A]	50.26 ± 5.31 [i]
CvY	203.05 ± 2.25 [b]	117.26 ± 7.96 [B]	85.47 ± 0.93 [ii]

Different letters indicate significantly different results ($p < 0.05$) for each column according to Fisher's test.

It is possible to underline the enhancement of bioactive compounds capable of antioxidant activity in mayonnaises obtained from the mixed system YPE + *C. vulgaris*. The most noticeable enrichment belongs to phenolic compounds, favorable to inhibit hydro-peroxides formation [75]. These bioactive compounds are positive not only for their antioxidant capacities but also for their antihypertensive potential. However, from the nutritional perspective, microalga in general still lack proper evidence of bioactive compounds' bioavailability after ingestion in the gastrointestinal tract [76].

A pertinent factor is connected with the viability of achieving greater antioxidant power if the amount of *C. vulgaris* incorporation increases [77]. In fact, previous studies approached antioxidant activity in foods with *C. vulgaris*, correlating the antioxidant potential with the total amount of incorporation. Increasing the biomass in food composition reveals positive effects in bacterial activity limitation and the promotion of antioxidant activity by a greater presence of phenolics, pigments, and vitamins [78]. This microalgae biomass and/or extracts represent an important alternative with potential application in both the pharmaceutical and food industries [79].

3.6. Color Evaluation

One of the main challenges for alternative Clean Label products is to acquire colors that resemble the traditional products. A constant effort to reduce and eliminate the components perceived as unnatural in the consumer's perspective is part of Clean Label's agenda. The usage of "artificial colorants" is nowadays associated with less healthy products, demanding the food industry to deeply study colorants of natural origin that can be incorporated into final products [80]. A comparison of Control mayonnaise with artificial colorant (undisclosed ingredient) and CvY mixed system with heterotrophic microalgae is present in Table 6. Regarding luminosity (L*), there are slight changes, mainly before a pH and salt reduction. This visual aspect can be related to an increase in the droplet size of the emulsions [67]. *C. vulgaris* incorporation also reveals a slight increase in green (a*) and blue (b*) tones. Even though small changes were detected by color tests, in terms of overall visual appearance differences detected by the human eye (ΔE), only a lowering pH depicts some barely noticeable deviation at close attention. Nevertheless, physical stability has the major impact, and if no instability is noticed, visual appearance slight changes are neglected by the consumers [81].

Table 6. CieLab color space coordinates (L*, a*, and b*) from each sample, plus color changes perception, and final pH.

Samples	L*	a*	b*	ΔE
Control	84.54 ± 1.42 [a]	−5.24 ± 0.09 [A]	24.93 ± 0.46 [i]	-
CvY	83.62 ± 0.72 [a]	−5.70 ± 1.42 [B]	24.19 ± 0.46 [ii]	1.26
CvY_ca0.1	83.18 ± 0.59 [ac]	−5.62 ± 0.04 [B]	24.04 ± 0.36 [ii]	0.48
CvY_ca0.4	81.36 ± 0.33 [b]	−5.99 ± 0.04 [C]	28.42 ± 0.43 [iii]	4.8
CvY_salt0.1	81.53 ± 1.17 [b]	−6.13 ± 0.09 [D]	25.97 ± 0.41 [iiii]	2.78
CvY_salt0.5	82.22 ± 0.28 [bc]	−6.14 ± 0.04 [D]	26.56 ± 0.31 [iiii]	2.789

Different letters indicate significantly different results ($p < 0.05$) for each column according to Tukey's test.

3.7. Sensorial Feedback and Purchase Intent

Consumer reception is subjective and influenced by external factors. Even so, the majority of consumers consider some important variables, such as the yield value, appearance, elasticity, consistency, texture, and flavor, when experiencing mayonnaise. The tasting panel compared the mayonnaises in Figure 4.

Figure 4. Target coriander mayonnaise (Control) is on the left, and YPE + Microalgae mixed system (CvY) analogue product is on the right.

At a second attempt, factors as viscosity perceived in the mouth infused with intensities of flavor notes sensory profile also reveal to contain a major impact [82]. In Figure 5, it is possible to understand how the tasting panel related Control and CvY in terms of organoleptic characteristics.

Most of the organoleptic profile unveils a balance and similar taste. Focusing on the aspect and color, the application of a heterotrophic microalgae better convinced the consumer since the lack of a green color avoided the initial shock and uncertainty [83]. In terms of flavor, global appreciation, and odor, the microalgae's distinct character was perceived and ranked below the egg yolk mayonnaise, which evidences that sensory acceptability manifests itself along with the amount of microalgae biomass total incorporation [58]. Ongoing trials of mixed incorporation percentages must be taken into account to balance the purchase intent present in Figure 6, based on the sensory experience.

Figure 5. Sensorial analysis preliminary assessment comparison between mayonnaise with egg (Control) and mayonnaise with a mixed system of YPE + Microalgae (CvY).

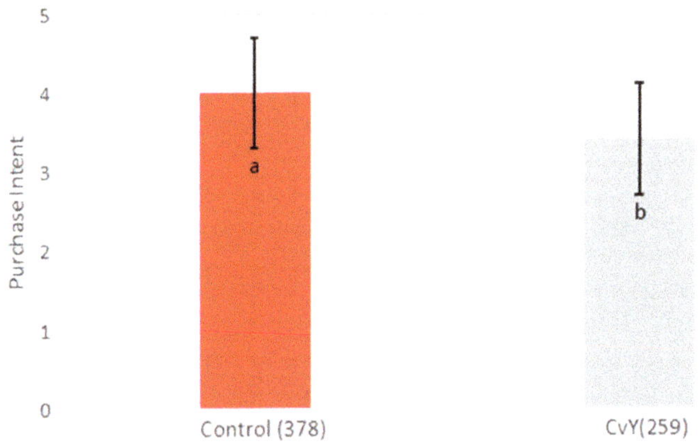

Figure 6. Purchase intent from the tasting panel after a blind sensorial analysis to compare mayonnaise with egg (Control) and mayonnaise with a mixed system of YPE + Microalgae (CvY). Different letters indicate significantly different results ($p < 0.05$) for each column according to Fisher's test.

Although upcoming generations of consumers tend to support plant-based foods in general, this positivism is still far away from a regular purchase intent. Despite the origin of the ingredients, most consumers expect foods to remain similar to their traditional counterparts. Both consumers and the food industry play a role in this matter since alternative products are usually linked to traditional products' similar responses. If alternative foods fail to reach previously emphasized expectations and turn out to offer an inferior experience, their sales performance will also be inferior [84].

4. Conclusions

Mixed systems composed of proteins with synergy to perform emulsifying roles, like yeast protein extract and white *Chlorella vulgaris*, can be used for the stabilization of commercial vegan and clean label mayonnaise, contributing to a new market offering.

Strategies to elaborate a balance consortium between alternative products to successfully mimic and achieve the consumers undistinguished appraisal around these alternative products become even closer to take a place in markets shelfs. Also, focusing on less disruptive organoleptic changes, as offering a similar color by opting for a heterotrophic microalga, seems to nourish a positive acceptance.

The studied systems show a certain robustness against variations in pH and salt concentration, which represents a positive effect to possible unexpected deviations in citric acid or salt amount during the production chain. However, excessively low pH values (around 3.6) should be avoided. In fact, the pH values of commercial emulsions are typically slightly higher. It was also found that increasing the salt concentration positively contributes to the increased degree of structuring of the emulsion, but the limits on salt incorporation in food are becoming more restricted, considering that the maximum recommended daily intake of NaCl is 5 g, according to the guidelines of the World Health Organization (WHO).

Further research is essential to evaluate the digestibility rate and underline the apparent health benefits of the consumption of mayonnaise produced with the mixed system (CvY). Moreover, stability and shelf life must also be taken into consideration. From this initial perspective, CvY systems depict a promising physical similarity, a boost in functional properties, and a good initial response by the tasting panel.

Author Contributions: Conceptualization, P.C., C.S., and A.R.; methodology, P.C., C.S., and A.R.; validation, N.K., and A.R.; formal analysis, P.C., and C.S.; investigation, P.C., and C.S.; resources, N.K., and A.R.; data curation, P.C., and C.S.; writing—original draft preparation, P.C.; writing—reviewing and editing, N.K., and A.R.; visualization, P.C.; supervision, N.K., and A.R.; project administration, N.K., and A.R.; funding acquisition, A.R. All authors have read and agreed to the published version of the manuscript.

Funding: This work was supported by (i) Mobilizing Agendas for Business Innovation&PRR I Notice No. 02/C05-i01/2022 project VIIAFOOD—Platform for Valorization, Industrialization and Commercial Innovation for Agro-Food (No. C644929456-00000040). WP3–A3.3.: MYA Protein Emulsions. (ii) FCT—Fundação para a Ciência e a Tecnologia, I.P., under the project UIDB/04129/2020 of LEAF-Linking Landscape, Environment.

Institutional Review Board Statement: The testing panel was run by the common standards used within the research center LEAF (Linking Landscape, Environment, Agriculture and Food) from Instituto Superior de Agronomia, Portugal. Volunteer participants were provided with an informed consent in accordance with the ethical standards of the local committee responsible for human experimentation and with The Code of Ethics of the World Medical Association (Declaration of Helsinki of 1975, as revised in 2013). In Portugal, we do not need approval from an Ethics Committee to conduct sensory analysis.

Informed Consent Statement: Informed consent was obtained from all subjects involved in the study.

Data Availability Statement: The original contributions presented in this study are included in the article. Further inquiries can be directed to the corresponding author.

Acknowledgments: VIIAFOOD—Plataforma de Valorização, Industrialização e Inovação para o AgroAlimentar project—ref. C644929456-00000040—funded by Agendas Mobilizadoras para a Inovação Empresarial—PRR I Aviso No 02/C05-i01/2022 program. MYA Protein Emulsions research project, funded by Casa Mendes Gonçalves. Proenol S.A. LEAF research unit.

Conflicts of Interest: Author Norton Komora was employed by the company Casa Mendes Gonçalves. The remaining authors declare that the research was conducted in the absence of any commercial or financial relationships that could be construed as a potential conflict of interest.

Appendix A

	Simple emulsion (65% oil; 1–4% protein; 34–31% water)				
		Rheology	Texture	Color	OSD
Trial 1	YPE				
	1%	-	-	+/-	-
	2%	-	-	+/-	-
	3%	-	-	+/-	+/-
	4%	+/-	+/-	+/-	+/-
	C. vulgaris				
	1%	-	-	+/-	-
	2%	-	-	+/-	-
	3%	-	+/-	+/-	+/-
	4%	+/-	+/-	+/-	+/
Main problems	YPE increase results in a stronger structure, *C. vulgaris* alone is not enough				
Main achievments	White color from YPE and heterotrophic microalgae resembles mayonnaise				

	Simple emulsion (65% oil; 4% mixed system; 31% water)				
		Rheology	Texture	Color	DSD
Trial 2	YPE + *C. vulgaris*				
	1%+ 3%	+	+	+/-	+/-
	2% + 2%	+/-	+/-	+/-	+
	3% + 1%	+/-	+/-	+/-	+
Main problems	YPE stabilizes emulsion but does not behave due to a stronger structure				
Main achievments	Proteins synergy in the mixed system mirror a similar behaviour				

	Substitution of egg fraction by mixed system (0.8% YPE + 1% *C. vulgaris*) in mayonnaise recipe (control)				
		Rheology	Texture	Color	DSD
Trial 3	YPE + *C. vulgaris*				
	0.8% + 1%	+	+	+/-	+/-
Main problems	Addition of acetic acid (vinegar) desintegrates the mixed system for higher percentages of YPE				
Main achievments	Physical behaviour and consumer acceptance validated				

Figure A1. Initial attempts to assess YPE and microalgae concentration to produce food emulsions under industry predetermined requirements.

References

1. Aschemann-Witzel, J.; Gantriis, R.F.; Fraga, P.; Perez-Cueto, F.J.A. Plant-based food and protein trend from a business perspective: Markets, consumers, and the challenges and opportunities in the future. *Crit. Rev. Food Sci. Nutr.* **2020**, *68*, 3119–3128. [CrossRef] [PubMed]
2. Weinrich, R.; Elshiewy, O. A cross-country analysis of how food-related lifestyles impact consumers' attitudes towards microalgae consumption. *Algal Res.* **2023**, *70*, 102999. [CrossRef]
3. Choi, H.W.; Ham, S.H.; Hahn, J.; Choi, Y.J. Developing plant-based mayonnaise using pea protein-xanthan gum conjugates: A maillard reaction approach. *LWT—Food Sci. Technol.* **2023**, *185*, 115137. [CrossRef]
4. Alu'datt, M.H.; Rababah, T.; Alhamad, M.N.; Ereijef, K.; Gammoh, S.; Kubow, S.; Tawalbeh, D. Preparation of mayonnaise from extracted plant protein isolates of chickpea, broad bean and lupin flour: Chemical, physiochemical, nutritional and therapeutic properties. *J. Food Sci. Technol.* **2017**, *54*, 1395–1405. [CrossRef] [PubMed]
5. Simões, S.; Carrera Sanchez, C.; Santos, A.J.; Figueira, D.; Prista, C.; Raymundo, A. Impact of Grass Pea Sweet Miso Incorporation in Vegan Emulsions: Rheological, Nutritional and Bioactive Properties. *Foods* **2023**, *12*, 1362. [CrossRef]
6. Vieira, M.; Simões, S.; Carrera-Sánchez, C.; Raymundo, A. Development of a Clean Label Mayonnaise Using Fruit Flour. *Foods* **2023**, *12*, 2111. [CrossRef]
7. Cabrita, M.; Simões, S.; Álvarez-Castillo, E.; Castelo-Branco, D.; Tasso, A.; Figueira, D.; Guerrero, A.; Raymundo, A. Development of innovative clean label emulsions stabilized by vegetable proteins. *Int. J. Food Sci. Technol.* **2023**, *58*, 406–422. [CrossRef]
8. Ribeiro, A.C.; Simões, S.; Perez-Gregorio, R.; Soares, S.; Figueira, D.; Branco, D.C.; Tasso, A.; Raymundo, A.; Mateus, N.; Freitas, V. Unravelling the effect of phenolic compounds in the design of a yeast protein-based emulsions. *Int. J. Food Sci. Technol.* **2023**, *58*, 3337–3348. [CrossRef]
9. Braga, A.R.C.; Nunes, M.C.; Raymundo, A. The Experimental Development of Emulsions Enriched and Stabilized by Recovering Matter from Spirulina Biomass: Valorization of Residue into a Sustainable Protein Source. *Molecules* **2023**, *28*, 6179. [CrossRef]
10. Aybar, M.; Simões, S.; Sales, J.R.; Santos, J.; Figueira, D.; Raymundo, A. Tenebrio molitor as a Clean Label Ingredient to Produce Nutritionally Enriched Food Emulsions. *Insects* **2023**, *14*, 147. [CrossRef]
11. Palacios, L.E.; Wang, T. Extraction of egg-yolk lecithin. *J. Am. Oil Chem. Soc.* **2005**, *82*, 565–569. [CrossRef]
12. Gu, L.; Yao, X.; McClements, D.J.; Liang, L.; Xiong, W.; Li, J.; Chang, C.; Su, Y.; Yang, Y. Fabrication, characterization, and application of catechin-dextran-egg white protein conjugates: Novel antioxidant emulsifiers. *LWT—Food Sci. Technol.* **2022**, *173*, 114268. [CrossRef]
13. Almonacid, S.; Simpson, R.; Teixeira, A. Heat transfer models for predicting Salmonella enteritidis in shell eggs through supply chain distribution. *J. Food Sci.* **2007**, *72*, 508–517. [CrossRef] [PubMed]
14. Ghorbani Gorji, S.; Smyth, H.E.; Sharma, M.; Fitzgerald, M. Lipid oxidation in mayonnaise and the role of natural antioxidants: A review. *Trends Food Sci. Technol.* **2016**, *56*, 88–102. [CrossRef]
15. Savani, P.; Puthiyedath, A.; Chandran, K.R.; George, S.; Prasad, P.S.; Annapure, U.S. Evaluation of the sensory properties and antioxidant activity of clean rosemary extracts for an effective replacement of EDTA in Mayonnaise. *Appl. Food Res.* **2023**, *3*, 100302. [CrossRef]
16. Tao, Z.; Yuan, H.; Liu, M.; Liu, Q.; Zhang, S.; Liu, H.; Jiang, Y.; Huang, D.; Wang, T. Yeast Extract: Characteristics, Production, Applications and Future Perspectives. *J. Microbiol. Biotechnol.* **2023**, *33*, 151–166. [CrossRef] [PubMed]
17. Ma, J.; Sun, Y.; Meng, D.; Zhou, Z.; Zhang, Y.; Yang, R. Yeast proteins: The novel and sustainable alternative protein in food applications. *Trends Food Sci. Technol.* **2023**, *135*, 190–201. [CrossRef]
18. Jach, M.E.; Serefko, A.; Ziaja, M.; Kieliszek, M. Yeast Protein as an Easily Accessible Food Source. *Metabolites* **2022**, *12*, 63. [CrossRef]

19. Barriga, J.A.T.; Cooper, D.G.; Idziak, E.S.; Cameron, D.R. Components of the bioemulsifier from *S. cerevisiae*. *Enzym. Microb. Technol.* **1999**, *25*, 96–102. [CrossRef]
20. Furtado, G.F.; Picone, C.S.F.; Cuellar, M.C.; Cunha, R.L. Breaking oil-in-water emulsions stabilized by yeast. *Colloids Surf. B Biointerfaces* **2015**, *128*, 568–576. [CrossRef] [PubMed]
21. Molino, A.; Iovine, A.; Casella, P.; Mehariya, S.; Chianese, S.; Cerbone, A.; Rimauro, J.; Musmarra, D. Microalgae characterization for consolidated and new application in human food, animal feed and nutraceuticals. *Int. J. Environ. Res. Public Health* **2018**, *15*, 2436. [CrossRef]
22. Gohara-Beirigo, A.K.; Matsudo, M.C.; Cezare-Gomes, E.A.; Monteiro de Carvalho, J.C.; Danesi, E.D.G. Microalgae trends toward functional staple food incorporation: Sustainable alternative for human health improvement. *Trends Food Sci. Technol.* **2022**, *125*, 185–199. [CrossRef]
23. Bito, T.; Okumura, E.; Fujishima, M.; Watanabe, F. Potential of Chlorella as a Dietary Supplement to Promote Human Health. *Nutrients* **2020**, *12*, 2524. [CrossRef] [PubMed]
24. Ruiz, J.; Wijffels, R.H.; Dominguez, M.; Barbosa, M.J. Heterotrophic vs autotrophic production of microalgae: Bringing some light into the everlasting cost controversy. *Algal Res.* **2022**, *64*, 102698. [CrossRef]
25. Cao, Y.; Miao, L. Consumer perception of clean food labels. *Br. Food J.* **2022**, *125*, 433–448. [CrossRef]
26. Asioli, D.; Aschemann-Witzel, J.; Caputo, V.; Vecchio, R.; Annunziata, A.; Næs, T.; Varela, P. Making sense of the "clean label" trends: A review of consumer food choice behavior and discussion of industry applications. *Food Res. Int.* **2017**, *99*, 58–71. [CrossRef] [PubMed]
27. Maruyama, K.; Sakashita, T.; Hagura, Y.; Suzuki, K. Relationship between Rheology, Particle Size and Texture of Mayonnaise. *Food Sci. Technol. Res.* **2007**, *13*, 1–6. [CrossRef]
28. Tass, M.A.; Bhat, A.A.; Riyaz, I.; Hakim, I.A. Russia-Ukraine War and Its Impact on Global Oil and Food Prices. *J. Adv. Zool.* **2024**, *45*, 1145–1151. [CrossRef]
29. Sánchez-Morales, D.; Martinez-Rodriguez, O.A.; Kyndt, J.; Martinez, A. Heterotrophic growth of microalgae: Metabolic aspects. *World J. Microbiol. Biotechnol.* **2015**, *31*, 1–9. [CrossRef]
30. Simões, A.; Miranda, M.; Cardoso, C.; Veiga, F.; Vitorino, C. Rheology by Design: A Regulatory Tutorial for Analytical Method Validation. *Pharmaceutics* **2020**, *12*, 820. [CrossRef] [PubMed]
31. Fischer, P.; Windhab, E.J. Rheology of food materials. *Curr. Opin. Colloid Interface Sci.* **2011**, *16*, 36–40. [CrossRef]
32. López-Castejón, M.L.; Bengoechea, C.; Espinosa, S.; Carrera, C. Characterization of ningprebiotic emulsions stabilized by inulin and β-lactoglobulin. *Food Hydrocoll.* **2019**, *87*, 382–393. [CrossRef]
33. Felix, M.; Guerrero, A.; Carrera-Sánchez, C. Optimization of Multiple W1/O/W2 Emulsions Processing for Suitable Stability and Encapsulation Efficiency. *Foods* **2022**, *11*, 1367. [CrossRef]
34. Khemiri, S.; Khelifi, S.; Messaoud, C.; Smaali, I. Bioprospecting of microalgae for a potential use as enzyme inhibitors, anti-ageing and prebiotic agents. *Biocatal. Agric. Biotechnol.* **2023**, *51*, 102759. [CrossRef]
35. Brand-Williams, W.; Cuvelier, M.E.; Berset, C. Use of a Free Radical Method to Evaluate Antioxidant Activity. *LWT—Food Sci. Technol.* **1995**, *28*, 25–30. [CrossRef]
36. Benzie, I.F.F.; Strain, J.J. The Ferric Reducing Ability of Plasma (FRAP) as a Measure of "Antioxidant Power: The FRAP Assay. *Anal. Biochem.* **1996**, *239*, 70–76. [CrossRef] [PubMed]
37. Benzie, I.; Devaki, M. The ferric reducing/antioxidant power (FRAP) assay for non-enzymatic antioxidant capacity: Concepts, procedures, limitations and applications. In *Measurement of Antioxidant Activity & Capacity: Recent Trends and Applications*; Wiley: Hoboken, NJ, USA, 2018; Volume 5, pp. 77–106.
38. Zhang, Q.; Zhang, J.; Shen, J.; Silva, A.; Dennis, D.A.; Barrow, C.J. A simple 96-well microplate method for estimation of total polyphenol content in seaweeds. *J. Appl. Phycol.* **2006**, *18*, 445–450. [CrossRef]
39. Kemp, S.E.; Hollowood, T.; Hort, J. *Sensory Evaluation: A Practical Handbook*; Wiley-Blackwell: Hoboken, NJ, USA, 2009.
40. Sriprablom, J.; Luangpituksa, P.; Wongkongkatep, J.; Pongtharangkul, T.; Suphantharika, M. Influence of pH and ionic strength on the physical and rheological properties and stability of whey protein stabilized o/w emulsions containing xanthan gum. *J. Food Eng.* **2019**, *242*, 141–152. [CrossRef]
41. Barnes, H.A.; Hutton, J.F.; Walters, K. Linear Viscoelasticity. In *An Introduction to Rheology*; Elsevier: Amsterdam, The Netherlands, 1989; pp. 46–52.
42. Schramm, G. The Relevance of Shear Rates on Rheological Data and on the Processibility of Visco-elastic Fluids. In *A Practical Approach to Rheology and Rheometry*, 2nd ed.; Gebrueder HAAKE GmbH: Karlsruhe, Germany, 1998; pp. 142–160.
43. Wang, Y.; Wang, S.; Li, R.; Wang, Y.; Xiang, Q.; Qiu, S.; Xu, W.; Bai, Y. Synergistic effect of corn fiber gum and chitosan in stabilization of oil in water emulsion. *LWT—Food Sci. Technol.* **2022**, *154*, 112592. [CrossRef]
44. Dickinson, E. Stabilising emulsion-based colloidal structures with mixed food ingredients. *J. Sci. Food Agric.* **2013**, *93*, 710–721. [CrossRef]

45. Lam, A.C.Y.; Karaca, A.C.; Tyler, R.T.; Nickerson, M.T. Pea protein isolates: Structure, extraction, and functionality. *Food Rev. Int.* **2018**, *34*, 126–147. [CrossRef]
46. Pastrana, A.J.; Gallegos, A.C.; Acosta, D.F.; Herrera, R.; Duque, J.F. Rheological behavior of quinoa, lentil, and rice flour mixtures under different pH conditions. *Food Hydrocoll.* **2025**, *158*, 110457. [CrossRef]
47. McClements, D.J. The future of food colloids: Next-generation nanoparticle delivery systems. *Curr. Opin. Colloid Interface Sci.* **2017**, *28*, 7–14. [CrossRef]
48. Dickinson, E. Colloids in Food: Ingredients, Structure, and Stability. *Annu. Rev. Food Sci. Technol.* **2015**, *6*, 211–233. [CrossRef] [PubMed]
49. Zhu, M.; Wang, H.; Sun, T.; Lv, J.; Wang, K.; Huan, S.; Li, Z.; Liu, Y.; Liu, S.; McClements, D.J.; et al. Plant-based Pickering emulsions using cellulose nanofibers and soy protein isolate: Stabilization and high environmental resistance. *Food Hydrocoll.* **2025**, *162*, 110943. [CrossRef]
50. Erazo, E.M.; Bosqui, K.; Rabelo, R.S.; Hubinger, M.D. Effect of pH and Pea Protein: Xanthan Gum Ratio on Emulsions with High Oil Content and High Internal Phase Emulsion Formation. *Molecules* **2021**, *26*, 5646. [CrossRef]
51. Zhang, X.; Qi, B.; Xie, F.; Hu, M.; Sun, Y.; Han, L.; Li, L.; Zhang, S.; Li, Y. Emulsion stability and dilatational rheological properties of soy/whey protein isolate complexes at the oil-water interface: Influence of pH. *Food Hydrocoll.* **2021**, *133*, 106391. [CrossRef]
52. Tu, Y.; Zhang, X.; Wang, L. Effect of salt treatment on the stabilization of Pickering emulsions prepared with rice bran protein. *Food Res. Int.* **2023**, *166*, 112537. [CrossRef]
53. Jurkowska, D.; Domian, E. The Effect of Heat- and Salt Treatment on the Stability and Rheological Properties of Chickpea Protein-Stabilized Emulsions. *Appl. Sci.* **2024**, *14*, 2698. [CrossRef]
54. Franco, J.M.; Partal, P.; Ruiz-Marquez, D.; Conde, B.; Gallegos, C. Influence of pH and Protein Thermal Treatment on the Rheology of Pea Protein-Stabilized Oil-in-Water Emulsions. *J. Am. Oil Chem. Soc.* **2000**, *77*, 975–984. [CrossRef]
55. Yue, M.; Huang, M.; Zhu, Z.; Huang, T.; Huang, M. Effect of ultrasound assisted emulsification in the production of Pickering emulsion formulated with chitosan self-assembled particles: Stability, macro, and micro rheological properties. *LWT—Food Sci. Technol.* **2022**, *154*, 112595. [CrossRef]
56. Yang, Q.; Qi, W.; Shao, Y.; Zhang, X.; Wu, F.; Zhang, Z. Stability and pH-Dependent Mechanism of Astaxanthin-Loaded Nanoemulsions Stabilized by Almond Protein Isolate. *Foods* **2024**, *13*, 4067. [CrossRef]
57. Quintana, J.M.; Califano, A.N.; Zaritzky, N.E.; Partal, P. Effect of Salt on the Rheological Properties of Low-in-Fat O/W Emulsions Stabilised with Polysaccharides. *Food Sci. Technol. Int.* **2002**, *8*, 213–221. [CrossRef]
58. Falcão, R.L.; Pinheiro, V.; Ribeiro, C.; Sousa, I.; Raymundo, A.; Nunes, M.C. Nutritional Improvement of Fresh Cheese with Microalga *Chlorella vulgaris*: Impact on Composition, Structure and Sensory Acceptance. *Food Technol. Biotechnol.* **2023**, *61*, 259–270. [CrossRef]
59. Chen, J.; Dickinson, E. Viscoelastic Properties of Protein-Stabilized Emulsions: Effect of Protein-Surfactant Interactions. *J. Agric. Food Chem.* **1998**, *46*, 91–97. [CrossRef] [PubMed]
60. Graça, C.; Mota, J.; Lima, A.; Ferreira, R.B.; Raymundo, A.; Sousa, I. Glycemic Response and Bioactive Properties of Gluten-Free Bread with Yoghurt or Curd-Cheese Addition. *Foods* **2020**, *9*, 1410. [CrossRef] [PubMed]
61. McClements, D.J. Interfacial Properties and Their Characterization. In *Food Emulsions: Principles, Practices, and Techniques*, 3rd ed.; CRC Press: Boca Raton, FL, USA, 2015; pp. 190–260.
62. Gu, Y.S.; Decker, E.A.; McClements, D.J. Influence of pH and carrageenan type on properties of β-lactoglobulin stabilized oil-in-water emulsions. *Food Hydrocoll.* **2005**, *19*, 83–91. [CrossRef]
63. McClements, D.J. Emulsion Stability. In *Food Emulsions: Principles, Practices, and Techniques*, 3rd ed.; CRC Press: Boca Raton, FL, USA, 2015; pp. 294–384.
64. Zhao, X.; Wu, T.; Xing, T.; Xu, X.; Zhou, G. Rheological and physical properties of O/W protein emulsions stabilized by isoelectric solubilization/precipitation isolated protein: The underlying effects of varying protein concentrations. *Food Hydrocoll.* **2019**, *95*, 580–589. [CrossRef]
65. Tesch, S.; Schubert, H. Influence of increasing viscosity of the aqueous phase on the short-term stability of protein stabilized emulsions. *J. Food Eng.* **2002**, *52*, 305–312. [CrossRef]
66. Abdolmaleki, K.; Mohammadifar, M.A.; Mohammadi, R.; Fadavi, G.; Meybodi, N.M. The effect of pH and salt on the stability and physicochemical properties of oil-in-water emulsions prepared with gum tragacanth. *Carbohydr. Polym.* **2016**, *140*, 342–348. [CrossRef] [PubMed]
67. Doğan, M.; Saraç, M.G.; Türker, D.A. Effect of salt on the inter-relationship between the morphological, emulsifying and interfacial rheological properties of O/W emulsions at oil/water interface. *J. Food Eng.* **2020**, *275*, 109871. [CrossRef]
68. French, D.J.; Fowler, J.; Taylor, P.; Clegg, P.S. Influence of salt concentration on the formation of Pickering emulsions. *Soft Matter* **2020**, *16*, 7342–7349. [CrossRef] [PubMed]
69. Martínez, I.; Riscardo, M.A.; Franco, J.M. Effect of salt content on the rheological properties of salad dressing-type emulsions stabilized by emulsifier blends. *J. Food Eng.* **2007**, *80*, 1272–1281. [CrossRef]

70. Shao, P.; Ma, H.; Zhu, J.; Qiu, Q. Impact of ionic strength on physicochemical stability of o/w emulsions stabilized by *Ulva fasciata* polysaccharide. *Food Hydrocoll.* **2017**, *69*, 202–209. [CrossRef]
71. Depree, J.A.; Savage, G.P. Physical and flavor stability of mayonnaise. *Trends Food Sci. Technol.* **2001**, *12*, 157–163. [CrossRef]
72. Li, C.-Y.; Kim, H.-W.; Li, H.; Lee, D.-C.; Rhee, H.-I. Antioxidative effect of purple corn extracts during storage of mayonnaise. *Food Chem.* **2014**, *152*, 592–596. [CrossRef] [PubMed]
73. Mattia, C.D.; Balestra, F.; Sacchetti, G.; Neri, L.; Mastrocola, D.; Pittia, P. Physical and structural properties of extra-virgin olive oil based mayonnaise. *LWT—Food Sci. Technol.* **2015**, *62*, 764–770. [CrossRef]
74. Cerro, D.A.; Maldonado, A.P.; Matiacevich, S.B. Comparative study of the physicochemical properties of a vegan dressing-type mayonnaise and traditional commercial mayonnaise. *Grasas Aceites* **2022**, *72*, e439. [CrossRef]
75. Alavi, N.; Golmakani, M. Improving oxidative stability of virgin olive oil by addition of microalga *Chlorella vulgaris* biomass. *J. Food Sci. Technol.* **2017**, *54*, 2464–2473. [CrossRef]
76. Cunha, S.A.; Coscueta, E.R.; Nova, P.; Silva, J.L.; Pintado, M.M. Bioactive Hydrolysates from *Chlorella vulgaris*: Optimal Process and Bioactive Properties. *Molecules* **2022**, *27*, 2505. [CrossRef] [PubMed]
77. Bengwayan, P.T.; Laygo, J.C.; Pacio, A.E.; Poyaoan, J.L.Z.; Rebugio, J.F.; Yuson, A.L.L. A Comparative Study on the Antioxidant Property of *Chlorella* (*Chlorella* sp.) Tablet and Glutathione Tablet. *E-Int. Sci. Res. J.* **2010**, *2*, 25–35.
78. Oliveira, S.; Pérez, M.D.T.; Sousa, I.; Raymundo, A. 3D-printed *Chlorella vulgaris* snacks: A contribution to a healthy diet. *Front. Food Sci. Technol.* **2023**, *3*, 1265828. [CrossRef]
79. Rodriguez-Garcia, I.; Guil-Guerrero, J.L. Evaluation of the antioxidant activity of three microalgal species for use as dietary supplements and in the preservation of foods. *Food Chem.* **2008**, *108*, 1023–1026. [CrossRef]
80. Gouveia, L.; Raymundo, A.; Batista, A.P.; Sousa, I.; Empis, J. *Chlorella vulgaris* and *Haematococcus pluviialis* biomass as colouring and antioxidant in food emulsions. *Eur. Food Res. Technol.* **2006**, *222*, 362–367. [CrossRef]
81. Davidov-Pardo, G.; Gumus, C.E.; McClements, D.J. Lutein-enriched emulsion-based delivery systems: Influence of pH and temperature on physical and chemical stability. *Food Chem.* **2016**, *196*, 821–827. [CrossRef]
82. Štern, P.; Valentová, H.; Pokorný, J. Rheological properties and sensory texture of Mayonnaise. *Eur. J. Lipid Sci. Technol.* **2001**, *103*, 23–28. [CrossRef]
83. Letras, P.; Oliveira, S.; Varela, J.; Nunes, M.C.; Raymundo, A. 3D printed gluten-free cereal snack with incorporation of *Spirulina* (*Arthrospira platensis*) and/or *Chlorella vulgaris*. *Algal Res.* **2022**, *68*, 102863. [CrossRef]
84. Jaeger, S.R.; Chheang, S.L.; Ares, G. Beyond plant-based alternatives to milk and meat: Product and individual variables influence purchase intention for plant-based yoghurt and eggs. *Food Qual. Prefer.* **2023**, *112*, 105019. [CrossRef]

foods

MDPI

Article

Salty Biscuits Enriched with Fresh and Dried Bee Pollen: Chemical, Technological, and Sensory Characterization

Alessandro Bianchi [1], Sonia Capparelli [1], Isabella Taglieri [1,2], Chiara Sanmartin [1,2,*], Laura Pistelli [1,2,3] and Francesca Venturi [1,2]

1 Department of Agriculture, Food and Environment, University of Pisa, Via del Borghetto 80, 56124 Pisa, Italy; alessandro.bianchi@phd.unipi.it (A.B.); sonia.capparelli@phd.unipi.it (S.C.); isabella.taglieri@unipi.it (I.T.); laura.pistelli@unipi.it (L.P.); francesca.venturi@unipi.it (F.V.)
2 Interdepartmental Research Centre "Nutraceuticals and Food for Health", University of Pisa, Via del Borghetto 80, 56124 Pisa, Italy
3 Interdepartmental Center of Agro-Environmental Research "Enrico Avanzi", University of Pisa, 56122 Pisa, Italy
* Correspondence: chiara.sanmartin@unipi.it

Abstract: Bee pollen is a potential functional food ingredient as it contains essential nutrients and a wide range of bioactive compounds. Among bakery products, sweet or salty biscuits are very popular, because they can be consumed quickly, have a long shelf life, and have a favorable taste and texture. The aim of this study was to investigate the effects of the enrichment of salty biscuits with bee pollen (fresh and dried) through their chemical-technological and sensory characteristics. The biscuit formulations were created by replacing the flour with an increasing amount (5% and 10%) of fresh (FP) and dried (DP) pollen. A formulation without pollen was used as the control (CB). To evaluate its potential as a fortification ingredient, pollen as well as salty biscuits were analyzed in terms of their chemical composition and sensory characteristics. In particular, biscuits with 5% fresh pollen (FPB5%) proved to be the formulation with the optimal combination of chemical-compositional and sensory characteristics. Given the increase in their antioxidant component, fortified biscuits can represent an interesting vehicle for phenolic compounds and carotenoids, with a characteristic sensory profile.

Keywords: fortification; antioxidant activity; carotenoids; functional foods; color; sensory quality; physicochemical properties; bakery products

1. Introduction

In the last few decades, the interest in food enriched with one or more micronutrients and/or nutrients has increased regardless of whether these are usually contained in the food [1–3]. In 2006, the constant growth of fortified foods led the European Union to issue EC regulation no. 1924/2006, which regulates the addition of vitamins and minerals to foods, reporting a list of those allowed to go together with their relative forms [2,4]. Many food ingredients (food additives or functional food additives) have been included in the formulation of baked goods to increase their diversity and quality, in terms of nutritional and sensory characteristics [5–7].

Sweet or salty biscuits are very popular bakery products, mainly because they can be consumed quickly, have a long shelf life, and have varied tastes and textures [8]. The essential ingredients are flour, oil, sugar and/or salt, water, and chemical leavening agents such as sodium bicarbonate. Regular consumption of high doses of certain ingredients (such as salt, sugar, and fat) may be considered unsafe, particularly when associated with highly processed foods [9,10]. So, in recent years, numerous studies have been conducted to improve the nutritional quality of such products [3,6,11].

Due to its nutritional properties, bee pollen is suitable for food supplementation because, in addition to being rich in nutrients, it contains important bioactive molecules, antioxidants, minerals, and vitamins [12–15]. These compounds can have beneficial effects on human health, preventing the onset of certain diseases [12,16]. Bee pollen contains important biologically active molecules, such as phenolic acid derivatives and flavonoid glycosides like quercetin and rutin [17–20]. Flavonoids are secondary plant metabolites known for their anti-inflammatory, anticancer, antiviral, antibacterial, and cardioprotective properties [21,22]. In addition, a fairly significant amount of vitamin B12, a nutrient not typically derived from plant sources, can be found in pollen [15,23,24], which is valuable particularly for vegetarian or gluten-free diets [16,25]. Moreover, bee pollen contains phytosterols, with beta-sitosterol contributing to its anti-teratogenic, anti-inflammatory, and immunostimulant effects [13,18,26].

Several studies [27–34] have demonstrated that adding bee pollen can improve the sensory properties of bread and sweet biscuits, particularly by improving their taste and rheological properties, which influence people's tactile sensations when chewing [34,35].

The unique techno-functional properties of bee pollen, such as hardness, chewiness, adhesiveness, gumminess, springiness, resilience, and cohesiveness [36], make it an excellent ingredient for incorporation into a wide range of food products [13,37]. Higher concentrations of bee pollen result in softer biscuits and bread [28,37,38]. Additionally, bee pollen's excellent emulsifying properties make it suitable for use as a natural emulsifier for the development of gluten-free bread [27,39].

A study [28] investigating the enrichment of gluten-free bread with multifloral bee pollen, in increasing percentages, demonstrated that this integration provided a well-leavened dough system. After baking, the enriched bread showed a significant improvement in rheological, technological-functional, and sensory properties compared to the control bread.

Another study [37] compared sweet biscuits enriched with monofloral bee pollen, such as rapeseed pollen (*Brassica napus* L.), phacelia (*Phacelia tanacetifolia* Benth.), and sunflower (*Helianthus annuus* L.) on the basis of their nutritional, physical, and sensory properties. The results confirmed that pollen can be used to improve the nutritional value of biscuits; however, it should be noted that the chemical composition of bee pollen, which varies according to environmental conditions, the plant species from which it is collected, and commercial processing and preservation techniques [14,40–42], greatly affects the overall acceptance of the final product [43]. The fortification of biscuits with bee pollen resulted in a statistically significant increase in the sugar, protein, ash, fiber, and polyphenol content, as well as the antioxidant potential of the final product [30,44]. In addition to the nutritional profile, the pollen positively affected the sensory profile of the biscuits, as they were characterized by a greater consistency and a darker surface than the control [30,44].

Sokmen et al. [45] showed that the fortification with multifloral bee pollen on reduced-fat biscuits significantly increased the overall sensory acceptability of the biscuits, their nutritional value, given the reduction in fat, as well as their total phenol content and antioxidant capacity values.

All the studies in the literature report that bee pollen is a natural ingredient that is well suited for fortifying foods such as baked goods [46] or yoghurt [47]. Not only does it enhance these products by providing antioxidant compounds with health benefits, but it also improves their sensory qualities, adding unique flavors, taste, and colors [37,48,49].

Nonetheless, further research is needed to understand the interactions between phenolic compounds and macronutrients in pollen-enriched bakery products and to improve the bio-accessibility of nutrients in these products [18,30,50].

However, to date, no published studies have investigated the application of bee pollen as a fortifying agent in salty biscuits or its effects on their technological-functional and sensory properties.

For this reason, this study aimed to address this gap by producing salty biscuits formulated with different levels (0%, 5%, and 10%) of fresh and dried multifloral bee pollen and evaluating their technological properties, physicochemical composition, and sensory profile both soon after cooking and over time (a timelapse of 60 days).

2. Materials and Methods

2.1. Raw Materials

The basic ingredients used to produce biscuits were purchased at a local supermarket (Coop Italia, Pisa, Italy), while the bee pollen used was donated by the Mancini agricultural company (Tirli, Grosseto, Italy) and subsequently frozen at −18 °C until further analysis.

The ingredients used for the biscuits' formulation were soft wheat flour type 00 (Coop Italia, Bologna, Italy), dry white wine (Tavernello, Caviro soc. coop. agricola, Faenza, Ravenna, Italy), extra virgin olive oil (Filippo Berio, Salov S.p.A, Lucca, Italy), sea salt (Coop Italia, Bologna, Italy), still water (Coop Italia, Bologna, Italy), and multifloral bee pollen.

2.2. Pollen-Drying Process

A total of 50 g of fresh pollen was dried at a temperature of 35 °C for 24 h in the Fruit Jerky Plus 6 dryer (Klarstein, Berlin, Germany) at the Department of Agricultural, Food, and Environment of the University of Pisa until reaching approximately 12% humidity.

2.3. Process of Making Salty Biscuits

The biscuit samples were prepared at room temperature (23 ± 2 °C) by replacing wheat flour with fresh (FP) or dried (DP) bee pollen, in the amounts of 5% and 10%. A control (CB) sample (pollen-free biscuits, 0%) was also prepared. To reduce bias, in samples prepared with DP, the needed amount of water was added to reach the same final fresh weight of 294 g.

The formulations used were reported in Table 1 and the samples were coded as follows:

- CB: control salty biscuits without pollen;
- FPB5%: salty biscuits with 5% fresh pollen;
- FPB10%: salty biscuits with 10% fresh pollen;
- DPB5%: salty biscuits with 5% dried pollen;
- DPB10%: salty biscuits with 10% dried pollen.

Table 1. Formulations adopted to produce about 300 g of salty biscuit dough.

Ingredients (g)	CB	FPB5%	FPB10%	DPB5%	DPB10%
Soft wheat flour type 00	184	174.8	165.6	174.8	165.6
Water	0	0	0	1.0	2.0
Multifloral bee pollen	0	9.2	18.4	8.2	16.4
Dry white wine	67	67	67	67	67
Extra virgin olive oil	40	40	40	40	40
Sea salt	3	3	3	3	3
Total fresh weight	294	294	294	294	294

All the ingredients were weighed in separate containers. The flour, salt, and olive oil were mixed until a crumbly mixture was obtained. The pollen was dissolved in the wine and added to the mixture. The dough was kneaded by hand until a homogeneous consistency was obtained. The dough was then shaped into biscuits approximately 3 × 3 × 0.5 cm in size.

The biscuits were baked in an electric convection oven (Fimar, OMNIA srl, Santarcangelo di Romagna, Rimini, Italy) at 160 °C for 40 min, subsequently cooled to room temperature, packaged in plastic bags (with an outer nylon layer and two plastic layers; Food Saver, Moncalieri, Torino, Italy) with an industrial packaging machine (450 GAS Lavezzini, Fiorenzuola d'Arda, Piacenza, Italy), and stored at 23 ± 2 °C. The entire biscuit-making process was repeated in triplicate, and three different batches were obtained for each formulation.

2.4. Technological and Chemical Composition of Bee Pollen and Biscuits

The dry matter (dm) was determined for a sample of approximately 5 g (bee pollen and biscuits) dried at 105 °C until constant weight was achieved, and the pH of biscuits was measured with a pH meter (pH 80+ DHS, XS Instrument, Modena, Italy) according to the AACC (American Association of Cereal Chemists) standard method, as previously reported [3].

Total titratable acidity (TTA) was determined in biscuits following Bianchi et al.'s study [51].

Water activity (a_w) of bee pollen and biscuits was measured by a HygroPalm HP23-AW-A equipment (Rotronic AG, Bassersdorf, Switzerland) [52].

Free acidity (FA) (% oleic acid) was determined in biscuits as previously reported [53]; 4 g of ground biscuits was added to 100 mL of 50% ethanol and stirred for 3 h at 22 °C. The solution was filtered, and 50 mL was then titrated with NaOH (0.02 N) until the indicator (phenolphthalein) changed.

Peroxide value (meq O_2/kg dm) was determined in biscuits as reported by Nanditha et al., 2009 [50].

For the evaluation of phytochemical compounds (total polyphenols, flavonoids, and antioxidant activity), 1 g of pollen (fresh and dried) and 5 g of biscuit sample were extracted at room temperature using 20 mL of a methanol/water solution (80% *v*/*v*) and sonicated for 30 min at a frequency of 2400 Mhz. The samples were centrifuged at 10,000× *g* rpm for 10 min at 23 °C, and the supernatants were collected and used for further analyses involving a spectrophotometer assay.

Total polyphenols were determined at 765 nm using the Folin–Ciocalteau method [54]. The results are expressed as mg gallic acid equivalent (GAE) per g dm.

Flavonoids were determined as reported by Bianchi et al. [55]. The absorbance was read at 510 nm, while the concentration of the samples was expressed as mg of catechin equivalents (CE) per g dm, comparing the measures to a standard curve of catechin.

Antioxidant activity was measured by three methods [56]: ABTS (2,2′-azino-bis [3-ethylbenzothiazoline-6-sulfonic acid]) by reading the absorbance at 734 nm, DPPH (2,2-diphenyl-1-picrylhydrazyl) by reading the absorbance at 515 nm, and FRAP (ferric ion reducing antioxidant power) by reading the absorbance at 593 nm. The results were expressed as µmol Trolox equivalents (TE) per g dm of sample, based on different standard curves of Trolox: 0–200 µmol L^{-1} for the DPPH assay, 0.2–1.5 mM range for ABTS, and 0–2.0 mM for the FRAP assay.

For carotenoid determination in bee pollen and biscuits, extraction (with saturated n-butanol) was performed following the methodology of Cabras [53], and concentration was determined at 435.8 nm expressed as mg of β-carotene per kg dm of biscuits or bee pollen, comparing the measurements to a standard curve of β-carotene.

All tests were performed in triplicate.

2.5. Colorimetric Analysis of Bee Pollen and Biscuits

The color of the bee pollen and biscuits was measured according to the CIE L*a*b* Color System using a tristimulus colorimeter (CLM-196 Benchtop, Eoptis, Trento, Italy). The analysis was performed on pollen and biscuit samples placed on an area of approximately 4.5 cm^2, and each sample was analyzed in triplicate.

The color was defined based on the chromatic coordinates, lightness (L*), green-red (a*), and blue-yellow components (b*). The cylindric coordinates, chroma (C*) and hue (h*), were also calculated as previously reported [57]. The color differences among samples (ΔE^*_{ab}) were calculated as previously reported [3] and are expressed in CIELAB units.

The whiteness (WI) and yellowness (YI) indices of the samples were calculated according to the equations provided by Bianchi et al. [3].

2.6. Sensory Evaluation of Biscuits

Sensory profiles of the biscuits were determined through quantitative descriptive analysis (QDA) from the "expert panel" of the University of Pisa.

QDA was conducted at room temperature, in a standard sensory laboratory (ISO 8589:2010 [58]) located in the Department of Agriculture, Food, and Environment (DAFE) by trained judges (10 assessors: 6 females and 4 males, aged between 23 and 63 years), who were primarily experts in bakery tasting. The assessors were selected based on their availability from a larger pool of judges who regularly collaborate with the DAFE at the University of Pisa. All judges underwent standardized training to enhance their ability to recognize, describe, and quantify tastes, odors, and texture properties in accordance with ISO 8586:2023 standard [59] and typically worked collaboratively.

Assessor records, including consent forms and personal data, were collected and maintained in compliance with national data protection laws. Sensory activities were conducted in consideration of ethical principles regarding the use of human subjects and adhered to health and safety regulations. The research obtained the approval of the bioethical committee of the University of Pisa (protocol n. 0088081/2023).

Prior to the tasting sessions of the experimental samples, a sub-group of the trained panelists participated in a consensus panel specifically designed to generate descriptors and their definitions.

A final set of 27 descriptive parameters was established through consensus among the panelists for the QDA sensory test. These included both quantitative descriptors (such as color intensity, hue (yellow-ochre), shape regularity, olfactory intensity, persistence, and specific sensory attributes like oily, grainy, floral, fruity, vegetal, frankness, toasted, chewing resistance, cohesiveness, hardness, friability, sweetness, saltiness, acidity, bitterness, astringency, umami, and aftertaste) and hedonic descriptors (such as visual attractiveness, olfactory pleasantness, taste pleasantness, and overall pleasantness). To facilitate the evaluation, an innovative sensory sheet specifically designed for tasting bee pollen biscuits was created (Figure S1).

The samples were presented in a different order at each tasting session, and 10 min intervals between each sample were set. Furthermore, a biscuit sample was randomly replicated to verify the performance of the panel at each tasting session. For evaluation, each assessor was provided with filtered water and asked to cleanse their palate between tastings.

Each attribute was evaluated on a 0–9 scale. All ratings were digitally acquired by the Input Sensory Soft 2.0 (ISS, Centro Studi Assaggiatori, Brescia, Italy). Finally, the overall hedonic index (HI) of the biscuits, which represents the overall acceptability of the product, was calculated based on the mean of the hedonic parameters, which was converted to a scale from 0 to 10, as previously reported [60]. The HI was calculated at time 0 (after cooling) and after 30 and 60 days of storage at 23 ± 2 °C in plastic bags to assess whether the acceptability of the product changed over time.

2.7. Statistical Analysis

Results were statistically analyzed by one-way analysis of variance (ANOVA) by the software CoStat 6.451 (CoHort Software, Pacific Grove, CA, USA), using Tukey's honestly significant difference test ($p < 0.05$).

The data processing of the sensory profile was carried out using the software Big Sensory Soft 2.0 (Centro Studi Assaggiatori, Brescia, Italy), and the statistical analyses were performed by interquartile two-way ANOVA, choosing samples and panelists as main factors. The correlation value between the quantitative descriptor and the hedonic one was calculated using Pearson correlation coefficient.

Finally, the graphic was created using the software JMP Pro 17 (SAS Institute, Cary, NC, USA).

3. Results and Discussion

3.1. Characterization of Dried and Fresh Bee Pollen

The physicochemical and technological characterization of fresh (FP) and dried (DP) bee pollen is reported in Table 2.

Table 2. Physicochemical and technological characterization of fresh (FP) and dried (DP) pollen.

Parameters	p-Value [1]	FP	DP
Dry matter (%)	***	72.93 ± 0.46 [b]	88.06 ± 0.12 [a]
Water activity (a_w)	***	0.713 ± 0.003 [a]	0.353 ± 0.002 [b]
Total polyphenols (mg GAE/g dm)	n.s.	25.97 ± 0.72	25.57 ± 1.15
Flavonoids (mg CE/g dm)	**	10.77 ± 0.30 [a]	8.45 ± 0.11 [b]
Carotenoids (mg/kg dm)	***	100.35 ± 0.43 [b]	117.54 ± 1.95 [a]
ABTS (μmol TE/g dm)	n.s	227.74 ± 3.41	225.56 ± 4.77
DPPH (μmol TE/g dm)	n.s.	19.03 ± 3.68	25.32 ± 7.11
FRAP (μmol TE/g dm)	n.s.	68.44 ± 0.53	72.44 ± 3.27
L*	n.s.	54.39 ± 2.71	52.68 ± 1.84
a*	n.s.	0.30 ± 0.86	1.24 ± 0.90
b*	n.s.	35.07 ± 4.41	35.65 ± 3.10
C*	n.s.	35.07 ± 4.42	35.68 ± 3.13
h*	n.s.	89.62 ± 1.26	88.06 ± 1.25
White index (WI)	n.s.	42.31 ± 0.80	40.67 ± 0.89
Yellow index (YI)	n.s.	91.89 ± 6.93	96.58 ± 5.49
ΔE^*_{ab}			2.0

Values are presented as the mean ± standard deviation (SD) of 3 samples. In the same row, different letters indicate significant difference among values. [1] Significance level: *** = $p < 0.001$; ** = $p < 0.01$; n.s. = not significant ($p \geq 0.05$). FP: fresh pollen; DP = dried pollen. ΔE^*_{ab} was calculated by taking fresh pollen as a reference to evaluate the difference after drying.

As expected, following the drying process, the water activity and flavonoid content decrease significantly. In contrast, the dry matter and carotenoids significantly increased in the dry pollen (DP). The total polyphenols and antiradical activity did not show significant variation during the drying process.

No significant differences were observed in the chromatic coordinates and colorimetric parameters between the fresh and dried pollen samples (Table 2). Both fresh and dried pollen grains have a color close to red (+a) and yellow (+b), characterized by a strong color intensity, as indicated by the colorimetric parameter (C*). The yellow index (YI) was higher than the white index (WI). As expected, the color difference (ΔE^*_{ab}) was found to be two, indicating no perceptible difference to the human eye [61].

3.2. Characterization of Biscuits

3.2.1. Chemical-Technological Characterization

The biscuits were analyzed by evaluating the following chemical-technological parameters, which provide information for defining the shelf life of the product (Table 3).

Table 3. Chemical-technological parameters of different salty biscuits.

Parameters	p-Value [1]	CB	FPB5%	FPB10%	DPB5%	DPB10%
Dry matter (%)	n.s.	96.93 ± 0.59	96.60 ± 0.36	96.02 ± 0.21	96.68 ± 0.95	97.75 ± 0.36
a_w	***	0.134 ± 0.030 c	0.208 ± 0.003 b	0.274 ± 0.006 a	0.119 ± 0.001 c	0.123 ± 0.001 c
pH	*	5.44 ± 0.01 ab	5.40 ± 0.04 b	5.52 ± 0.04 a	5.43 ± 0.02 ab	5.50 ± 0.01 b
TTA (meq/g dm)	n.s.	13.04 ± 0.37	13.76 ± 1.04	11.88 ± 0.99	12.76 ± 0.49	12.02 ± 0.49
FA (% oleic acid dm)	**	0.06 ± 0.01 b	0.10 ± 0.01 a	0.11 ± 0.01 a	0.09 ± 0.01 a	0.11 ± 0.01 a
PV (meq O_2/kg dm)	n.s.	2.79 ± 0.06	2.43 ± 0.22	2.86 ± 0.37	2.33 ± 0.37	2.76 ± 0.29

Values are presented as the mean ± standard deviation (SD) of 3 samples. In the same row, different letters indicate significant difference among values. [1] Significance level: *** = $p < 0.001$; ** = $p < 0.01$; * = $p < 0.05$; n.s. = not significant ($p \geq 0.05$). CB: control salty biscuits without pollen; FPB5%: salty biscuits with 5% fresh pollen; FPB10%: salty biscuits with 10% fresh pollen; DPB5%: salty biscuits with 5% dried pollen; DPB10%: salty biscuits with 10% dried pollen.

The water activity does not seem to be affected by fortification; the values are close to the range defined as optimal for this type of baked product (a_w = 0.30) [62]. As observed, 5% and 10% pollen fortification, as well as pollen drying, did not have any significant effect on the parameters of titratable acidity (TTA), dry matter, pH, and peroxide values (Table 3).

As expected, however, the water activity was influenced by the drying process of the pollen, with a decrease in value (Table 3).

The peroxide value (PV) falls within the legal limits, which for dry biscuits is less than 10 meq O_2/kg dm [63]. This suggests the good oxidative stability of the salty biscuits.

In contrast, there is a significant variation in the value of the free acidity (FA) of the biscuits likely due to the lipids present in pollen [12,64,65]. However, these values are low and should not affect the oxidative stability of the product.

3.2.2. Phytochemical Characterization

As expected, the addition of fresh or dried pollen resulted in notable differences in the phytochemical components compared to the control sample (Table 4).

In particular, the content of total phenolic compounds increased significantly in the enriched biscuit, compared to the control sample, in line with the 5–10% fortification. The increase is particularly marked for polyphenols, ranging from the value of 0.67 mg GAE/g dm in the control sample (CB) to 2.77 mg GAE/g dm in FPB10% and to 3.06 mg GAE/g in BPB10% (Table 4). These results fall within the range previously reported by other authors [29,45].

Flavonoids are also affected by this increase and the differences are statistically significant ($p < 0.01$). This increase is more pronounced in biscuits with fresh pollen (FP) compared to those with dried pollen (DP).

Table 4. Phytochemical parameters of different salty biscuits.

Parameters	p-Value [1]	CB	FPB5%	FPB10%	DPB5%	DPB10%
Total polyphenols (mg GAE/g dm)	***	0.67 ± 0.09 d	1.83 ± 0.14 c	2.77 ± 0.15 b	2.03 ± 0.20 c	3.06 ± 0.07 a
Flavonoids (mg CE/g dm)	**	0.20 ± 0.01 d	0.28 ± 0.01 bc	0.60 ± 0.04 a	0.29 ± 0.04 bc	0.42 ± 0.07 b
Carotenoids (mg β-carotene/kg dm)	***	3.96 ± 0.04 e	11.36 ± 0.64 d	18.96 ± 0.22 b	12.50 ± 0.33 c	20.53 ± 0.30 a
ABTS (μmol TE/g dm)	***	2.44 ± 0.17 d	5.22 ± 0.30 c	7.29 ± 0.16 b	5.92 ± 0.46 c	8.75 ± 0.70 a
DPPH (μmol TE/g dm)	***	0.94 ± 0.21 d	2.29 ± 0.07 c	3.26 ± 0.09 b	3.63 ± 0.24 b	5.06 ± 0.28 a
FRAP(μmol TE/g dm)	***	1.71 ± 0.10 d	6.28 ± 0.16 c	9.05 ± 0.64 b	10.73 ± 0.40 b	15.54 ± 0.43 a

Values are presented as the mean ± standard deviation (SD) of 3 samples. In the same row, different letters indicate significant difference among values. [1] Significance level: *** = $p < 0.001$; ** = $p < 0.01$. CB: control salty biscuits without pollen; FPB5%: salty biscuits with 5% fresh pollen; FPB10%: salty biscuits with 10% fresh pollen; DPB5%: salty biscuits with 5% dried pollen; DPB10%: salty biscuits with 10% dried pollen.

Similarly, the results of the antioxidant capacity (ABTS, DPPH, and FRAP) showed an increasing trend together with the percentage of fresh or dried pollen added to the product [45]. The increase is particularly marked for DPPH and FRAP, with values rising from 0.94 μmol TE/g dm and 1.71 μmol TE/g dm, respectively, in CB to 7.29 μmol TE/g and 9.05 μmol TE/g (FBP10%) and 8.75 μmol TE/g and 15.54 μmol TE/g (DBP10%) (Table 4).

Regarding carotenoids, the content significantly increased as the quantity of pollen increased, going from 3.69 mg β-carotene/kg dm (CB) to 11.36 mg β-carotene/kg dm (FBP5%) and 26.48 mg β-carotene/kg dm (FBP10%) for the biscuits with fresh pollen and from 12.50 mg β-carotene/kg dm (DBP5%) to 15.69 mg β-carotene/kg dm (DBP10%) for those enriched with dried pollen. It is also possible to assert that with the same percentage of fortification, following cooking, the content of these molecules increases [45,66]. The increase could be explained both by the hydrolysis of bound molecules and the presence of fats (e.g., oil) in the dough, which protects carotenoids, improves their stability, and limits oxidation. Proteins and phenolic compounds in pollen can further stabilize carotenoids during cooking.

3.2.3. Color Characterization

The addition of pollen made the biscuits different from a chromatic point of view. Analyzing the color of the biscuits, it was observed that the addition of fresh or dried pollen significantly darkened the surface of the biscuits compared to the control (Table 5).

Table 5. Color parameters of different salty biscuits.

Parameters	p-Value [1]	CB	FPB5%	FPB10%	DPB5%	DPB10%
L*	***	70.70 ± 1.43 a	62.71 ± 1.80 b	57.90 ± 1.03 c	54.63 ± 1.10 d	47.11 ± 0.81 e
a*	***	5.10 ± 1.30 d	6.98 ± 0.75 c	8.02 ± 0.24 c	9.81 ± 0.09 b	11.43 ± 0.36 a
b*	***	31.68 ± 1.15 d	39.49 ± 0.38 a	40.63 ± 0.65 a	36.64 ± 0.06 b	34.68 ± 1.33 c
C*	***	32.11 ± 0.93 c	40.11 ± 0.42 a	41.41 ± 0.65 a	37.93 ± 0.07 b	36.52 ± 1.37 b
h*	***	80.81 ± 2.63 a	79.98 ± 1.05 ab	78.83 ± 0.34 b	75.02 ± 0.13 c	71.75 ± 0.20 d
White index (WI)	***	56.51 ± 0.28 a	45.22 ± 1.34 b	40.94 ± 0.41 c	40.86 ± 0.85 c	35.72 ± 0.44 d
Yellow index (YI)	***	64.02 ± 1.03 d	90.02 ± 2.69 c	100.25 ± 0.80 ab	95.86 ± 1.91 b	105.15 ± 2.74 a

Values are presented as the mean ± standard deviation (SD) of 3 samples. In the same row, different letters indicate significant difference among values. [1] Significance level: *** = $p < 0.001$. CB: control salty biscuits without pollen; FPB5%: salty biscuits with 5% fresh pollen; FPB10%: salty biscuits with 10% fresh pollen; DPB5%: salty biscuits with 5% dried pollen; DPB10%: salty biscuits with 10% dried pollen.

Since pollen is darker than wheat flour, the brightness (L*) of the 5% and 10% fortified biscuits was significantly lower than the CB; it is hypothesized that the products of the Maillard reaction increase with the increase in the concentration of pollen due to the high amount of amino acids [31,67].

As the quantity of pollen added increased, a significant increase in the red (+a) and yellow (+b) (Table 5) in the biscuits was observed, presumably associated with the main pollen pigments, such as carotenoids and flavonoids [19,24]. Furthermore, the results related to the chromatic value (C*) suggest that the fortification with pollen significantly increases the color intensity of the biscuits, compared to the control. As expected, there was a decrease in the white index (WI) and an increase in the yellow index (YI), respectively (Table 5).

Furthermore, the color differences (ΔE^*_{ab}) among the samples were calculated (Table 6).

Table 6. Color differences (ΔE^*_{ab}) among the different salty biscuits.

ΔE^*_{ab}	CB	FPB5%	FPB10%	DPB5%	DPB10%
CB		11.33	15.88	17.46	24.60
FPB5%			5.05	9.02	16.92
FPB10%				5.45	12.78
DPB5%					7.94
DPB10%					

CB: control salty biscuits without pollen; FPB5%: salty biscuits with 5% fresh pollen; FPB10%: salty biscuits with 10% fresh pollen; DPB5%: salty biscuits with 5% dried pollen; DPB10%: salty biscuits with 10% dried pollen.

The results indicate that the 5% (FPB5% and DPB5%) and 10% (FPB10% and DPB10%) fortified biscuits have very different colors compared to the control ($\Delta E^*_{ab} > 12$).

There was also a difference in color between the samples with 5% and 10% fortification (Figure S2), but it was significantly higher in the samples with dried pollen ($6 < \Delta E^*_{ab} < 12$) compared to the fresh ones ($3 < \Delta E^*_{ab} < 6$). The color was a fundamental parameter that significantly influenced the consumers' perceptions and decisions. A uniform golden color was often associated with well-baked and appetizing biscuits. If the color was too dark, it may have suggested that the biscuits were burnt or had a bitter taste [68].

3.2.4. Sensory Characterization

From the visual descriptors (Figure 1a), it can be observed that the addition of pollen, fresh and dried, resulted in significant visual differences compared to the control sample, in line with what was expected based on the colorimetric evaluations. In fact, there were evident significant differences in terms of color intensity and hue (from yellow to ochre), which increased as the percentage of pollen increased.

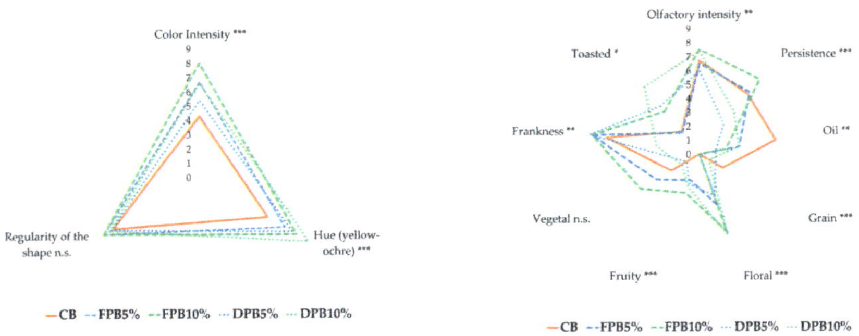

(a) (b)

Figure 1. Sensory evaluation: (**a**); Visual descriptors; (**b**) Olfactory descriptors. Significance level: *** = $p < 0.001$; ** = $p < 0.01$; * = $p < 0.05$; n.s. = not significant ($p \geq 0.05$). CB: control salty biscuits without pollen; FPB5%: salty biscuits with 5% fresh pollen; FPB10%: salty biscuits with 10% fresh pollen; DPB5%: salty biscuits with 5% dried pollen; DPB10%: salty biscuits with 10% dried pollen.

Almost all the olfactory descriptors analyzed (Figure 1b) showed significant differences. Enrichment with pollen was associated with high values of frankness, intensity, and olfactory persistence. The use of pollen determines a higher perception of floral, fruity, and vegetal aromatic notes; the CB sample, on the other hand, is distinguished by a clearer perception of the notes of oil and wheat. Toasted notes also emerge for the DPB10% sample.

From the taste descriptors (Figure 2a), the enrichment with pollen makes the biscuits more astringent, with higher hints of bitterness that resulted in a more complex taste profile, compared to the control sample, which is perceived as sweeter.

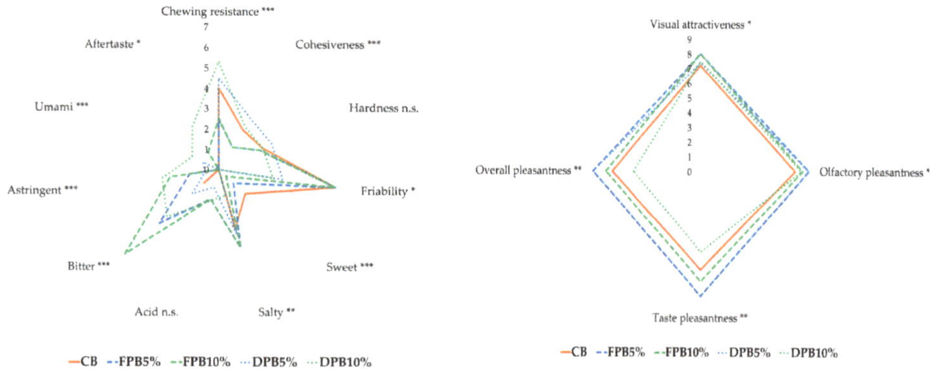

(a) (b)

Figure 2. Sensory evaluation: (**a**) Taste descriptors; (**b**) Hedonic descriptor. Significance level: *** = $p < 0.001$; ** = $p < 0.01$; * = $p < 0.05$; n.s. = not significant ($p \geq 0.05$). CB: control salty biscuits without pollen; FPB5%: salty biscuits with 5% fresh pollen; FPB10%: salty biscuits with 10% fresh pollen; DPB5%: salty biscuits with 5% dried pollen; DPB10%: salty biscuits with 10% dried pollen.

As regards the evaluation of rheological properties, the greatest differences are perceived for chewing resistance and cohesiveness. In chewing resistance, the highest scores are shown by fortified biscuits with dried pollen and control samples, and the same is true for cohesiveness. On the other hand, biscuits enriched with fresh pollen are characterized by greater crumbliness. However, the addition of pollen does not seem to significantly affect the hardness of the product.

The hedonic descriptors (visual attractiveness, olfactory pleasantness, taste pleasantness, and overall pleasantness) were all significant (Figure 2b), especially in regards to taste and the level of global satisfaction, for which DPB10% showed the lowest value.

In general, the sensory profile of the fortified biscuits appears to be much broader and they received higher scores for positive taste and smell descriptors than the control, which was the least aromatically favorable.

In addition, it is essential to be aware of which descriptors contribute most to the characterization of the samples. Figure S3 shows the weight assumed by each quantitative and hedonic descriptor on a scale from zero (minimum) to one (maximum).

As can be seen, visual descriptors, such as the intensity of color and hue, as well as olfactory descriptors such as the perception of oily, floral, toasted, and frankness notes, contribute to a greater extent to the characterization of biscuits, while taste descriptors include resistance to chewing, crumbliness, and a hint of bitterness. As expected, the hedonic parameters (visual attractiveness, olfactory pleasantness, gustatory pleasantness, and overall pleasantness) contribute equally to the characterization of the product.

Figure 3 shows the HI calculated for the biscuits analyzed. The hedonic quality expressed by a product represents a fundamental value determining its success.

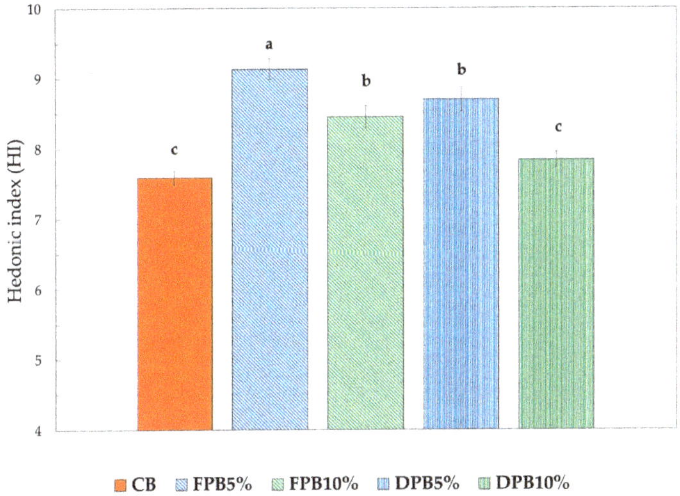

Figure 3. Hedonic index at time 0. Different letters indicate significant difference among values ($p < 0.05$). CB: control salty biscuits without pollen; FPB5%: salty biscuits with 5% fresh pollen; FPB10%: salty biscuits with 10% fresh pollen; DPB5%: salty biscuits with 5% dried pollen; DPB10%: salty biscuits with 10% dried pollen.

It can be seen that the FPB5% group obtained the highest hedonic index. In general, it can be seen that the formulations with fresh pollen showed higher values than the biscuits with pollen dried at the same percentage, as well as control samples. However, all formulations showed overall hedonic indices greater than the acceptability index (equal to six) [51,60].

While the acceptability of a product is important, what is even more critical is identifying which features to highlight or downplay in order to improve the product. This enables targeted product innovation, optimizing the use of resources for maximum effectiveness. The hedonic generators in Figure 4 show to what extent the descriptor contributed to the pleasantness highlighted by the panel.

Characters that negatively affect hedonic quality are highlighted in red, and those that increase hedonic quality are highlighted in blue. For the statistical technique used, values over 0.30, both positive and negative, must be considered; over −0.5 and +0.5, the correlation is significant.

The most consistent contribution to the hedonic index is due to the hedonic descriptors themselves: taste pleasantness (+0.97), olfactory pleasantness (+0.92), global (+0.95), and visual attractiveness (+0.82), followed by the acidity (+0.52), which is probably correlated with the sensory perception of the pollen itself. Oil (−0.53) and aftertaste (−0.51), on the other hand, negatively affect the hedonic quality of the product. Other descriptors that, albeit to a lesser extent, have contributed positively to the hedonic quality of the product were frankness (the absence of olfactory defects), fruity and floral aromatic notes, hue, and friability.

Shelf-life refers to the length of time between the production of a food and the loss of those characteristics that make it acceptable to the consumer [69]. In the case of dry biscuits, the limit used to determine this acceptability is often based on the level of sensory quality [60].

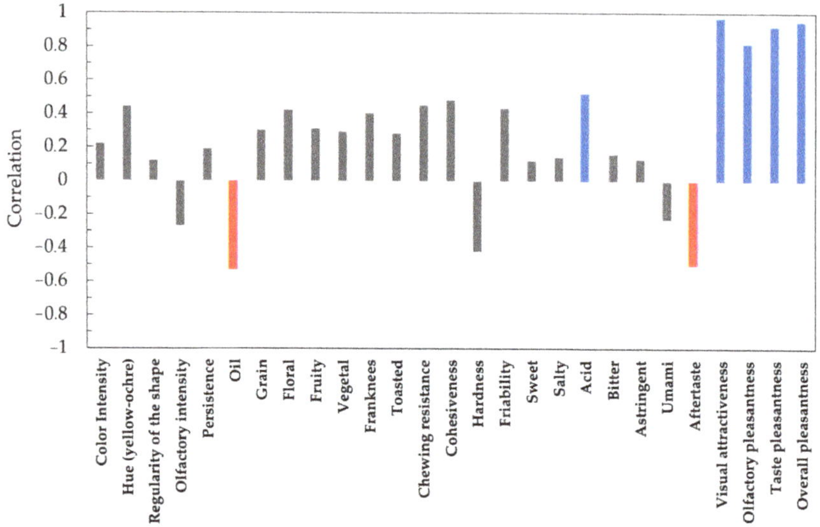

Figure 4. Correlation value for the quantitative and hedonic descriptors. Red color: significant negative effect; Blue color: significant positive effect; Grey color: no significant effect.

During the present study, sensory quality was therefore evaluated after 30 and 60 days of storage at a temperature of 23 ± 2 °C inside a plastic film bag. The data collected were synthesized through the hedonic index to evaluate their changes over time (Figure 5a,b). It can be observed that in the period of time considered, the fortified biscuits, both with fresh and dried pollen, have consistently maintained their sensory properties, and none of the samples evaluated received an overall evaluation below the limit of acceptability.

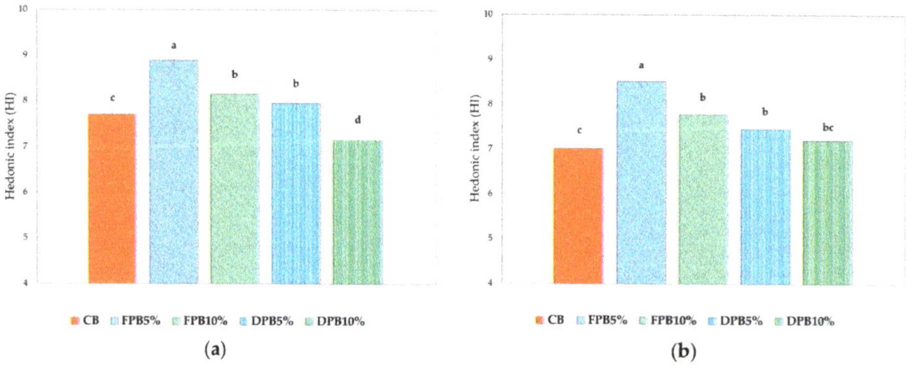

Figure 5. Hedonic index (HI): (**a**) After 30 days of storage; (**b**) After 60 days of storage. Different letters indicate significant difference among values ($p < 0.05$). CB: control salty biscuits without pollen; FPB5%: salty biscuits with 5% fresh pollen; FPB10%: salty biscuits with 10% fresh pollen; DPB5%: salty biscuits with 5% dried pollen; DPB10%: salty biscuits with 10% dried pollen.

4. Conclusions

The process of the fortification of the biscuits with bee pollen enabled us to obtain a high-quality finished product with an improved nutritional and sensory profile, enriched in bioactive compounds that increase its nutraceutical profile.

Based on the statistical analysis, adding 10% bee pollen appears to be the most effective for enriching biscuits with phenolic and carotenoid compounds, thereby enhancing their antioxidant activity. This supplementation level provides maximum nutritional benefits. However, even a 5% addition of pollen significantly improved these chemical and compositional parameters. Overall, biscuits fortified with 5% fresh pollen emerged as the optimal formulation, balancing chemical, technological, and sensory properties, including distinctive flavors and colors. This suggests that fortifying biscuits with pollen could greatly expand the availability of health-promoting snacks and broaden the potential applications of bee pollen in the food industry.

Furthermore, based on the data obtained, it could be suggested that the choice between fresh or dried pollen did not significantly affect the analyzed parameters of the biscuits. This could be explained by the mild drying treatment, applied to the pollen, which resulted in a final moisture content of 12%, above the established commercial limits of 2–4%.

Finally, it was observed that, despite the short observation time, the fortified biscuits maintained their sensory properties consistently over time.

5. Future Prospects

As a future goal, it would be valuable to implement a more targeted drying treatment for pollen to better understand its impact on fortified biscuits and, ultimately, to extend this approach to gluten-free products. In addition, conducting stability tests on the enriched biscuits, integrated with the sensory evaluation and chemical characterization of the samples together with microbiological analyses over time, could help determine their shelf-life. This approach would enable the development of a product that combines chemical, sensory, and nutritional qualities and a long shelf-life.

Supplementary Materials: The following supporting information can be downloaded at: https://www.mdpi.com/article/10.3390/foods14030527/s1, Figure S1: Sensory sheet used by the judges; Figure S2: Images of the biscuits after cooking. CB: control salty biscuits without pollen, FPB5%: salty biscuits with 5% fresh pollen, FPB10%: salty biscuits with 10% fresh pollen, DPB5%: salty biscuits with 5% dried pollen, DPB10%: salty biscuits with 10% dried pollen; Figure S3: Final weight assumed by each descriptor.

Author Contributions: Conceptualization, A.B., L.P., C.S. and F.V.; methodology, A.B., S.C. and C.S.; software, A.B.; validation, A.B., C.S. and F.V.; formal analysis, A.B., S.C., I.T. and C.S.; investigation, A.B., S.C., C.S., L.P. and F.V.; resources, C.S., L.P. and F.V.; data curation, A.B., S.C. and I.T.; writing—original draft preparation, A.B., S.C., I.T., C.S., L.P. and F.V.; writing—review and editing, A.B., S.C., I.T., C.S., L.P. and F.V.; visualization, A.B., C.S., L.P. and F.V.; supervision, C.S., L.P. and F.V.; project administration, L.P.; funding acquisition, L.P. All authors have read and agreed to the published version of the manuscript.

Funding: This research was funded by Regione Toscana, ("Beta" project) Reg. UE 1305/2013—PSR 2014–2022, Misura 19.2 "Progetti di rigenerazione delle Comunità" GAL F.A.R. Maremma, CUP: G94H22001370002.

Institutional Review Board Statement: The research obtained the approval of the Ethics Committee of the University of Pisa (protocol n. 0088081/2023). The research was conducted according to the ethical guidelines, and informed consent was obtained from all participants.

Informed Consent Statement: Informed consent was obtained from all subjects involved in the study.

Data Availability Statement: The original contributions presented in the study are included in the article/Supplementary Material; further inquiries can be directed to the corresponding author.

Acknowledgments: We would like to thank the Mancini agricultural company (Tirli, Grosseto, Italy), which provided the bee pollen.

Conflicts of Interest: The authors declare no conflicts of interest.

References

1. Pointke, M.; Pawelzik, E. Plant-Based Alternative Products: Are They Healthy Alternatives? Micro- and Macronutrients and Nutritional Scoring. *Nutrients* **2022**, *14*, 601. [CrossRef] [PubMed]
2. Lutter, C.K.; Dewey, K.G. Proposed Nutrient Composition for Fortified Complementary Foods. *J. Nutr.* **2003**, *133*, 3011S–3020S. [CrossRef]
3. Bianchi, A.; Venturi, F.; Zinnai, A.; Taglieri, I.; Najar, B.; Macaluso, M.; Merlani, G.; Angelini, L.G.; Tavarini, S.; Clemente, C.; et al. Valorization of an Old Variety of *Triticum aestivum*: A Study of Its Suitability for Breadmaking Focusing on Sensory and Nutritional Quality. *Foods* **2023**, *12*, 1351. [CrossRef] [PubMed]
4. Anadón, A.; Ares, I.; Martínez-Larrañaga, M.-R.; Martínez, M.-A. Chapter 73—Evaluation and regulation of food supplements: European perspective. In *Nutraceuticals*, 2nd ed.; Gupta, R.C., Lall, R., Srivastava, A.B.T.-N., Eds.; Academic Press: Cambridge, MA, USA, 2021; pp. 1241–1271, ISBN 978-0-12-821038-3.
5. Fan, L.; Zhang, S.; Yu, L.; Ma, L. Evaluation of antioxidant property and quality of breads containing *Auricularia auricula* polysaccharide flour. *Food Chem.* **2007**, *101*, 1158–1163. [CrossRef]
6. Chakraborty, N.; Chakraborty, R. Innovative Ingredients Fortified Nutrients Enriched Biscuits and Cookies: Quality and Sensory Analyses. *Curr. Nutr. Food Sci.* **2023**, *19*, 615–641. [CrossRef]
7. Sahi, S.S. 14—Applications of natural ingredients in baked goods. In *Natural Food Additives, Ingredients and Flavourings*; Baines, D., Seal, R., Eds.; Woodhead Publishing: Sawston, UK, 2012; pp. 318–332. ISBN 978-1-84569-811-9.
8. Galić, K.; Ćurić, D.; Gabrić, D. Shelf Life of Packaged Bakery Goods—A Review. *Crit. Rev. Food Sci. Nutr.* **2009**, *49*, 405–426. [CrossRef] [PubMed]
9. Lustig, R.H. Ultraprocessed Food: Addictive, Toxic, and Ready for Regulation. *Nutrients* **2020**, *12*, 3401. [CrossRef]
10. Fuhrman, J. The Hidden Dangers of Fast and Processed Food. *Am. J. Lifestyle Med.* **2018**, *12*, 375–381. [CrossRef] [PubMed]
11. Arepally, D.; Reddy, R.S.; Goswami, T.K.; Datta, A.K. Biscuit baking: A review. *LWT* **2020**, *131*, 109726. [CrossRef]
12. Thakur, M.; Nanda, V. Composition and functionality of bee pollen: A review. *Trends Food Sci. Technol.* **2020**, *98*, 82–106. [CrossRef]
13. Kalaycıoğlu, Z.; Kanbur, E.D.; Kolaylı, S.; Erim, F.B. Antioxidant activities, aliphatic organic acid and sugar contents of Anatolian bee bread: Characterization by principal component analysis. *Eur. Food Res. Technol.* **2023**, *249*, 1351–1361. [CrossRef]
14. Dong, J.; Yang, Y.; Wang, X.; Zhang, H. Fatty acid profiles of 20 species of monofloral bee pollen from China. *J. Apic. Res.* **2015**, *54*, 503–511. [CrossRef]
15. Ecem Bayram, N. Vitamin, mineral, polyphenol, amino acid profile of bee pollen from *Rhododendron ponticum* (source of "mad honey"): Nutritional and palynological approach. *J. Food Meas. Charact.* **2021**, *15*, 2659–2666. [CrossRef]
16. Kieliszek, M.; Piwowarek, K.; Kot, A.M.; Błażejak, S.; Chlebowska-Śmigiel, A.; Wolska, I. Pollen and bee bread as new health-oriented products: A review. *Trends Food Sci. Technol.* **2018**, *71*, 170–180. [CrossRef]
17. Mosić, M.; Trifković, J.; Vovk, I.; Gašić, U.; Tešić, Ž.; Šikoparija, B.; Milojković-Opsenica, D. Phenolic Composition Influences the Health-Promoting Potential of Bee-Pollen. *Biomolecules* **2019**, *9*, 783. [CrossRef] [PubMed]
18. Rzepecka-Stojko, A.; Stojko, J.; Kurek-Górecka, A.; Górecki, M.; Kabała-Dzik, A.; Kubina, R.; Moździerz, A.; Buszman, E. Polyphenols from Bee Pollen: Structure, Absorption, Metabolism and Biological Activity. *Molecules* **2015**, *20*, 21732–21749. [CrossRef] [PubMed]
19. Bleha, R.; Shevtsova, T.V.; Živčáková, M.; Korbářová, A.; Ježková, M.; Saloň, I.; Brindza, J.; Synytsya, A. Spectroscopic Discrimination of Bee Pollen by Composition, Color, and Botanical Origin. *Foods* **2021**, *10*, 1682. [CrossRef] [PubMed]
20. Giampieri, F.; Quiles, J.L.; Cianciosi, D.; Forbes-Hernández, T.Y.; Orantes-Bermejo, F.J.; Alvarez-Suarez, J.M.; Battino, M. Bee Products: An Emblematic Example of Underutilized Sources of Bioactive Compounds. *J. Agric. Food Chem.* **2022**, *70*, 6833–6848. [CrossRef]
21. Ullah, A.; Munir, S.; Badshah, S.L.; Khan, N.; Ghani, L.; Poulson, B.G.; Emwas, A.-H.; Jaremko, M. Important Flavonoids and Their Role as a Therapeutic Agent. *Molecules* **2020**, *25*, 5243. [CrossRef]
22. Mutha, R.E.; Tatiya, A.U.; Surana, S.J. Flavonoids as natural phenolic compounds and their role in therapeutics: An overview. *Future J. Pharm. Sci.* **2021**, *7*, 25. [CrossRef]
23. de Arruda, V.A.S.; Pereira, A.A.S.; de Freitas, A.S.; Barth, O.M.; de Almeida-Muradian, L.B. Dried bee pollen: B complex vitamins, physicochemical and botanical composition. *J. Food Compos. Anal.* **2013**, *29*, 100–105. [CrossRef]
24. Mărgăoan, R.; Cornea-Cipcigan, M. Carotenoids and Vitamins of Pollen. In *Pollen Chemistry & Biotechnology*; Ecem Bayram, N., Kostic, A.Ž., Can Gercek, Y., Eds.; Springer International Publishing: Cham, Switzerland, 2023; pp. 147–177. ISBN 978-3-031-47563-4.
25. Andrewski, E.; Cheng, K.; Vanderpool, C. Nutritional Deficiencies in Vegetarian, Gluten-Free, and Ketogenic Diets. *Pediatr. Rev.* **2022**, *43*, 61–70. [CrossRef] [PubMed]
26. Trautwein, E.A.; Demonty, I. Phytosterols: Natural compounds with established and emerging health benefits. *Oléagineux Corps Gras Lipides* **2007**, *14*, 259–266. [CrossRef]

27. Kostić, A.Ž.; Milinčić, D.D.; Barać, M.B.; Ali Shariati, M.; Tešić, Ž.L.; Pešić, M.B. The Application of Pollen as a Functional Food and Feed Ingredient—The Present and Perspectives. *Biomolecules* **2020**, *10*, 84. [CrossRef] [PubMed]
28. Conte, P.; Del Caro, A.; Balestra, F.; Piga, A.; Fadda, C. Bee pollen as a functional ingredient in gluten-free bread: A physical-chemical, technological and sensory approach. *LWT* **2018**, *90*, 1–7. [CrossRef]
29. Conte, P.; Del Caro, A.; Urgeghe, P.P.; Petretto, G.L.; Montanari, L.; Piga, A.; Fadda, C. Nutritional and aroma improvement of gluten-free bread: Is bee pollen effective? *LWT* **2020**, *118*, 108711. [CrossRef]
30. Krystyjan, M.; Gumul, D.; Ziobro, R.; Korus, A. The fortification of biscuits with bee pollen and its effect on physicochemical and antioxidant properties in biscuits. *LWT-Food Sci. Technol.* **2015**, *63*, 640–646. [CrossRef]
31. Dundar, A.N. Total phenolic and antioxidant bioaccessibilities of cookies enriched with bee pollen. *J. Food Process. Preserv.* **2022**, *46*, e16085. [CrossRef]
32. Camacho-Bernal, G.I.; Cruz-Cansino, N.D.; Ramírez-Moreno, E.; Delgado-Olivares, L.; Zafra-Rojas, Q.Y.; Castañeda-Ovando, A.; Suárez-Jacobo, Á. Addition of Bee Products in Diverse Food Sources: Functional and Physicochemical Properties. *Appl. Sci.* **2021**, *11*, 8156. [CrossRef]
33. Jabłońska, M.; Karpińska-Tymoszczyk, M.; Surma, M.; Narwojsz, A.; Reszka, M.; Błaszczak, W.; Sawicki, T. Enrichment of shortcrust pastry cookies with bee products: Polyphenol profile, in vitro bioactive potential, heat-induced compounds content, colour parameters and sensory changes. *Sci. Rep.* **2024**, *14*, 23652. [CrossRef] [PubMed]
34. Yan, S.; Wan, Y.; Wang, F.; Xue, X.; Wu, L. Fortification of bread with bee pollen, and its effects on quality attributes and antioxidant activity. *Int. Food Res. J.* **2021**, *28*, 517–526. [CrossRef]
35. Anis, U.; Fatharani, A.; Yuwita, F.; Gusriani, I. Physical Characteristics of Substituted Biscuits with Bee Pollen. *Agritropica J. Agric. Sci.* **2024**, *7*, 28–33. [CrossRef]
36. Martinez-Velasco, J.D.; Filomena-Ambrosio, A.; Garzón-Castro, C.L. Technological tools for the measurement of sensory characteristics in food: A review. *F1000Research* **2024**, *12*, 340. [CrossRef]
37. Végh, R.; Csóka, M.; Stefanovits-Bányai, É.; Juhász, R.; Sipos, L. Biscuits Enriched with Monofloral Bee Pollens: Nutritional Properties, Techno-Functional Parameters, Sensory Profile, and Consumer Preference. *Foods* **2023**, *12*, 18. [CrossRef] [PubMed]
38. Aldarf, R.; Nadaf, D.M.; Sultaneh, D.A. Effect of Bee Pollen Addition on the Chemical Composition, Antioxidant Properties Rheological and Sensory Properties of Biscuits. *Asian J. Adv. Res.* **2023**, *6*, 9–19.
39. Kostić, A.Ž.; Barać, M.B.; Stanojević, S.P.; Milojković-Opsenica, D.M.; Tešić, Ž.L.; Šikoparija, B.; Radišić, P.; Prentović, M.; Pešić, M.B. Physicochemical composition and techno-functional properties of bee pollen collected in Serbia. *LWT-Food Sci. Technol.* **2015**, *62*, 301–309. [CrossRef]
40. Domínguez-Valhondo, D.; Bohoyo Gil, D.; Hernández, M.T.; González-Gómez, D. Influence of the commercial processing and floral origin on bioactive and nutritional properties of honeybee-collected pollen. *Int. J. Food Sci. Technol.* **2011**, *46*, 2204–2211. [CrossRef]
41. Almeida-Muradian, L.B.; Pamplona, L.C.; Coimbra, S.; Barth, O.M. Chemical composition and botanical evaluation of dried bee pollen pellets. *J. Food Compos. Anal.* **2005**, *18*, 105–111. [CrossRef]
42. Capparelli, S.; Pieracci, Y.; Sagona, S.; Flamini, G.; Coppola, F.; Sanmartin, C.; Taglieri, I.; Venturi, F.; Felicioli, A.; Pistelli, L. The volatile and sensory profiles of Tuscan bee pollens stored at different temperatures. *Nat. Prod. Res.* **2024**, 1–8. [CrossRef]
43. Negrao, A.F.; Orsi, R.O. Harvesting season and botanical origin interferes in production and nutritional composition of bee pollen. *An. Acad. Bras. Cienc.* **2018**, *90*, 325–332. [CrossRef]
44. Mahmoud, F.; Latif, S.; Mohamed, S.; Saleh, S. Nutritional Quality and Sensory Attributes of Biscuits Fortified with Bee Pollen. *Aswan Univ. J. Sci. Technol.* **2022**, *2*, 1–13. [CrossRef]
45. Sokmen, O.; Ozdemir, S.; Dundar, A.N.; Cinar, A. Quality properties and bioactive compounds of reduced-fat cookies with bee pollen. *Int. J. Gastron. Food Sci.* **2022**, *29*, 100557. [CrossRef]
46. Kaur, J.; Rasane, P.; Kumar, V.; Nanda, V.; Bhadariya, V.; Kaur, S.; Singh, J. Exploring the Health Benefits of Bee Pollen and Its Viability as a Functional Food Ingredient. *Rev. Agric. Sci.* **2024**, *12*, 65–78. [CrossRef] [PubMed]
47. Karabagias, I.K.; Karabagias, V.K.; Gatzias, I.; Riganakos, K.A. Bio-Functional Properties of Bee Pollen: The Case of "Bee Pollen Yoghurt". *Coatings* **2018**, *8*, 423. [CrossRef]
48. Sipos, L.; Végh, R.; Bodor, Z.; Zaukuu, J.-L.Z.; Hitka, G.; Bázár, G.; Kovacs, Z. Classification of Bee Pollen and Prediction of Sensory and Colorimetric Attributes—A Sensometric Fusion Approach by e-Nose, e-Tongue and NIR. *Sensors* **2020**, *20*, 6768. [CrossRef]
49. Cornara, L.; Biagi, M.; Xiao, J.; Burlando, B. Therapeutic Properties of Bioactive Compounds from Different Honeybee Products. *Front. Pharmacol.* **2017**, *8*, 412. [CrossRef] [PubMed]
50. Nanditha, B.R.; Jena, B.S.; Prabhasankar, P. Influence of natural antioxidants and their carry-through property in biscuit processing. *J. Sci. Food Agric.* **2009**, *89*, 288–298. [CrossRef]
51. Bianchi, A.; Taglieri, I.; Zinnai, A.; Macaluso, M.; Sanmartin, C.; Venturi, F. Effect of Argon as Filling Gas of the Storage Atmosphere on the Shelf-Life of Sourdough Bread—Case Study on PDO Tuscan Bread. *Foods* **2022**, *11*, 3470. [CrossRef]

52. Park, D.; Lee, H.J.; Kumar Sethukali, A.; Yim, D.-G.; Park, S.; Jo, C. Effects of Temperature on the Microbial Growth and Quality of Unsealed Dry Pet Food during Storage. *Food Sci. Anim. Resour.* **2024**. [CrossRef]
53. Cabras, P.; Tuberoso, I.G.C. *Analisi dei Prodotti Alimentari*, 2nd ed.; Piccin Nuova Libraria S.p.A: Padova, Italy, 2014; ISBN 8829923117.
54. Monacci, E.; Sanmartin, C.; Bianchi, A.; Pettinelli, S.; Taglieri, I.; Mencarelli, F. Plastic film packaging for the postharvest quality of fresh hop inflorescence (*Humulus lupulus*) cv. Cascade. *Postharvest Biol. Technol.* **2023**, *206*, 112575. [CrossRef]
55. Bianchi, A.; Sanmartin, C.; Taglieri, I.; Macaluso, M.; Venturi, F.; Napoli, M.; Mancini, M.; Fabbri, C.; Zinnai, A. Effect of Fertilization Regime of Common Wheat (*Triticum aestivum*) on Flour Quality and Shelf-Life of PDO Tuscan Bread. *Foods* **2023**, *12*, 2672. [CrossRef] [PubMed]
56. Sanmartin, C.; Taglieri, I.; Bianchi, A.; Parichanon, P.; Puccinelli, M.; Pardossi, A.; Venturi, F. Effects of Temperature and Packaging Atmosphere on Shelf Life, Biochemical, and Sensory Attributes of Glasswort (*Salicornia europaea* L.) Grown Hydroponically at Different Salinity Levels. *Foods* **2024**, *13*, 3260. [CrossRef] [PubMed]
57. Modesti, M.; Shmuleviz, R.; Macaluso, M.; Bianchi, A.; Venturi, F.; Brizzolara, S.; Zinnai, A.; Tonutti, P. Pre-processing Cooling of Harvested Grapes Induces Changes in Berry Composition and Metabolism, and Affects Quality and Aroma Traits of the Resulting Wine. *Front. Nutr.* **2021**, *8*, 728510. [CrossRef] [PubMed]
58. *ISO 8589:2010*; Sensory Analysis—General Guidance for the Design of test Rooms. International Organization for Standardization: Geneva, Switzerland, 2010.
59. *ISO 8586:2023*; Sensory Analysis—Selection and Training of Sensory Assessors. International Organization for Standardization: Geneva, Switzerland, 2023.
60. Bianchi, A.; Venturi, F.; Palermo, C.; Taglieri, I.; Angelini, G.L.; Tavarini, S.; Sanmartin, C. Primary and secondary shelf-life of bread as a function of formulation and MAP conditions: Focus on physical-chemical and sensory markers. *Food Packag. Shelf Life* **2024**, *41*, 101241. [CrossRef]
61. Martínez, J.A.; Melgosa, M.; Pérez, M.M.; Hita, E.; Negueruela, A.I. Note. Visual and Instrumental Color Evaluation in Red Wines. *Food Sci. Technol. Int.* **2001**, *7*, 439–444. [CrossRef]
62. Beuchat, L.R.; Komitopoulou, E.; Beckers, H.; Betts, R.O.Y.P.; Bourdichon, F.; Fanning, S.; Joosten, H.A.N.M.; Ter Kuile, B.H. Low--Water Activity Foods: Increased Concern as Vehicles of Foodborne Pathogens. *J. Food Prot.* **2013**, *76*, 150–172. [CrossRef] [PubMed]
63. Kaur, P.; Choudhary, M.; Sharma, S. Fatty acid profiling and oxidative stability of biscuits available in the market of the city of Ludhiana, India. *Sci. Rep.* **2023**, *13*, 21791. [CrossRef]
64. Manning, R. Fatty acids in pollen: A review of their importance for honey bees. *Bee World* **2001**, *82*, 60–75. [CrossRef]
65. Milinčić, D.D.; Kostić, A.Ž.; Stanojević, S.P.; Pešić, M.B. Techno-Functional Properties of Pollen. In *Pollen Chemistry & Biotechnology*; Ecem Bayram, N., Kostic, A.Ž., Can Gercek, Y., Eds.; Springer International Publishing: Cham, Switzerland, 2023; pp. 291–318. ISBN 978-3-031-47563-4.
66. Mladenović Drinić, S.D.; Vukadinović, J.Z.; Srdić, J.; Milašinović Šeremešić, M.S.; Anđelković, V.B. Effect of Cooking on the Content of Carotenoids and Tocopherols in Sweet Corn. *Food Feed Res.* **2021**, *48*, 119–129. [CrossRef]
67. Starowicz, M.; Zieliński, H. How Maillard Reaction Influences Sensorial Properties (Color, Flavor and Texture) of Food Products? *Food Rev. Int.* **2019**, *35*, 707–725. [CrossRef]
68. Heenan, S.P.; Hamid, N.; Dufour, J.-P.; Harvey, W.; Delahunty, C.M. Consumer freshness perceptions of breads, biscuits and cakes. *Food Qual. Prefer.* **2009**, *20*, 380–390. [CrossRef]
69. Taormina, P.J. Purposes and Principles of Shelf Life Determination. In *Food Safety and Quality-Based Shelf Life of Perishable Foods*; Taormina, P.J., Hardin, M.D., Eds.; Springer International Publishing: Cham, Switzerland, 2021; pp. 1–26, ISBN 978-3-030-54375-4.

foods

MDPI

Article

Sustainable Protein Sources: Functional Analysis of *Tenebrio molitor* Hydrolysates and Attitudes of Consumers in Poland and Spain Toward Insect-Based Foods

Marcelina Maciejewska [1,*], Anna Dąbrowska [1] and Marina Cano-Lamadrid [2]

1 Department of Development Functional Food Products, Wrocław University of Environmental and Life Sciences, 50-366 Wrocław, Poland; anna.dabrowska@upwr.edu.pl
2 Department of Agri-Food Technology, Miguel Hernández University of Elche, Crta. de Beniel, Km. 3,2, 03312 Orihuela, Spain; marina.canol@umh.es
* Correspondence: marcelina.maciejewska@upwr.edu.pl

Abstract: This study explores the potential of *Tenebrio molitor* protein hydrolysates as functional food ingredients, evaluating their bioactivity and consumer acceptance of the incorporation of edible insects into food across Poland and Spain. By aligning technical advancements with consumer preferences, this research bridges the gap between laboratory innovation and market feasibility, contributing to the development of sustainable functional foods. The study optimized the process of enzyme hydrolysis using serine protease from *Cucurbita ficifolia*, thereby enhancing DPPH scavenging capacity increased from 3.15 ± 0.53 to 8.17 ± 0.62 μM Trolox/mL and ABTS decolorization capacity increased from 4.29 ± 0.01 to 10.29 ± 0.01 μM Trolox/mL after 5 h of hydrolysis. Consumer surveys incorporating the Food Neophobia, Insect Phobia, and Entomophagy Scales revealed demographic and cultural influences on entomophagy acceptance. Among respondents, 27.1% in Poland and 25.7% in Spain had previously consumed insect-based products, while Polish participants showed a higher willingness to adopt insect-enriched foods. The study confirmed that hydrolysis enhances the antioxidant activity of *T. molitor* protein hydrolysates and that demographic and cultural factors significantly influence consumer acceptance of insect-based foods.

Keywords: edible insects; *Tenebrio molitor*; *Cucurbita ficifolia*; protein hydrolysis; consumer acceptance; functional foods

1. Introduction

The expanding global population causes a constantly growing demand for protein supply, the production of which is likely affecting the environment in severely negative ways. For example, CO_2 emission based on cattle alone can lead to the release of 7.1 gigatons annually. By the year 2050, an increase of 75% is projected in meat consumption, which is a major contributor to environmental degradation. Therefore, it is crucial to reconsider the current food practices to ensure a more sustainable future. Recommendations to address climate change and environmental issues include reducing meat consumption, improving agricultural efficiency, and exploring sustainable protein sources. Environmentally friendly sources of protein, such as soybeans, legumes, nuts, seeds, whole grains, algae biomass, and insects are gaining prominence. Insects, as an alternative protein source, are promising in addressing the environmental issues associated with meat production as they have the potential to reduce greenhouse gas emissions [1–3].

Insects have long been a food source across various cultures, particularly in Africa Asia, Latin America, and Australia, although Western societies view insect consumption with skepticism. Cultural perceptions of disgust are a barrier, requiring strategies to promote awareness of the nutritional and environmental benefits insects offer [4,5]. Studies show varying levels of food neophobia and neophilia across Europe. While food neophobia may act as a barrier to the acceptance of insect-based foods, food neophilia can equally drive curiosity and openness to try novel and unfamiliar products, creating the potential for the gradual adoption of such foods [6]. As Ghosh et al. point out, the acceptance or rejection of edible insects is influenced by multiple factors, including cultural traditions, sensory attributes (such as taste and texture), and the availability of insects as a food source. They emphasize that strategies to promote the acceptance of insect-based foods must address both cultural perceptions and sensory experiences, highlighting the importance of positive exposure to such foods to overcome barriers of disgust and unfamiliarity [7].

The European Food Safety Authority (EFSA) has approved several insect species, including *Tenebrio molitor* (yellow mealworm), as novel foods. However, protein hydrolysates derived from mealworms have not yet been authorized for human consumption. Known for its rich protein profile and essential amino acids like leucine and lysine, *T. molitor* is considered one of the most promising insect-based foods [8,9]. Research into food neophobia underscores the importance of understanding cultural differences when introducing insects as a novel food source [10,11].

Tenebrio molitor, or yellow mealworm, offers a high protein content (41–66%) and a mild flavor profile, including nutty and umami notes. It has a well-established farming system, primarily for fish and bird feed, which could support its expanded use in human diets [12,13]. As a source of bioactive peptides, mealworms also show health benefits such as antioxidant and antimicrobial effects, making them suitable for sustainable, nutritious foods [14,15]. Additionally, Ghosh et al. highlighted that *Tenebrio molitor* has a high-fat content (34.5%) and a protein profile that meets all essential amino acid requirements recommended by FAO/WHO/UNU [16]. This combination of high protein quality and lipid content positions mealworms as a nutritionally dense and versatile ingredient for future food systems. In order to fully benefit from the nutritional potential of *Tenebrio molitor*, in particular, its high protein and bioactive peptide content, it is essential to use methods that optimize its functional properties.

Enzymatic hydrolysis is an efficient and gentle process that extracts bioactive peptides from protein sources, retaining nutritional value and enhancing functional properties. This method allows for control over product composition, including amino acid profiles and molecular weights, making it ideal for developing protein-rich diets [17]. The demand for new, cost-effective enzymes has led to the exploration of unconventional enzymes, including those from *Cucurbita ficifolia*.

Serine protease from *C. ficifolia* is a promising unconventional enzyme due to its specific proteolytic properties, which target proteins like casein, gluten, and ovalbumin. When applied to whey proteins and αs-casein in cow's milk, this protease has been shown to significantly reduce immunoreactivity to IgE and IgG, suggesting potential applications in allergy prevention, especially for infants at high allergy risk [17–21].

The unique properties of serine protease from *C. ficifolia* support its application in developing hydrolysates and enhancing the nutritional profiles of various protein sources. This enzyme, with its effective and affordable production, also contributes to expanding enzyme options in food production and diversifying protein hydrolysates, which include free amino acids, oligopeptides, and bioactive peptides with antioxidant activities.

This study evaluates the potential of edible insects as a sustainable alternative protein source. The bioactivity of *Tenebrio molitor* proteins is investigated following enzymatic hydrolysis using non-commercial serine proteases derived from *Cucurbita ficifolia*. In this study, bioactivity refers to the biological properties of *Tenebrio molitor* protein hydrolysates, particularly their antioxidant potential, which was evaluated using DPPH-free radical scavenging and ABTS cation decolorization assays. These properties are indicative of the hydrolysates' functional potential for use in health-promoting food products. Additionally, consumer acceptance of new food products, particularly edible insects, is assessed through a survey of Polish and Spanish consumers. The survey employed the Food Neophobia Scale (FNS), the Insect Phobia Scale (INS), and the Entomophagy Scale (ENS). Moreover, the survey evaluated consumers' willingness to incorporate insects into other food matrices and their preferences based on the types and sensory properties of insect-enriched products.

2. Materials and Methods

2.1. Materials

The larvae of yellow mealworm (*Tenebrio molitor*) were sourced from TENEBRIA Ltd. (Lubawa, Poland). The larvae were frozen at −80 °C for 24 h, followed by sublimation drying for an additional 24 h at 40 °C under 1.650 mbar pressure. After drying, lipid extraction was performed using supercritical fluid extraction (SFE) with carbon dioxide at 30 MPa pressure and 60 °C temperature. This process resulted in defatted yellow mealworm flour, which was stored at −80 °C until further use. The serine protease enzyme was extracted from Asian pumpkin (*Cucurbita ficifolia*) following the methodology described by Dryjański et al. [22]. All reagents and chemicals used in this study were of analytical grade and were procured from standard suppliers.

2.2. Experimental Design

The study comprised two interconnected phases. In the initial phase, the focus was on optimizing the enzymatic hydrolysis of proteins from defatted yellow mealworm flour to enhance their functional properties, particularly their antioxidant activity. In the subsequent phase, consumer attitudes and preferences toward insect-based foods were evaluated through a comprehensive survey among Polish and Spanish participants.

2.3. Hydrolysis of Proteins

The proteolytic activity against casein was quantified in accordance with the methodologies delineated by Chrzanowska and Koaczkowska [23]. A 0.5% (w/v) protein solution was prepared by suspending defatted yellow mealworm flour in distilled water, with the pH adjusted to 4.5. The suspension was then homogenized using a blender and subjected to centrifugation at 8000 rpm to separate the plasma. This process was repeated after subjecting the suspension to a freezing and drying cycle to ensure consistent protein extraction. Enzymatic hydrolysis was then carried out using serine protease at a dose of 300 U/mg of hydrolyzed protein in a 0.1 M Tris-HCl buffer (pH 8.0). The hydrolysis reactions were carried out at temperatures of 37 °C for durations of 1, 3, and 5 h, with the reaction being terminated by thermal inactivation at 90 °C. The protein content was determined using the Lowry method [24]. The degree of hydrolysis (DH) was then quantified based on the content of free amino groups (FAG), with trinitrobenzene sulfonic acid (TNBS) being used as described by Snyder and Sobocinski and Kuchroo and Ramilly [25,26]. Subsequently, the hydrolyzed protein fractions were subjected to SDS-PAGE under denaturing conditions [27].

2.4. Determination of Antioxidant Activity

Antioxidant activity was assessed using two standard methods: DPPH radical scavenging activity and ABTS cation decolorization assay. The DPPH assay was conducted in accordance with the methodology described by Kedare and Singh [28], while the ABTS assay adhered to the procedure outlined by Re et al. [29]. Both methods were applied to hydrolysates obtained after 1, 3, and 5 h of hydrolysis, as well as to fully hydrolyzed samples, to determine their antioxidant potential at different stages of hydrolysis.

2.5. Consumer Survey

An English questionnaire was designed by the authors to address the objectives of this study. It was then translated into Polish and Spanish to ensure linguistic consistency and cultural relevance for the respondents in both countries. Google Forms was used as the platform for data collection. Ethical approval was obtained from the UMH Institutional Review Board prior to the distribution of the questionnaire (ref: DTA.MCL.2240123). The questionnaire received a total of 140 responses (n = 140), with 70 respondents from each of the targeted countries (Poland and Spain). It was structured into sections, covering demographics, consumer behavior, and consumer preferences. Participants' responses were evaluated on a hedonic scale from 1 to 7, with 1 indicating complete disagreement and 7 indicating complete agreement with the statements [30–34]. The consumer survey was conducted to understand consumer attitudes and preferences, ensuring that edible insect-based products align with market needs. Psychometric tools provided insights into factors influencing acceptance, bridging the gap between innovation and consumer demand. The consumer survey focused on the general acceptance of insect-based foods to explore cultural and demographic influences. Specific questions about *T. molitor* protein hydrolysates were avoided to ensure reliability, as respondents were unlikely to have prior familiarity. This foundational approach informs future studies on hydrolysate applications.

2.6. Statistical Analyses

The statistical analysis of the study was conducted using Excel and RStudio R 4.4.2. The analysis began with a one-way analysis of variance (ANOVA) to identify statistically significant differences in the means of the data. This was followed by a Tukey's post hoc test for multiple comparisons. All laboratory experiments were carried out in triplicate (n = 3), and a p-value threshold was used for significance ($p < 0.05$). The resulting data are expressed as the mean ± standard deviation (SD). The consumer survey data analysis followed the methodological framework described by Fernández-Ruiz et al., which included reliability assessments and mixed model ANOVA to analyze relationships between food neophobia and consumer preferences. Laboratory analyses, including protein hydrolysis and antioxidant activity determinations, were conducted in accordance with the methodological framework established by Szołtysik et al., thereby ensuring the reliability and replicability of the results [35,36].

3. Results and Discussion

3.1. Determination of Antioxidant Activity of Hydrolysates

The results presented in Figure 1 were obtained from both methods and indicate that an increase in the degree of hydrolysis occurs with an increase in hydrolysis time. The highest values (26.38%) were obtained after 5 h of degradation. As expected, the degree of hydrolysis increased proportionally with hydrolysis time, consistent with well-established biochemical principles. Consequently, this finding substantiates the efficacy of *Cucurbita ficifolia* protease in hydrolyzing defatted yellow mealworm protein within the experimental parameters. The enzymatic hydrolysis process led to a progressive increase in the degree of hydrolysis [%] as shown in Figure 2. Notably, the highest free amino group (FAG) concentration was observed after a five-hour reaction, reaching 8602.84 [μmol Gly/g protein], with the total hydrolysis score culminating at 32,608.1 [μmol Gly/g protein]. This result highlights the efficiency of the hydrolysis process over time, demonstrating a direct correlation between the reaction duration and protein degradation, as evidenced by the measurable increase in FAG levels.

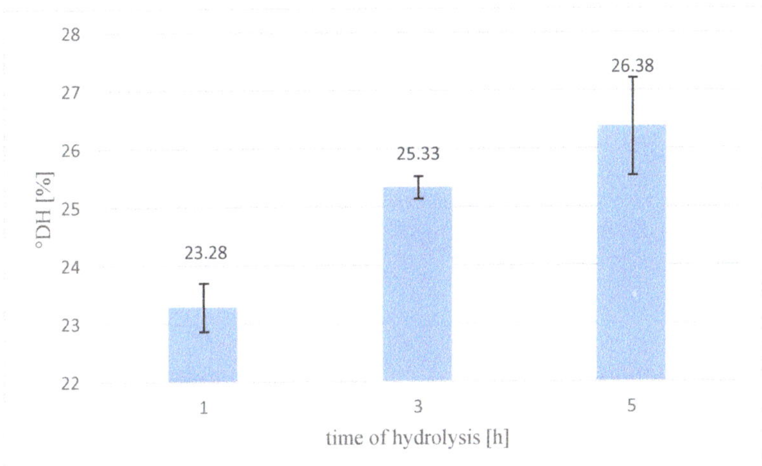

Figure 1. Degree of hydrolysis [%] obtained with the use of serine protease from Asian pumpkin at a dose of 300 U/mL protein.

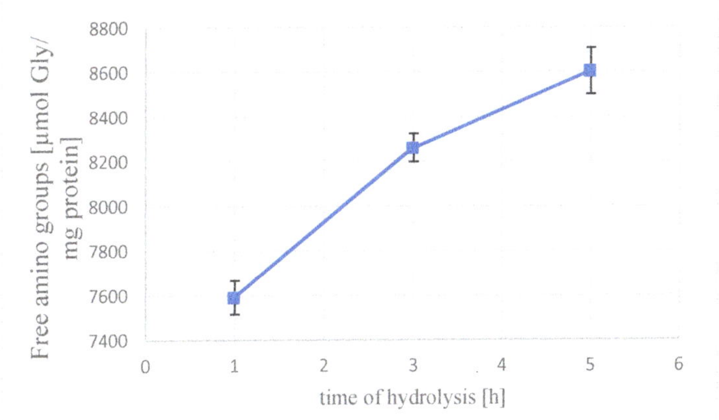

Figure 2. Increase in free amino groups [µmol Gly/mg protein] in hydrolysate.

SDS-PAGE analysis under denaturing conditions confirmed protein degradation over time. The control sample showed intact high molecular weight insect proteins. After 1 h of hydrolysis, the protein fraction partially degraded, forming peptides of 25 to 35–40 kDa, indicated by additional lower molecular weight bands and reduced intensity of original bands. Further hydrolysis (3 and 5 h) showed similar degradation. The deepest degradation occurred within the first hour, aligning with the increase in the degree of hydrolysis and FAG levels, demonstrating the efficiency of the hydrolysis process.

Following enzymatic hydrolysis with serine protease, the antioxidant activity of the hydrolysates was analyzed to evaluate the functional impact of the process. The results, presented in Table 1, highlight the effectiveness of the hydrolysis process in enhancing antioxidant activity, measured through DPPH scavenging and ABTS decolorization, with extended hydrolysis times. The 1 h hydrolysate shows moderate antioxidant activity, which further improves in the 3 h and 5 h samples. The highest antioxidant activity is observed in the fully hydrolyzed sample, suggesting that prolonged hydrolysis enhances the release of antioxidant peptides. This pattern aligns with previous studies showing that insect protein hydrolysates exhibit significant antioxidant potential. Studies highlight that hydrolysates, particularly those produced with alcalase enzymes, show enhanced antioxidant and hepatoprotective effects compared to raw powder. These enhancements render hydrolysates highly suitable for functional food applications, where superior solubility, digestibility, and bioactivity are paramount. Previous studies have demonstrated the antioxidant potential of insect protein hydrolysates, which is attributed to the release of bioactive peptides during enzymatic hydrolysis. These peptides have shown significant antioxidant and hepatoprotective effects, highlighting their suitability for functional food applications [37,38].

Table 1. Bioactivity values: antioxidant activity as the ability to scavenge DPPH free radicals and decolorize ABTS cations [μM Trolox/mL].

Sample	DPPH	ABTS
1H	3.15 ± 0.53	4.29 ± 0.01
3H	3.48 ± 0.07	4.99 ± 0.32
5H	3.65 ± 0.01	5.15 ± 0.02
total hydrolysis	8.17 ± 0.62	10.29 ± 0.01

3.2. Consumer Survey

Table 2 provides the demographic breakdown of the consumer survey among Spanish and Polish respondents. The sample consists of more females in both groups. Most Polish respondents fall within the 18–25 age range, while the Spanish cohort is more evenly distributed across age groups, particularly in the 26–35 and 46–55 brackets. These demographic differences influence perceptions of entomophagy. Men are generally more inclined to try unprocessed insects, whereas processed insect products show a more balanced acceptance across genders [5]. The younger Polish demographic, with a higher percentage of females, suggests that targeting younger female consumers might be an effective approach. In contrast, the broader age range in Spain indicates a wider target demographic.

Studies by Mustapa and Kallas and Ordoñez López et al. indicate that food preferences tend to solidify with age, leading to higher levels of food neophobia among middle-aged respondents due to established eating habits [39,40]. Education levels also differ significantly, with more Polish respondents having only a high school education and an equal number holding bachelor's degrees, while Spanish respondents show higher levels of postgraduate education (32.9% Ph.D. and 30% master's degrees). Most Polish consumers live in large cities, whereas Spanish consumers are more dispersed, with nearly 25% residing in villages and over 20% in small cities. The higher educational levels among Spanish participants may increase their openness to novel foods, such as insect-based products, due to greater awareness of their benefits [5].

Insect-based foods show generally low engagement among Polish and Spanish consumers, as shown in Table 3, with only a small portion having consciously tried insects or insect-enriched products. This limited exposure reflects a broader trend of food neophobia and cultural hesitance toward novel foods, particularly insect-based products. Such barriers are well-documented in research, which emphasizes that unfamiliarity and cultural perceptions strongly influence consumer acceptance. Addressing these factors through awareness campaigns focused on the nutritional and environmental benefits of insect-based foods could help mitigate neophobic attitudes and increase willingness to try these products [41].

Table 2. Demographic section results among Spanish and Polish consumers (n = 140).

Characteristic	Breakdown	Polish [%]	Spanish [%]
Gender	Female	70.0	65.7
	Male	30.0	34.3
Age	18–25	62.9	21.4
	26–35	22.9	28.6
	36–45	2.9	12.9
	46–55	7.1	15.7
	56–65	1.4	15.7
	>65	2.9	5.7
Education level	Less than High School	0.0	1.4
	High School	41.4	5.7
	Bachelor's degree	41.4	28.6
	Master's degree	14.3	30.0
	Ph.D.	1.4	32.9
	I prefer not to answer	1.4	1.4
City size	Village	21.4	24.3
	City up to 50 thousand residents	8.6	21.4
	City of 50–100 thousand residents	12.9	12.9
	City over 100 thousand residents	55.7	37.1
	I do not know	1.4	4.3

Table 3. Previous experience with the consumption of insects among Polish and Spanish consumers [n, %].

Questions		Poland		Spain	
		Yes	No	Yes	No
Have you ever conscientiously tasted insects as food?	[n]	19	51	18	52
	[%]	27.1	72.9	25.7	74.3
Have you ever conscientiously tasted food enriched with insects?	[n]	17	53	20	50
	[%]	24.3	75.7	28.6	71.4

Consumer willingness to try insect-based products, influenced by various factors, is illustrated in Table 4, with Spanish respondents generally showing a higher baseline acceptance than Polish respondents. Both groups show increased willingness when informed about the nutritional benefits and assured of the safety of insect-based products, highlighting the importance of educational efforts. Familiarity and knowledge about edible insects, as indicated by prior studies, play a critical role in consumer openness. Emphasizing nutritional and safety aspects in targeted marketing could therefore boost acceptance, particularly in Poland, where these assurances seem to significantly impact consumer willingness [32,42].

Table 4. Willingness to consume insects among Polish and Spanish consumers [n, %].

Answers		Poland	Spain
Yes, of course	[n]	8	18
	[%]	11.4	25.7
Only if I were aware of beneficial nutritional value of those products	[n]	20	14
	[%]	28.6	20.0
Only if I would be sure about the safety of those products	[n]	20	15
	[%]	28.6	22.0
Only if there wouldn't be noticeable change in taste of products	[n]	15	10
	[%]	21.4	14.0
No, never	[n]	7	13
	[%]	10	18

Table 5 illustrates the distribution of attitudes toward entomophagy, with Spanish respondents showing a higher prevalence of neophilic attitudes compared to Polish respondents, while most individuals in both groups display neutral attitudes. This pattern aligns with existing research, which identifies food neophobia as a barrier to the acceptance of insect-based foods. Studies indicate that neophobic tendencies, often linked to factors such as age and education, reduce the willingness to try novel foods, including insect-based options, especially among older and less-educated individuals. The high internal consistency of the FNS, INS, and ENS scales confirms their reliability in assessing these attitudes. Targeted educational and marketing strategies that emphasize the safety, nutritional benefits, and environmental advantages of insect-based products could effectively reduce food and insect phobia, thereby enhancing acceptance [40].

Table 5. Results of Food Neophobia, Insect Phobia, and Entomophagy Scales [n, %].

Food Neophobia Scale Results (FNS)			Insect Phobia Scale (INS)			Entomophagy Scale (ENS)		
	Poland	Spain		Poland	Spain		Poland	Spain
Neophilia	25.71	32.86	Insectophilia	12.86	24.29	Entomophilia	32.86	24.29
Neutral	71.43	61.43	Neutral	58.57	50.00	Neutral	33.00	51.43
Neophobia	2.86	5.71	Insectophobia	28.57	25.71	Entomophobia	20.00	24.29
Cronbach's Alpha	0.78	0.79	Cronbach's Alpha	0.89	0.81	Cronbach's Alpha	0.91	0.90

Differences in attitudes between Polish and Spanish consumers regarding the incorporation of insects into other food matrices, especially in animal products, are presented in Table 6. Polish respondents generally show a stronger acceptance, finding the idea of consuming insect-fed animal products more natural and beneficial for environmental sustainability. In contrast, Spanish respondents exhibit lower acceptance, with some viewing insect-fed animal products as less palatable or even unappealing. Both groups, however, recognize the potential environmental benefits of insect incorporation in the food chain. These findings suggest that framing insect-fed products within an environmental and sustainability context could help increase consumer openness, especially in populations that currently display more reservations.

Table 6. Mean scores and standard deviation [mean ± SD] of each question for incorporation of insects on a food matrix, where 1—completely disagree and 7—completely agree.

Questions	Poland	Spain
It seems natural to me that animals feed on insects	6.26 ± 1.08 [a]	5.73 ± 1.53 [b]
The idea of consuming meat/milk from animals fed on insects is not disgusting	5.54 ± 1.65 [a]	4.66 ± 2.14 [b]
I believe that consuming meat/milk from animals fed with insects-based feed is good for the environment	5.50 ± 1.39 [a]	5.20 ± 1.64 [a]
I believe there are no other ways to improve environmental sustainability in animal production than introducing insects	4.83 ± 1.59 [a]	4.10 ± 1.83 [b]
I would be willing to consume meat/milk from animals fed on insect meal as long as the food is not expensive	5.40 ± 1.54 [a]	5.00 ± 2.01 [a]

[a, b]: Diffrent letters indicate statistically significant differences between means at $p < 0.05$—Tukey's post hoc test.

Figure 3 illustrates a higher overall acceptance among Polish consumers for incorporating insects into food matrices compared to Spanish consumers. The stronger acceptance in Poland suggests a greater openness to insect-based innovations, likely influenced by younger demographics and a more urbanized population. Spanish consumers' reluctance may stem from cultural factors such as unfamiliarity with insect-based products and potential disgust associated with insects as food sources. Addressing these cultural and psychological barriers through targeted marketing and educational campaigns could foster greater acceptance by highlighting the benefits of insect-based food production, such as sustainability and reduced environmental impact.

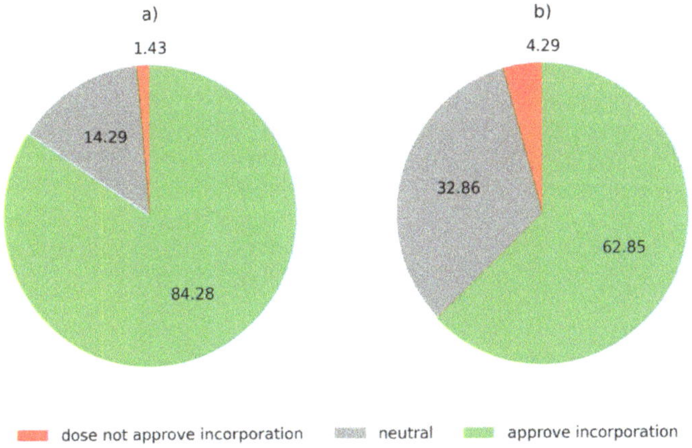

Figure 3. Spanish and Polish consumer acceptance of insect's incorporation in food matrix [%]: (**a**) Polish consumer responses. (**b**) Spanish consumer responses.

A comparative analysis of consumer preferences for products enhanced with insect powder among Polish and Spanish respondents (Figure 4) reveals significant differences and shared trends. Both groups show a clear preference for salty snacks and dry cereal products, with Polish consumers exhibiting a more pronounced inclination toward these categories. Additionally, Polish respondents demonstrate greater interest in sweet snacks and ready-to-eat meals, which may reflect differences in taste, food traditions, and food availability. In contrast, Spanish respondents indicate greater reluctance toward insect-enhanced products, highlighting cultural and demographic influences on acceptance.

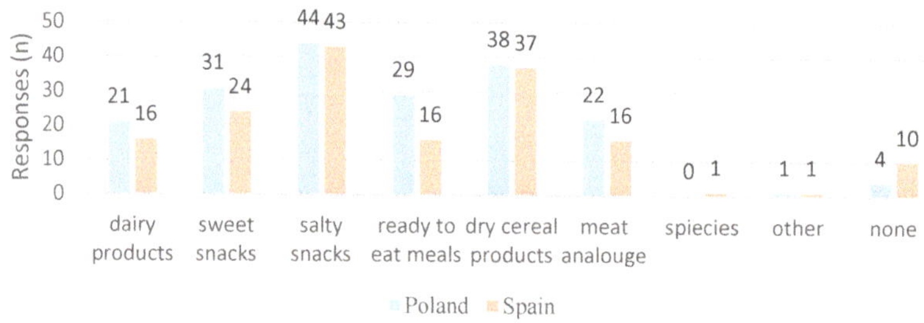

Figure 4. Consumer preferences about products enhanced with insect powder.

The relevance of different sensory aspects for consumers regarding insect-enhanced food products is considered. It is found that texture, smell, and taste are of greater importance than color (Figure 5).

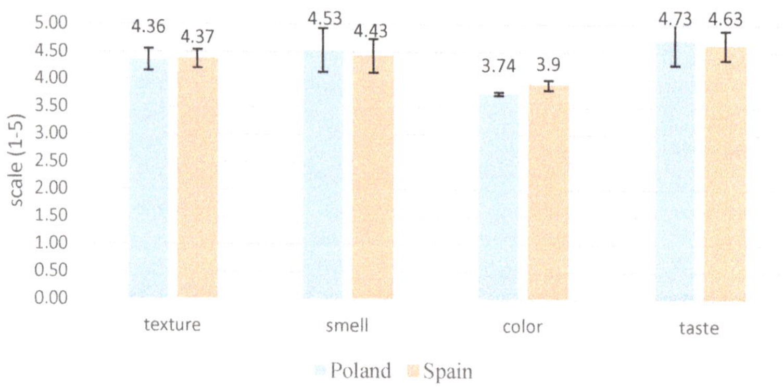

Figure 5. Consumer Perceptions of Sensory Aspects in Insect-Enhanced Foods (0—Completely Irrelevant, 5—Highly Relevant).

When evaluating the relevance of sensory aspects for insect-enhanced food products, both Polish and Spanish consumers prioritize texture, smell, and taste over color. This aligns with the findings of the study by Martins et al., which emphasize the importance of sensory experiences in consumer acceptance of novel foods, particularly insect-based products [42].

4. Conclusions

This study investigated the functional properties of *Tenebrio molitor* protein hydrolysates, with a particular emphasis on their antioxidant activity and the consumer acceptance of insect-based foods. The null hypothesis—that enzymatic hydrolysis using Cucurbita ficifolia protease would enhance the bioactivity of yellow mealworm protein hydrolysates and that consumer acceptance would vary by demographic factors—was confirmed. The hydrolysis process was found to significantly enhance antioxidant activity, as evidenced by DPPH and ABTS assays. Additionally, the study revealed notable differences in consumer attitudes between Polish and Spanish populations.

From a scientific perspective, the findings demonstrate the effectiveness of *Cucurbita ficifolia* protease in producing bioactive peptides with enhanced antioxidant properties. This contributes to the growing body of research on non-commercial proteases and underscores the potential of yellow mealworm protein as a sustainable and functional ingredient for health-promoting foods. The observed correlation between hydrolysis efficiency and bioactivity supports the applicability of bioactive peptides in addressing oxidative stress, making them valuable for functional food formulations.

The consumer survey revealed that acceptance of insect-based foods is influenced by cultural and demographic factors. Younger and urbanized Polish consumers showed greater openness, while Spanish participants demonstrated cautious interest, likely shaped by established food habits and higher levels of education. These findings suggest that tailored marketing and educational strategies are essential to overcoming barriers to acceptance. For instance, emphasizing the sustainability and nutritional benefits of insect-based foods can appeal to environmentally conscious consumers, while addressing sensory concerns can mitigate reluctance.

In terms of practical applications, the study underscores the potential of mealworm hydrolysates as adaptable ingredients for incorporation into diverse food matrices. Their enhanced bioactivity, when combined with insights into consumer attitudes, provides a strong foundation for the development of innovative insect-based products that are aligned with health and sustainability trends. Future research should prioritize the establishment of links between consumer perceptions and specific product characteristics, such as hydrolysates, through targeted surveys and sensory evaluations. This integrative approach serves to fill the gap between technical advancements and market demands, providing a foundation for the successful adoption of insect-based functional foods.

Author Contributions: Conceptualization, M.M., A.D. and M.C.-L.; Data curation, M.M.; Formal analysis, M.M.; Investigation, M.M.; Methodology, M.M., A.D. and M.C.-L.; Project administration, M.M., A.D. and M.C.-L.; Supervision, A.D. and M.C.-L.; Validation, M.M., A.D. and M.C.-L.; Visualization, M.M.; Writing—original draft, M.M.; Writing—review & editing, A.D. and M.C.-L. All authors have read and agreed to the published version of the manuscript.

Funding: This research received no external funding.

Institutional Review Board Statement: The study was conducted in accordance with the Declaration of Helsinki and approved by the Institutional Review Board of Miguel Hernández University of Elche (protocol code DTA.MCL.2240123, approved prior to data collection).

Informed Consent Statement: Informed consent was obtained from all subjects involved in the study.

Data Availability Statement: The original contributions presented in the study are included in the article. Should further inquiries necessitate additional information, the corresponding author can be contacted directly.

Acknowledgments: This work was conducted as part of Marcelina Maciejewska's Master's Thesis at the University of Environmental Sciences in Wroclaw and the Universidad Miguel Hernández in Elche. Marcelina Maciejewska is very thankful to Anna Dąbrowska and Marina Cano Lamadrid, for their invaluable guidance, support, and encouragement throughout this research. Their expertise and insights were instrumental in the completion of this work.

Conflicts of Interest: The authors declare no conflicts of interest.

References

1. Dzhalyukova, Z.M.; Fursov, V.A.; Rodionova, V.I.; Shvachkina, L.A.; Zolotareva, T.A. Socio-economic factors of the edible insects' market development. *Entomol. Appl. Sci. Lett.* **2022**, *9*, 33–40. [CrossRef]
2. Sobczak, P.; Grochowicz, J.; Łusiak, P.; Żukiewicz-Sobczak, W. Development of alternative protein sources in terms of a sustainable system. *Sustainability* **2023**, *15*, 12111. [CrossRef]
3. Ordoñez-Araque, R.; Egas-Montenegro, E. Edible insects: A food alternative for the sustainable development of the planet. *Int. J. Gastron. Food Sci.* **2021**, *23*, 100304. [CrossRef]
4. Imathiu, S. Benefits and food safety concerns associated with consumption of edible insects. *NFS J.* **2020**, *18*, 1–11. [CrossRef]
5. Kröger, T.; Dupont, J.; Büsing, L.; Fiebelkorn, F. Acceptance of insect-based food products in western societies: A systematic review. *Front. Nutr.* **2022**, *8*, 759885. [CrossRef]

6. Cano-Lamadrid, M.; Martínez-Zamora, L.; Castillejo, N.; Cattaneo, C.; Pagliarini, E.; Artés-Hernández, F. How does the phytochemical composition of sprouts and microgreens from Brassica vegetables affect the sensory profile and consumer acceptability? *Postharvest Biol. Technol.* **2023**, *203*, 112411. [CrossRef]
7. Ghosh, S.; Jung, C.; Meyer-Rochow, V.B. What governs selection and acceptance of edible insect species? In *Edible Insects in Sustainable Food Systems*; Halloran, A., Flore, R., Vantomme, P., Roos, N., Eds.; Springer: Cham, Switzerland, 2018; pp. 333–355. [CrossRef]
8. European Commission. Commission Implementing Regulation (EU) 2022/169 of 8 February 2022 authorising the placing on the market of frozen, dried and powder forms of yellow mealworm (*Tenebrio molitor* larva) as a novel food under Regulation (EU) 2015/2283 of the European Parliament and of the Council and amending Commission Implementing Regulation (EU) 2017/2470. *Off. J. Eur. Union* **2022**, *L28*, 10–16. Available online: http://data.europa.eu/eli/reg_impl/2022/169/oj (accessed on 9 February 2022).
9. Gkinali, A.A.; Matsakidou, A.; Vasileiou, E.; Paraskevopoulou, A. Potentiality of *Tenebrio molitor* larva-based ingredients for the food industry: A review. *Trends Food Sci. Technol.* **2022**, *119*, 495–507. [CrossRef]
10. Rabadán, A.; Bernabéu, R. A systematic review of studies using the food neophobia scale: Conclusions from thirty years of studies. *Food Qual. Prefer.* **2021**, *93*, 104241. [CrossRef]
11. Clemente-Villalba, J.; Cano-Lamadrid, M.; Issa-Issa, H.; Hurtado, P.; Hernández, F.; Carbonell-Barrachina, Á.A.; López-Lluch, D. Comparison on sensory profile, volatile composition, and consumers' acceptance for PDO or non-PDO tigernut (*Cyperus esculentus* L.) milk. *LWT* **2021**, *140*, 110606. [CrossRef]
12. Ghaly, A.E.; Alkoaik, F.N. The yellow mealworm as a novel source of protein. *Am. J. Agric. Biol. Sci.* **2009**, *4*, 319–331. [CrossRef]
13. Zielińska, E.; Baraniak, B.; Karaś, M.; Rybczyńska, K.; Jakubczyk, A. Selected species of edible insects as a source of nutrient composition. *Food Res. Int.* **2015**, *77*, 460–466. [CrossRef]
14. Fitzgerald, R.J.; Murray, B.A. Bioactive peptides and lactic fermentations. *Int. J. Dairy Technol.* **2006**, *59*, 118–125. [CrossRef]
15. Haque, E.; Chand, R.; Kapila, S. Biofunctional properties of bioactive peptides of milk origin. *Food Rev. Int.* **2008**, *25*, 28–43. [CrossRef]
16. Ghosh, S.; Lee, S.M.; Jung, C.; Meyer-Rochow, V.B. Nutritional composition of five commercial edible insects in South Korea. *J. Asia Pac. Entomol.* **2017**, *20*, 686–694. [CrossRef]
17. Akbarian, M.; Khani, A.; Eghbalpour, S.; Uversky, V.N. Bioactive peptides: Synthesis, sources, applications, and proposed mechanisms of action. *Int. J. Mol. Sci.* **2022**, *23*, 1445. [CrossRef]
18. Mohammad, A.W.; Kumar, A.G.; Basha, R.K. Optimization of enzymatic hydrolysis of tilapia (*Oreochromis* Spp.) scale gelatine. *Int. Aquat. Res.* **2015**, *7*, 27–39. [CrossRef]
19. Babij, K.; Bajzert, J.; Dąbrowska, A.; Szołtysik, M.; Zambrowicz, A.; Lubec, G.; Stefaniak, T.; Willak-Janc, E.; Chrzanowska, J. Hydrolysis with *Cucurbita ficifolia* serine protease reduces antigenic response to bovine whey protein concentrate and αs-casein. *Amino Acids* **2015**, *47*, 2335–2343. [CrossRef]
20. Binod, P.; Palkhiwala, P.; Gaikaiwari, R.; Nampoothiri, K.M.; Duggal, A.; Dey, K.; Pandey, A. Industrial enzymes-present status and future perspectives for India. *Indian J. Biotechnol.* **2013**, *12*, 346–355. Available online: http://nopr.niscpr.res.in/handle/123456789/17451 (accessed on 11 November 2024).
21. Pokora, M.; Szołtysik, M.; Dąbrowska, A.; Chrzanowska, J.; Trziszka, T. Wykorzystanie niekomercyjnych enzymów proteolitycznych do hydrolizy białek jaja. *Acta Sci. Pol.* **2010**, *17*, 45–54.
22. Dryjanski, M.; Otlewski, J.; Polanowski, A.; Wilusz, T. Serine Proteinase from *Cucurbita ficifolia* Seed; Purification, Properties, Substrate Specificity and Action on Native Squash Trypsin Inhibitor (CMTI I). *Biol. Chem.* **1990**, *371*, 889–896. [CrossRef]
23. Chrzanowska, J.; Kołaczkowska, M. Production of extracellular proteolytic enzymes by *Beauveria bassiana*. *Acta Mycol.* **1998**, *33*, 277–285. [CrossRef]
24. Lowry, O.H.; Rosebrough, N.J.; Farr, A.L.; Randall, R.J. Protein measurement with the Folin phenol reagent. *J. Biol. Chem.* **1951**, *193*, 265–275. [CrossRef]
25. Snyder, S.L.; Sobocinski, P.Z. An improved 2, 4, 6-trinitrobenzenesulfonic acid method for the determination of amines. *Anal. Biochem.* **1975**, *64*, 284–288. [CrossRef] [PubMed]
26. Kuchroo, C.N.; Rahilly, J.; Fox, P.F. Assessment of proteolysis in cheese by reaction with trinitrobenzene sulphonic acid. *Ir. J. Food Sci. Technol.* **1983**, *7*, 129–133. Available online: https://www.jstor.org/stable/25558072 (accessed on 17 November 2024).
27. Laemmli, U.K. Cleavage of structural proteins during the assembly of the head of bacteriophage T4. *Nature* **1970**, *227*, 680–685. [CrossRef] [PubMed]
28. Kedare, S.B.; Singh, R.P. Genesis and development of DPPH method of antioxidant assay. *J. Food Sci. Technol.* **2011**, *48*, 412–422. [CrossRef] [PubMed]
29. Re, R.; Pellegrini, N.; Proteggente, A.; Pannala, A.; Yang, M.; Rice-Evans, C. Antioxidant activity applying an improved ABTS radical cation decolorization assay. *Free Radic. Biol. Med.* **1999**, *26*, 1231–1237. [CrossRef] [PubMed]

30. Cantalapiedra, F.; Juan-García, A.; Juan, C. Perception of food safety associated with entomophagy among higher-education students: Exploring insects as a novel food source. *Foods* **2023**, *12*, 4427. [CrossRef] [PubMed]
31. Moruzzo, R.; Mancini, S.; Boncinelli, F.; Riccioli, F. Exploring the acceptance of entomophagy: A survey of Italian consumers. *Insects* **2021**, *12*, 123. [CrossRef]
32. Szakály, Z.; Kovács, B.; Soós, M.; Kiss, M.; Balsa-Budai, N. Adaptation and validation of the food neophobia scale: The case of Hungary. *Foods* **2021**, *10*, 1766. [CrossRef] [PubMed]
33. Kasza, G.; Izsó, T.; Szakos, D.; Nugraha, W.S.; Tamimi, M.H.; Süth, M. Insects as food: Changes in consumers' acceptance of entomophagy in Hungary between 2016 and 2021. *Appetite* **2023**, *188*, 106770. [CrossRef] [PubMed]
34. Ordoñez-López, M.F.; Ghnimi, S.; Liu, C. Willingness to consume insect-based food in France: Determinants and consumer perspectives. *LWT* **2023**, *185*, 115179. [CrossRef]
35. Fernández-Ruiz, V.; Claret, A.; Chaya, C. Testing a Spanish-version of the food neophobia scale. *Food Qual. Prefer.* **2013**, *28*, 222–225. [CrossRef]
36. Szołtysik, M.; Kucharska, A.Z.; Dąbrowska, A.; Zięba, T.; Bobak, Ł.; Chrzanowska, J. Effect of two combined functional additives on yoghurt properties. *Foods* **2021**, *10*, 1159. [CrossRef] [PubMed]
37. Quah, Y.; Tong, S.R.; Bojarska, J.; Giller, K.; Tan, S.A.; Ziora, Z.M.; Esatbeyoglu, T.; Chai, T.-T. Bioactive peptide discovery from edible insects for potential applications in human health and agriculture. *Molecules* **2023**, *28*, 1233. [CrossRef] [PubMed]
38. Cho, H.R.; Lee, S.O. Novel hepatoprotective peptides derived from protein hydrolysates of mealworm (*Tenebrio molitor*). *Food Res. Int.* **2020**, *133*, 109194. [CrossRef] [PubMed]
39. Mustapa, M.A.C.; Kallas, Z. Towards more sustainable animal-feed alternatives: A survey on Spanish consumers' willingness to consume animal products fed with insects. *Sustain. Prod. Consum.* **2023**, *41*, 9–20. [CrossRef]
40. Guiné, R.P.F.; Florença, S.G.; Costa, C.A.; Correia, P.M.R.; Cruz-Lopes, L.; Esteves, B.; Ferreira, M.; Fragata, A.; Cardoso, A.P.; Campos, S.; et al. Edible Insects: Perceptions of Marketing, Economic, and Social Aspects among Citizens of Different Countries. *Foods* **2023**, *12*, 4229. [CrossRef]
41. Jeżewska-Zychowicz, M.; Plichta, M.; Drywień, M.E.; Hamulka, J. Food neophobia among adults: Differences in dietary patterns, food choice motives, and food labels reading in Poles. *Nutrients* **2021**, *13*, 1590. [CrossRef]
42. Martins, O.M.; Bucea-Manea-Țoniș, R.; Coelho, A.S.; Simion, V.E. Sensory perception nudge: Insect-based food consumer behavior. *Sustainability* **2022**, *14*, 11541. [CrossRef]

foods

MDPI

Review

Therapeutic Potential of White Kidney Beans (*Phaseolus vulgaris*) for Weight Management

Hassan Muzaffar [1,2,†], Muhammad Jehangir [3,†], Jiayue Hu [4], Yiyang Yu [1], Mingzhou Yu [1,*] and Yonghong Hu [1,*]

[1] College of Food Science and Light Industry, Nanjing Tech University, Nanjing 211816, China; hassan14665@gmail.com (H.M.); yuyy@njtech.edu.cn (Y.Y.)
[2] School of Food Science and Engineering, South China University of Technology, Guangzhou 510640, China
[3] Jiangsu Provincial University Key Laboratory of Intelligent Medical Sensing Materials and Devices, Institute of Chemical Biology and Functional Molecules, School of Chemistry and Molecular Engineering, Nanjing Tech University, Nanjing 211816, China; jehangirm62@gmail.com
[4] College of Biotechnology and Pharmaceutical Engineering, Nanjing Tech University, Nanjing 211816, China; 202361218090@njtech.edu.cn
* Correspondence: ymz@njtech.edu.cn (M.Y.); chemywg@126.com (Y.H.)
† These authors contributed equally to this work.

Abstract

The escalating global prevalence of obesity underscores the need for effective and sustainable nutritional interventions. Functional foods, especially white kidney beans (*Phaseolus vulgaris*), show a promising avenue to link fundamental biochemical insights with clinically feasible interventions, supporting their potential as an adjunct dietary strategy for managing and preventing obesity. This review critically examines the mechanistic roles of white kidney bean in weight regulation, which includes suppression of starch digestion, attenuation of postprandial glycemia, modulation of appetite and satiety, and hypolipidemic effects. Clinical and preclinical evidence supports the potential of white kidney bean as a nutraceutical for metabolic health, demonstrating consistent reductions in body fat mass, glycemic excursion, and overall weight. Nevertheless, significant limitations persist, including heterogeneity in trial designs, absence of dose standardization, and inadequate long-term safety assessments. Furthermore, this review addresses food fortification, advancements in supplement formulation, and cooking techniques that enhance both consumer acceptability and the bioactivity of white kidney bean (WKB), along with the significance of regulatory standards to ensure safety and quality. Future research should integrate clinical, molecular and food technology methods to improve the translation of experimental findings into precision nutritional strategies for obesity management.

Keywords: *Phaseolus vulgaris*; white kidney bean; anti-obesity; starch inhibition; fat metabolism; protein hydrolysates; alpha-amylase inhibitor (α-AI); functional food

1. Introduction

Obesity and overweight represent major contemporary health concerns. Obesity ranks as the fifth leading risk factor for global mortality [1]. According to the World Health Organization, over 0.83 billion individuals were affected by obesity in 2020, with projections indicating an increase to more than 1.53 billion by 2035 [2]. Obesity results from an energy imbalance between intake and expenditure, often attributable to insufficient physical activity and consumption of diets high in fats and sugars but low in fiber and micronutrients, such as vitamins and minerals [3].

Obesity is associated with several diseases, including type 2 diabetes, cardiovascular disease, and various cancers [4–7]. Irrespective of genetic predisposition, an unhealthy lifestyle combined with obesity can precipitate type 2 diabetes [8,9]. The primary strategies for managing excess weight encompass diet and physical activity [10]. Numerous dietary regimens and nutritional approaches featuring varying proportions of proteins, fats, and carbohydrates have been proposed for weight loss [11,12]. In recent years, restricting carbohydrate intake particularly processed carbohydrates, has emerged as an prominent weight loss strategy [13]. Given that carbohydrates constitute a primary caloric source in most diets, inhibiting or delaying their digestion is regarded as an effective strategy for weight management [14]. Recently, white kidney beans has garnered substantial interest for its capacity to inhibit carbohydrate digestion and facilitate weight management [15].

White kidney beans (*Phaseolus vulgaris* L.) are native to the Mesoamerican and Andean regions [16]. They are abundant in complex carbohydrates, protein and bioactive compounds that promote metabolic health [3]. White kidney beans, which are rich in potassium (K) and magnesium (Mg), have been reported to confer benefits in individuals with conditions, such as atherosclerosis and hyperlipidemia [17]. *Phaseolamin*, a naturally occurring α-amylase inhibitor found in beans, impedes starch hydrolysis and retards carbohydrate digestion [18,19]. This process reduces caloric absorption, lowers the glycemic index of meals, and facilitates blood glucose regulation [20–22]. WKBE also enhances metabolic equilibrium and beneficially modulates the gut microbiota [23,24]. Furthermore, it mitigates fat accumulation and hyperglycemia while promoting weight loss [25]. A recent study indicates that α-amylase inhibitors derived from white kidney beans exhibit no adverse effects on the human body and can serve as safe dietary agents [26]. It also effectively regulates energy intake and appetite, thereby demonstrating promising outcomes in clinical studies involving obese patients, as illustrated in Figure 1 [27]. Although numerous studies have documented the nutritional properties and therapeutic importance of white kidney beans (*Phaseolus vulgaris*), a comprehensive synthesis of their impact on weight management remains scarce.

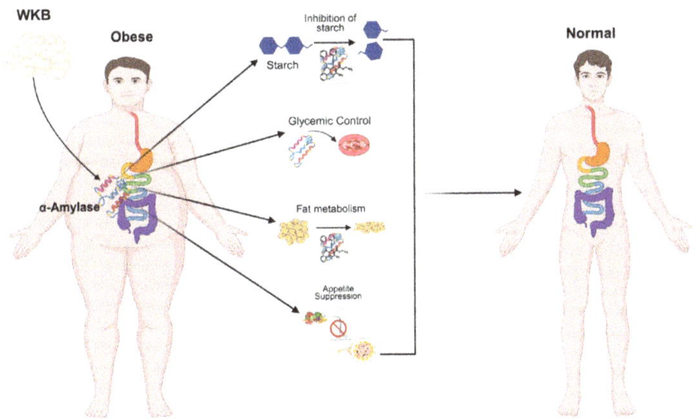

Figure 1. The Therapeutic Role of White Kidney Beans in Obesity Control.

This review summarizes the current scientific literature on the pivotal role of white kidney bean extract for the prevention and treatment of obesity and related metabolic diseases. WKBE confers metabolic benefits extending beyond carbohydrates catabolism. Its elevated protein and fiber content promote satiety and appetite suppression. Studies in both animal and human models have demonstrated that WKBE improves lipid metabolism and reduces fat accumulation. Existing evidence from current and prior research supports WKBE as a valuable supplement for obesity and metabolic health management.

2. Methodology

2.1. Study Design

The manuscript presents a narrative review that provides a comprehensive critical synthesis and integration of recent scientific literature regarding the anti-obesity and metabolic effects of white kidney bean (*Phaseolus vulgaris*) extracts. The approach was chosen to organize information from diverse study models, including in vitro enzyme assays, phytochemical analysis and human clinical trials.

2.2. Literature Search Strategy

A systematic literature search was conducted across multiple electronic databases to encompass the contemporary scientific literature. Searches were conducted in key databases, including Google Scholar, PubMed, Xmol, Elict and Scispace with the final update in July 2025. The search strategy employed a combination of keywords, including "white kidney bean extract", "*Phaseolus vulgaris*", "phaseolamin", "weight loss", "weight management", "obesity", "alpha-amylase inhibitor", "clinical trial", "RCT", "dose standardization", "safety", "consumption", etc.

3. Nutritional Profile and Functional Properties of Pulses

Pulses exhibit a distinctive nutritional profile, featuring high protein content, low fat levels, and substantial dietary fiber. They are also abundant in complex carbohydrates, resistant starch, and essential vitamins, including folate (vitamin B9) [28]. Moreover, pulses contain a variety of non-nutritive bioactive compounds, including enzyme inhibitors, phytates, saponins, lectins, phenolic compounds, and oligosaccharides. These compounds contribute to their health-promoting effects and play a role in reducing the risk of non-communicable diseases linked to dietary and lifestyle changes [29]. Pulses serve as an excellent source of essential nutrients, comprising approximately 21–25% crude protein, 60–65% carbohydrates, and key micronutrients including magnesium, iron, potassium, and B vitamins [30]. They provide both soluble and insoluble dietary fiber, which contributes to weight management and cholesterol regulation. In addition, the bioactive compounds present in pulses exhibit anti-inflammatory and anticancer properties. Daily consumption of 100–200 g of pulses can fulfill recommended intakes of minerals and iron, rendering them a valuable element of a balanced diet [31]. Pulses represent a suitable and cost-effective source of protein, carbohydrates, vitamins, and minerals for obesity prevention, diabetes management, and chronic disease risk reduction [32]. They satisfy the increasing demand for dietary proteins, alleviate protein-energy malnutrition, improve soil fertility, and diminish greenhouse gas emissions [33]. Pulses are incorporated into traditional and innovative food products, including pasta, snacks, breakfast foods, and soups to enhance nutritional quality and technological attributes [34]. The gluten-free nature and abundance of resistant starch and fiber render them suitable for low-calorie and gluten-free diets. Techno-functional properties such as fat and water absorption capacity, foaming stability, and viscosity depend on their composition [35].

White kidney beans (*Phaseolus vulgaris* L.) are a widely consumed pulse and a vital global source of nutrients and dietary protein [36]. They display diverse physicochemical, cooking, hydration, and nutritional properties [37]. Although certain studies have investigated their functional properties, the digestibility and techno-functional attributes of diverse accessions remain largely unexamined [38].

White kidney beans are nutrient-dense, providing a balanced profile of complex carbohydrates, dietary fiber, and proteins. Their carbohydrate fraction, primarily comprising complex polysaccharides, facilitates slower digestion and absorption, thereby reducing postprandial glucose and insulin levels and potentially aiding weight loss. Dietary fiber supports gastrointestinal health and extends satiety, thereby contributing to weight management. Their high protein content and superior digestibility render white kidney beans a valuable staple in vegetarian and vegan diets [39].

4. Chemical Composition of White Kidney Beans

The chemical composition of white kidney beans (*Phaseolus vulgaris*) encompasses a broad spectrum of macronutrients and micronutrients, rendering them a valuable nutritional element [40]. These beans primarily consist of carbohydrates and proteins, with minor lipid content, along with essential minerals and amino acids [41].

4.1. Macronutrients

White kidney beans (*Phaseolus vulgaris*) are nutrient-rich legumes with a macronutrient profile predominantly comprising proteins and carbohydrates, accompanied by a minor fat fraction. This profile positions them as a staple plant-based functional food.

4.1.1. Proteins

White kidney beans contain substantial protein levels, typically ranging from 20% to 33% of dry weight [42]. Globulins predominate, primarily phaseolin and vicilin, constituting 50–70% of total protein. Albumins comprise approximately 20% of the protein fraction, thereby augmenting nutritional quality [43]. These proteins are abundant in essential amino acids yet, akin to other legumes, relatively deficient in methionine. White kidney beans offer a favorable ratio of essential amino acids, ranging from 0.29% to 0.36%, underscoring their value as a plant-based protein source [44]. Processing methods, such as cooking and roasting, can enhance the protein content and enhance digestibility [45]. Moreover, isolated proteins from white kidney beans exhibits outstanding solubility and functional properties, rendering them suitable for diverse food applications. These isolated proteins interact with starch to form complexes that retard starch digestion, thereby lowering the glycemic index and promoting weight loss [46,47].

4.1.2. Carbohydrates

Carbohydrates represent the predominant macronutrient in white kidney beans, constituting 45–60% of dry weight [48]. Starch is the primary carbohydrate, comprising 70.9–83.1% of the total, with a high proportion of resistant starch (RS) and slowly digestible starch (SDS) [49]. This composition suggests that an extensive portion of the starch is not directly broken down in the digestive tract, which helps with weight loss and regulates blood sugar levels after consumption [50]. They also contain a variety of non-starch polysaccharides, including mannose, galactose, arabinose and galacturonic acid. White kidney beans exhibit low sugar content, rendering them appropriate for low-sugar diets [51].

4.1.3. Lipids

White kidney bean (WKB) contains minimal fat, ranging from 1% to 2.8%. The majority comprises polyunsaturated fatty acids, particularly linolenic and α-linolenic acids, which support cardiovascular health. Their low-fat profile renders white kidney beans ideal for weight management and cardiovascular disease [43,44].

4.2. Micronutrients

White kidney beans (*Phaseolus vulgaris*) contain a variety of micronutrients, including essential minerals and vitamins, which underscore their nutritional and health-promoting attributes [17].

4.2.1. Minerals

White kidney beans are rich in calcium, magnesium, potassium, phosphorus and zinc. For example, the average amounts recorded for kidney beans are approximately 8.2 mg iron, 407 mg phosphorus, and 143 mg calcium per 100 g of dry weight, with substantial potassium and magnesium content [52]. These minerals play an important role in various physiological activities, including bone health (phosphorus and calcium), oxygen transport (iron), and potassium and magnesium for muscle function and the nervous system, as well as zinc for immune support [53]. The mineral content can be modulated through agronomic practices; for example, biofortification with zinc sulfate and iron has been reported to elevate levels in kidney beans, thereby enhancing nutritional value [54]. Moreover, application of natural fertilizers, such as zeolite and sheep manure can increase concentrations of potassium, iron, and zinc [55].

4.2.2. Vitamins

White kidney beans are a suitable source of various types of B vitamins. Including B1 thiamine, B2 riboflavin, B3 niacin, and B6 pyridoxine, which are necessary for the central nervous system and energy metabolism [56]. They also contain vitamin E (predominantly γ-tocopherol), a fat-soluble antioxidant that assists in protecting cells from damage caused by oxidative stress. Moreover, they are not rich in fat-soluble vitamins such as vitamins A, D, K, and C like other legumes [43].

4.2.3. Bioactive Compounds

White kidney beans (*Phaseolus vulgaris*) contain large amounts of flavonoids and phenolic compounds. The overall quantity of flavonoid content ranges from 0.19 to 7.05 mg/g, while the phenolic compounds vary from 0.25 to 0.37 mg rutin equivalent (RE)/g dry weight. These chemical compounds can exhibit antioxidant activities [57]. Recent studies haved identified various distinct bioactive compounds in *Phaseolus vulgaris*, such as eugenin, trans-ferulic acid ethyl ester, glutamyl isoleucine α-L-glutamyl-L-leucine and denatonium as shown in Figure 2; however, their precise proportions remain undetermined [58]. Isolation and characterization of these bioactive compounds from white kidney beans confirm the presence of diverse compounds with multiple health benefits [59]. Relative to flavonoids and phenolic compounds, white kidney beans contain lesser amounts of tannins, saponins and coumarins, which show antifungal and antibacterial properties [60].

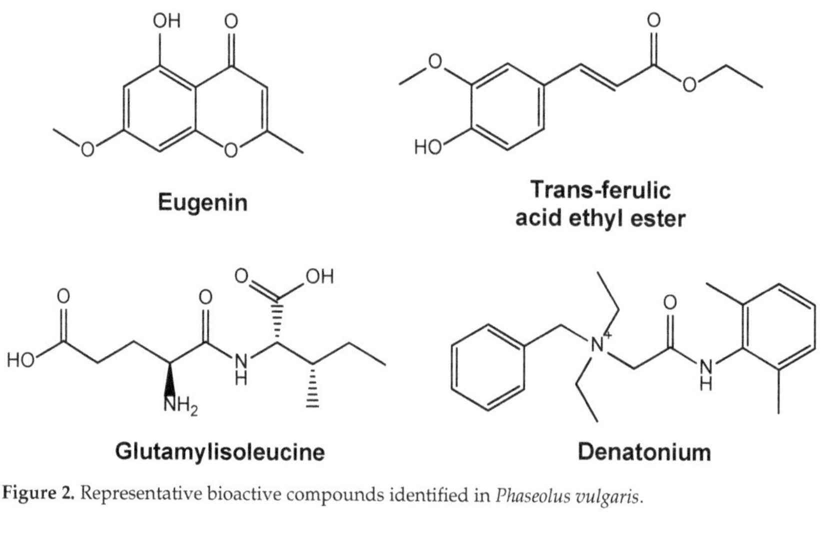

Figure 2. Representative bioactive compounds identified in *Phaseolus vulgaris*.

5. Mechanism of Weight Loss

White kidney beans (*Phaseolus vulgaris*) have attracted considerable interest not only for their culinary versatility but also for their potential in weight management. A meta-analysis by Davkova et al. (2024) demonstrated that *Phaseolus vulgaris* supplementation exerts a significant effect on body weight and fat mass [61]. Upon ingestion, carbohydrates must be hydrolyzed and converted into monosaccharides for intestinal absorption [50]. The enzymes responsible for converting complex carbohydrates into monosaccharides are amylases and β-glucosidases. These enzymes are indispensable for this process [62]. Initially, amylase hydrolyzes complex starch into oligosaccharides. Subsequently, β-glucosidases further hydrolyze oligosaccharides into monosaccharides [63]. Studies have identified three α-amylase inhibitors in white kidney beans: α-AIL, α-A1, and α-A12. These inhibitors effectively attenuate amylase activity, thereby facilitating weight reduction [64].

5.1. α-Amylase Inhibition and Glycemic Control

Inhibition of carbohydrate digestion represents a primary mechanism by which white kidney bean extract (WKBE) facilitates weight management. This is primarily attributable to its α-amylase inhibitory activity. This activity attenuates the hydrolysis of complex starches into absorbable monosaccharides. Consequently, carbohydrate absorption is delayed, resulting in attenuated postprandial glucose responses and reduced caloric intake, as illustrated in Figure 3 [65,66]. Preclinical studies in animal models corroborate this mechanism. Rodents administered WKBE exhibited reduced carbohydrate absorption, attenuated postprandial glycemia, and decreased body weight [23,67]. Long-term supplementation in animals also yielded reductions in visceral fat mass and improvements in metabolic flexibility, further substantiating the association between inhibited carbohydrate metabolism and body weight regulation [68]. Beyond inhibiting carbohydrate digestion, WKBE promotes weight management by stabilizing postprandial blood glucose and insulin levels, thereby improving overall glycemic control. This is particularly pertinent for individuals with metabolic disorders, such as diabetes [69–71]. Notably, in vitro gut model studies demonstrated α-amylase inhibition, whereas in vivo experiments failed to confirm significant enzyme inhibition [65]. This suggests that direct active-site inhibition may not represent the sole mechanism of α-amylase inhibition. Alternatively, WKBE may exert indirect effects, such as delayed starch hydrolysis, partial enzyme inhibition, or modulation of gut microbiota activity.

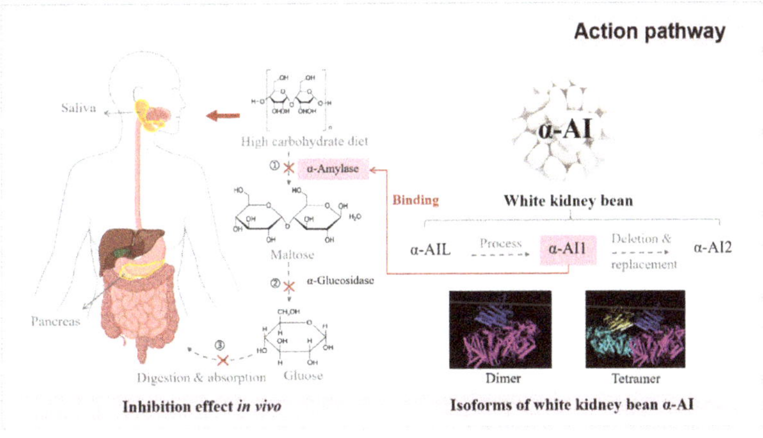

Figure 3. Insight into α-Amylase inhibition and glycemic control mechanisms of WKB. Reproduced with permission [66], Copyright 2024, Elsevier Ltd.

Overall, these findings indicate that WKBE exerts anti-obesity effects by diminishing carbohydrate absorption—via α-amylase inhibition and glycemic control—through modulation of gut microbiota and enzyme activity.

5.2. Gut Microbiota and Gastrointestinal Effects

WKBE can also alter gut microbiota composition. In a randomized, placebo-controlled crossover trial, Houghton et al. (2023) [72] assessed gut microbiota composition, gastrointestinal inflammation, and stool characteristics in 20 healthy individuals following seven days of white kidney bean (WKB) consumption. WKBE administration elevated the relative abundance of Firmicutes while reducing that of Bacteroidetes as shown in Figure 4 [72]. However, no significant differences were observed in digestive symptoms, stool frequency, or fecal calprotectin between the placebo and WKBE groups, indicating minimal effects on gastrointestinal health [73].

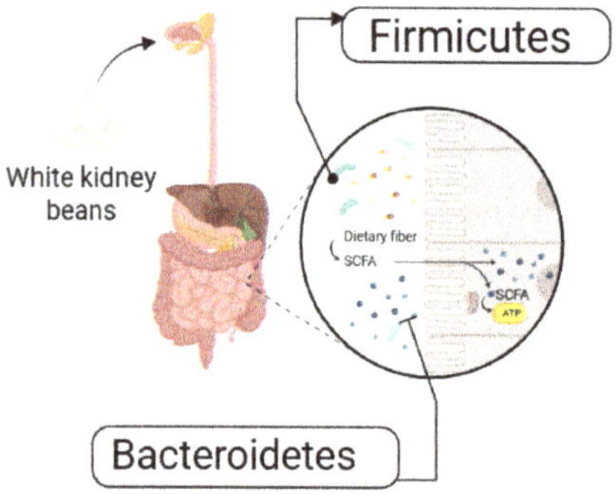

Figure 4. Gut Microbiota regulation by WKB.

5.3. Appetite Suppression and Satiety

Alongside its effects on blood sugar regulation and carbohydrate metabolism, WKBE may also play a role in body weight management by controlling satiety and overall energy intake. This effect is partially attributable to its elevated protein and fiber content, which delay gastric emptying and promote satiety [74,75]. Moreover, α-amylase inhibition prolongs the release of glucose from starch, thereby stabilizing blood glucose levels and mitigating appetite fluctuations [76].

This satiety-promoting effect is corroborated by both human and animal studies. In rodents, supplementation with WKBE reduced food intake and body weight gain relative to controls. These results were linked to reduced ghrelin (the hunger hormone) and elevated peptide YY and GLP-1, hormones that regulate satiety [77]. These findings suggest that WKBE influences both behavioral and physiological dimensions of appetite, thereby supporting weight loss through diminished energy intake [78].

5.4. Fat Metabolism and Lipid Profile

An additional mechanism by which WKBE supports weight management involves improving lipid profiles and modulating fat metabolism. WKB contains bioactive compounds that may reduce lipid absorption in the intestine and promote fat utilization as an energy source [79]. Studies have shown that white kidney beans contain compounds capable of modulating lipid absorption and ameliorating lipid profiles. This is critical for both weight loss and cardiovascular health [64]. WKBE facilitates weight management through enhanced lipid profiles and modulated fat metabolism. WKB harbors bioactive compounds that may attenuate intestinal lipid absorption and enhance fat oxidation as an energy substrate [80]. This mechanism is pertinent to both weight loss and cardiovascular health.

Preclinical studies provide further mechanistic insights. Rodents fed a high-fat diet supplemented with WKBE demonstrated lower LDL cholesterol, reduced serum triglycerides and improved hepatic lipid metabolism. Histological investigations also exhibited reduced lipid accumulation in the adipose tissue and liver, consistent with decreased fat storage [81]. These studies confirm that WKBE reduces lipid absorption, enhances lipid catabolism, and mitigating fat deposition.

Both human and preclinical studies provide evidence of WKBE's role in ameliorating lipid profiles, modulating lipid metabolism, and contributing to its overall anti-obesity effects [82].

6. Clinical Efficacy and Evidence Synthesis

Clinical trials have demonstrated that white kidney bean extract (WKBE) attenuates postprandial blood glucose levels and facilitates weight management. For instance, Barrett et al. (2011) [83] examined four crossover trials as shown in Table 1. These trials revealed that WKBE significantly reduced the postprandial glucose area under the curve (AUC) by up to 66% and accelerated return to baseline glucose levels, indicative of diminished carbohydrate absorption. A dose–response relationship was evident, with 1500 mg proving more efficacious than 750 mg, and a 3000 mg powdered formulation reducing postprandial hyperglycemia by 34.1% [84]. Furthermore, a multi-ingredient formulation containing 500 mg of the extract mitigated postprandial glucose and insulin excursions in individuals with elevated fasting blood glucose. Multiple studies confirm that WKBE supplementation promotes weight loss. Over periods of 35–84 days, participants receiving WKBE exhibited significantly greater weight loss (1.9–3.2 kg), along with reductions in waist circumference and body fat, relative to placebo groups [74,83,85,86]. This effect is attributable to enhanced satiety and diminished appetite, resulting in improved portion control and reduced food intake [70,87]. In addition to promoting weight loss, WKBE supplementation improves cardiovascular risk markers by lowering triglyceride and cholesterol levels while increasing HDL cholesterol [70]. These findings suggest that WKBE may concurrently facilitate weight loss and mitigate cardiovascular risk.

Table 1. Clinical Outcomes of WKBE Supplementation.

Trial/Study (Year)	Design	Dose/Duration	Main Outcomes	Efficacy	References
Tzeng et al. (2022)	Clinical/Human	Not specified	Improve insulin sensitivity, reduce fasting glucose, fructosamine and HbA1c.	Promote glycemic control.	[84]
Barrett et al. (2011) Trial 1	11 adults, crossover	1500 mg, single meal	Decrease glucose AUC by 66%, leading to quicker return to baseline.	Dose-dependent effect.	[83]
Barrett et al. (2011) Trial 2	7 adults, crossover	750 mg, single meal	Reduce glucose AUC by 28%.	Lower dose, lower impact.	
Barrett et al. (2011) Trial 3	13 adults, crossover	1500–3000 mg, single meal	(Capsule form) No impact, 3000 mg powder causes hyperglycemia by 34.1%.	Precise formulation is necessary.	
Barrett et al. (2011) Trial 4	13 individuals, FBG > 126 mg/dL.	500 mg coleus/chitosan	Reduced glucose and insulin levels after 30–60 min compared to placebo.	A product developed by combining.	
60-day RCT (n = 101)	Obese adults	1000 mg × 3/day for 60 days	Loss of weight of 1.9 kg, as well as waist circumference.	Placebo: weight loss of 0.4 kg.	[88]
Wang et al. (2020)	Obese adults	35 days, 2400 mg/day	Reduced weight by 2.2 kg, body fat and BMI.	Placebo-controlled.	[85]
84-day RCT (n = 62)	Obese adults	Dosage not specified	3.2 kg weight, 2.8% of fat reduction, 3.7 cm of waist circumference.	Controls are unchanged.	[70]

Table 1. *Cont.*

Trial/Study (Year)	Design	Dose/Duration	Main Outcomes	Efficacy	References
Jäger et al. (2024)	Double-blind, placebo-controlled	Dosage not specified	Increase fullness, decrease appetite and overindulging and improve portion management.	Modulation of appetite.	
60-day RCT	Obese adults	Dosage not specified	Reduced weight and intake of food enhanced satiety vs. placebo	Consistent with other research findings.	[89]
35-day RCT	Obese adults	2400 mg/day	Lower triglycerides, cholesterol, fat mass, and body weight.	Cardiometabolic advantages.	[85]
84-day RCT (n = 62)	Obese adults	Dosage not specified	Reduction in weight, BMI, waist circumference and improvement in HDL.	Cardiometabolic benefit.	[70]

7. Transformative Processing and Bioactivity Enhancement

White kidney beans are abundant in proteins, bioactive compounds and phenolics; however, their limited bioavailability and antinutritional factors may compromise functional and nutritional benefits. The digestibility, bioactivity, and health benefits of white kidney beans can be enhanced through transformative processing techniques, ranging from conventional boiling to advanced technologies.

7.1. Enzymatic Hydrolysis

Protein fractions from white kidney beans are typically modified via enzymatic and chemical hydrolysis to augment their bioactive and functional properties for food and nutraceutical applications [90]. Enzymatic hydrolysis employs common proteases, such as pepsin, papain, alcalase, and flavourzyme as shown in Figure 5. These enzymes cleave peptide bonds, yielding protein hydrolysates that improve foaming, emulsifying, and solubility properties [91–93]. For instance, pepsin hydrolysis achieved a degree of hydrolysis (DH) of 8.6% (60 min), thereby enhancing solubility, emulsifying, and foaming properties [94]. Enzymatic hydrolysis generates bioactive peptides exhibiting antioxidant, antimicrobial, and antihypertensive activities. Furthermore, papain- or alcalase-derived hydrolysates display pronounced antioxidant and antibacterial properties and inhibit enzymes associated with diabetes and hypertension [95].

7.2. Chemical Hydrolysis

Chemical hydrolysis produces smaller peptides and amino acids from proteins using acids or bases. Although essential for complete protein degradation and amino acid analysis, chemical hydrolysis lacks the selectivity of enzymatic methods [96]. Proteins extracted from white kidney beans using citric acid and sodium hydroxide can subsequently be fractionated and analyzed; these methods alter protein fractions, charge properties, and solubility, thereby influencing functional characteristics [97]. Nevertheless, acid hydrolysis remains a standard method for determining amino acid composition in white kidney bean proteins, as it completely degrades proteins into constituent amino acids for analysis [96]. Chemical hydrolysis, however, has limitations; it can destroy certain amino acids, such as tryptophan, and generate undesirable byproducts, thereby reducing its suitability for producing food-grade protein hydrolysates [98]. Although efficient for analytical purposes, chemical hydrolysis is infrequently employed for producing bioactive peptides in nutraceutical applications owing to its lack of specificity [99].

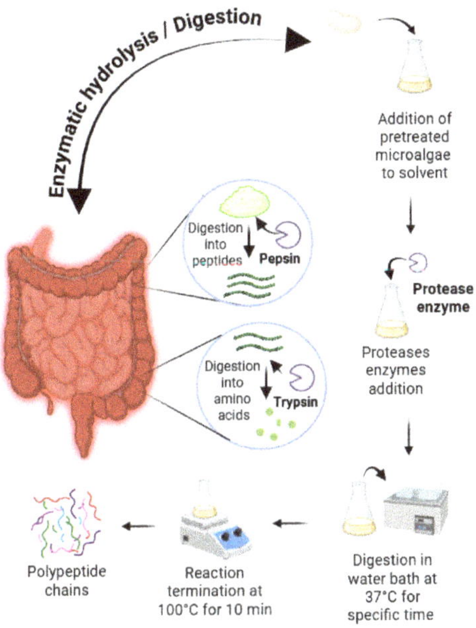

Figure 5. Enzymatic and chemical hydrolysis of WKB.

7.3. Protein Functionalization and Modification Techniques

Techniques such as heat treatment, ultrasound, and pH-shifting can effectively modify the surface properties of white kidney bean protein concentrates. These modifications significantly enhance foaming capacity, solubility, and surface hydrophobicity—all critical for applications in baked goods and emulsified foods. Heat treatment and pH-shifting can improve foaming capacity while yielding the highest solubility and emulsifying stability [100]. Moreover, solid-state fermentation can enhance functional properties (emulsifying, foaming, and oil- and water-holding capacities) and mineral bioavailability, thereby promoting the incorporation of kidney bean flour into fortified foods [101].

7.4. Applications for Ingredient Formulation

Incorporating modified white kidney bean components into baked goods, pasta, and meat products fortifies these items, enhancing nutritional quality and functional properties, including texture, stability, and yield [92]. Employing chemical, enzymatic, physical, and fermentation methods functionalizes and modifies white kidney beans, significantly enhancing their value as food ingredients. These methods augment bioactivity, functional attributes, and nutritional value, rendering them suitable for diverse innovative and health-promoting food formulations [102].

8. Functional Aspects, Processing, and Consumer Aspects of WKB

WKB exhibits substantial chemical and nutritional properties, including abundant phaseolin protein, which comprises 20–30% of its content [42,103]. These beans are low in fat, with a high proportion of unsaturated fatty acids, appealing to consumers with cardiovascular disorders. Additionally, fiber and antioxidant compounds support digestive health in patients with chronic conditions [104]. The high resistant starch content in WKB contributes to a low glycemic index, rendering it promising for patients with hyperglycemia, which often precipitates overweight [105]. These nutritional and functional attributes are significantly influenced by WKB processing and extraction techniques.

8.1. Correlation Between Processing, Functional Aspects, and Consumer Acceptability of WKB

Biological processes substantially affect the nutritional profile of WKB. For instance, germination, fermentation, soaking, cooking, and autoclaving enhance nutrient bioavailability and digestibility while diminishing antinutritional factors [30,43,104]. Conversely, extraction and modification methods—including physical, chemical, and biological approaches—are applied to improve the solubility and emulsifying properties of WKB [90,103]. These attributes broaden their applicability as functional foods for consumers.

8.2. Correlation Between Functional and Consumer Aspects

WKB, with its substantial bioactive potential, is utilized to develop low-glycemic-index foods owing to its α-amylase inhibitor, which regulates blood glucose levels and mitigates overweight, particularly in diabetic patients [66]. The protein and fiber contents of WKB render it promising for plant-based products, such as meat analogs and dairy substitutes [106]. A primary consumer consideration is economic acceptability, as WKB serves as an inexpensive and sustainable protein source. This supports food security and yields economic benefits, particularly in developing regions [104].

9. Comparative Analysis of Other Protein Sources

9.1. Comparison with Animal-Based Proteins

Plant-derived proteins are increasingly replacing animal proteins in the United States. The market for plant-based meat has rapidly grown, and it is estimated to increase at a rate of 19.3% yearly between 2022 and 2030 [107]. The overall risk of mortality is directly related to red meat (beef) and, more specifically, to processed meat [108]. One potential reason for the observed association between beef consumption and a reduced risk of chronic illness is the high content of saturated fats, carcinogens generated during cooking, as well as heme iron-derived molecules, such as heterocyclic amines and N-nitroso compounds [45]. Heterocyclic amines (HAAS/HCAS) are generated when high-protein diets like red meat are cooked at high temperatures. These chemical compounds are classified as strong carcinogens and mutagens that can cause oxidative stress, inflammation, and DNA adduct formation. They may adversely damage DNA, disrupt normal cellular function and enhance the promotion of tumor growth in a number of organs, including the breast, liver, colon and others. Besides these effects, certain HAAS/HCAS) to protein aggregation and mitochondrial dysfunction, which can lead to neurodegenerative diseases such as Alzheimer's and Parkinson's diseases [109]. HCAs possess the ability to disrupt lipid metabolism, and cause renal and intestinal dysfunction, which may contribute to liver illnesses. N-nitroso compounds (NOCs), such as nitrosamines, are created endogenously in the stomach from dietary precursors during food processing. NOCs are extremely carcinogenic, genotoxic and mutagenic. In animal model studies, more than 90% of research indicated that NOCs are highly carcinogenic, leading to benign tumors in the gastrointestinal system. Diets focused mainly on plants have been linked to improved health [110]. Most beans and pulses are rich in resistant starch, dietary fiber, and essential micronutrients (e.g., iron, magnesium, copper, phosphorus, potassium, manganese, and B vitamins) [111]. Furthermore, they have a notable protein content of 21–25% by weight. The commonly consumed dry beans, Phaseolus vulgaris, have an average of 0.5 g of total fat per half-cup.

According to projections, common beans or Phaseolus vulgaris are a staple food for 500 million people due to their protein richness (22–27% of seed weight) and carbohydrates (39–47% of seed weight). A 100 g serving of dry common beans can provide 9–25 g of protein, meeting over 20% of an average adult's daily requirement. The protein from beans meets the basic human protein needs as recommended by the FAO and WHO. Furthermore, several studies have shown the anti-inflammatory, antioxidant, anti-obesity, anti-carcinogenic, anti-diabetic, and cardioprotective potential of Phaseolus vulgaris extract (PVE) [112].

9.2. Comparison with Other Plant-Based Proteins

WKBs serve as an inexpensive source of proteins, vitamins, resistant starch, minerals, polyphenols, and dietary fiber. Legumes, when combined with cereal, typically provide a good source of protein (20–40%) as shown in Table 2 [113]. High protein content, widespread acceptance, and affordability of beans render them the most suitable source for protein isolate production. Among the most extensively grown and eaten legumes worldwide are beans and peas [114]. Peas (*Pisum sativum* L.) are the second-highest leguminous crop farmed on more than 25 million acres globally. Field peas contain higher levels of lysine and tryptophan than cereal grains and have lower trypsin inhibitor content than soybeans. According to Sai. UT et al. (2009) [115], navy beans exhibited the highest levels of ash, fat, and crude fiber, while red kidney beans showed the highest carbohydrate content. Adzuki bean showed the most vigorous trypsin-inhibitory activity. Globulin polypeptides were the predominant polypeptides in both the entire protein and the protein isolate, with molecular weights of 45–55 kDa. Every seed's protein isolate contained at least one major glycopeptide, and the whole protein contained more than two. At pH 4–5, all protein extracts show minimal protein solubility. Regarding functional properties, kidney and navy beans demonstrated superior foamability and foam stability compared to adzuki beans. Adzuki beans exhibited the highest oil absorption capacity, whereas no significant difference in water absorption was observed between adzuki and kidney beans as shown in Table 2.

Table 2. Comparison of white kidney with plant and animal-based protein sources.

Category	Protein Source	Nutritional Composition	Health Benefits	Limitations	Ref
Legume Protein	White kidney bean	*Phaseolus vulgaris* contains around 22–27% protein. It also contains starch resistance, dietary fiber, minerals (K, Mg, Fe), and bioactive compounds like polyphenols, saponins, and α-amylase inhibitors.	WKB shows its potential role for anti-obesity, anti-diabetic, anti-inflammatory, antioxidant, cardio-protective, and hypolipidemic effects.	WKB contains some anti-nutritional components such as phytates, lectins, and trypsin inhibitors when insufficiently processed.	[116–118]
Animal-based Proteins	Fish, Beef, and dietary	It contains Protein, critical amino acids, heme iron, and B12.	It supplies essential nutrients, improves muscle growth, and whey and casein enhance satiety and lean mass.	It shows some serious adverse effects on human health. Red meat is connected to an increased cardiovascular risk and cancer, due to its high saturated fat content.	[119,120]
Plant-based Protein	Soy Protein	It includes 40% which is rich in essential amino acids and isoflavones.	It reduces cholesterol levels. Improves bone strength, cardio-protective, and helps weight management.	It contains phytoestrogen, the causative factor of allergy.	[121,122]
	Pea Protein (*Pisum sativum* L.)	It contains 20–25% protein, rich in tryptophan and lysine, and is less efficient than a trypsin inhibitor.	It enhances blood pressure and controls glucose level, boosts muscle protein synthesis, and increases satiety.	Pea protein shows low digestibility as compared to animal-based protein and low methionine.	[123–125]
	Other Legumes, including (Red Kidney, Navy, and adzuki Beans	Red kidney beans contain a high amount of carbohydrates, while navy beans have more fiber, and adzuki beans show strong trypsin-inhibitory activity.	Exhibits anti-carcinogenic properties, antioxidants, anti-hypertensive, and supports gastrointestinal health.	With the presence of tannins and trypsin inhibitors, their protein solubility is a challenge.	[125–128]

10. Safety and Side Effects

10.1. Allergenic Potential

The utilization of raw or inadequately processed kidney beans is constrained by antinutritional factors, stemming from various non-nutritional compounds that can interfere with normal metabolism, inhibit nutrient digestion and absorption, and initiate allergic reactions [129]. Consumption of raw or undercooked kidney beans can induce acute gastrointestinal symptoms attributable to phytohemagglutinin; this risk is mitigated by adequate heat treatment [130]. Bento et al. (2021) [131] highlighted that antinutritional factors—a diverse array of substances also known as non-nutritional factors or plant bioactive components, exert protective effects against various malignancies and coronary atherosclerotic disease. Antinutritional components, including toxic alkaloids and proteins, adversely affect animal and human health by reducing nutrient absorption from kidney beans (KBs) and their products [131].

Most antinutritional constituents in KBs comprise lectin, trypsin inhibitor, saponin, phytic acid, and other compounds capable of chelating ionic cofactors, directly inhibiting protease action, or forming irreversible complexes that impede protein digestion [95]. Methods for inactivating non-nutritional components in KBs and producing KB products with negligible levels of these constituents are shown in Figure 6 [129]. The non-nutritional components included in KBs are diverse in size and form, and each has its own physiological impact. Phytates and tannins bind proteins and minerals, thereby reducing their bioavailability and digestibility. Digestion of protein and starch is impacted by trypsin inhibitors, which bind proteases and diminish their activity. Lectins trigger allergic responses by binding to carbohydrate moieties on cell surfaces. These antinutritional factors can impair nutrient absorption in humans and animals, elicit allergic reactions, disrupt normal metabolism, and adversely affect the skin, respiratory, and gastrointestinal systems; in severe cases, they may cause food poisoning. Consequently, their presence restricts human growth and KB consumption by compromising the quality and nutritional value of KB products [132].

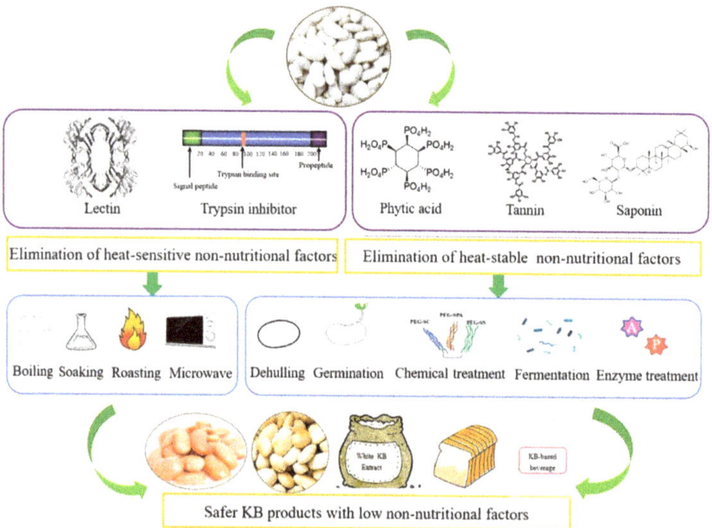

Figure 6. Methods for inactivating non-nutritional components in kidney beans, adopted with permission from [129].

10.2. Digestive Issues

The α-amylase inhibitor inhibits starch digestion by blocking access to the enzyme's active site. The inhibitory efficiency of α-AI isoforms is influenced by multiple parameters, including temperature, incubation duration, pH, and the presence of specific ions. The optimal temperature range is 22–37 °C, with an optimum pH of 4.5–5.5 for the inhibitor. Boiling beans for five minutes completely inactivate the inhibitor, while at 0 °C, its activity ceases. According to three separate studies, optimal incubation periods are 10, 40, and 100 min, varying test conditions—a pH of 4.5 for the shorter incubation periods and a pH of 6.9 for the longer ones, presumed to be the reason for the variations in incubation lengths [133]. Due to lectins and other antinutritional factors, consuming raw or inadequately processed KBs can decrease protein digestibility, impair the utilization and absorption of other nutrients, cause symptoms such as weight loss, loss of appetite, dyspnea, and, in severe cases, poisoning.

Trypsin inhibitors, known as anti-metabolic proteins, bind to digestive enzymes such as trypsin and chymotrypsin, blocking protein hydrolysis and the absorption of amino acids [134]. Tannins, water-soluble phenolic metabolites, are divided into three groups called hydrolysable tannins (as ellagitannins and gallotannins), complex tannins (like proanthocyanidins), and condensed tannins. Condensed tannins can precipitate and bind protein fractions, reducing their availability and making them hard to absorb and digest [135]. Saponins, known as tri-terpenoids or steroidal glycosides, present in KBs can hinder digestive or metabolic pathways, concurrently adhering to vital nutrients such as zinc, iron, and vitamin E [136].

11. Formulations and Delivery Methods

11.1. Supplements

Patients with diabetes can regulate their blood sugar levels by consuming foods with a low glycemic index (GI). The findings of Ma et al. (2018) demonstrated that WKB contains an alpha-amylase inhibitor that reduces the estimated GI (eGI) of gluten-rich diets by suppressing the activity of mammalian amylase [137]. The US Food and Drug Administration consider amylase inhibitors from white kidney bean extract under the Generally Recognized As Safe (GRAS) principle. Therefore, interest has grown in the widespread use of white kidney bean extracts for the preparation of low-GI meals [138]. There are still challenges with the industrial techniques of creating low-GI meals [139]. Despite the use of commercial white kidney bean extracts, such as *Phaseolus vulgaris*, to reduce postprandial glucose levels [140]. However, a suitable way to utilize WKBE in the food industry is to consider the variables (such as Temperature, pH, incubation time, etc.) that influence the bean inhibitory activities. Because of its nutritious value and ease of preparation, instant porridge is gaining popularity worldwide [141]. On the other hand, fast porridges' comparatively high GI is unfavorable to the control of glucose [142]. WKBE suppresses amylase activity and offers a way to lower the GI of different kinds of fast porridge [143]. Research demonstrated that fermented bean-based drinks are manufactured by combining several techniques. They are more nutritious and beneficial to health [144]. The fermentation of bean-based drinks with lactic acid resulted in nutrient-dense nondairy products, compared with those made from dairy. Lactic acid bacteria enhanced the digestibility of fermented beverages by reducing the oligosaccharides that aggregate in the digestive tract. Overall, the fermented-based drinks are suitable for bacterial cells to survive both the fermentation process and the cold storage time. They might be described as functional products that act as probiotic vessels [145].

11.2. Fortification

The method of adding an exogenous substance to basic food components to enhance their pro-health properties is termed dietary fortification. Despite this, the model is commonly used to improve both nutritional and nutraceutical qualities of food [145]. A previous study demonstrated that adding phenolic compounds to the food system significantly modified these areas due to the interaction with the food matrix constituents [146]. The interaction between phenolic compounds and the dietary matrix has been reported to negatively influence antioxidant activity and the in vitro bioaccessibility of phenolic compounds [147]. It is feasible to use the combination index to assess the food-matrix interaction as well as the affinity of phenolic compounds for the food matrix. The phenolic-food matrix interaction changed during the simulated digestion, as shown by the comparison of the combination for the sample obtained before and after in vitro digestion. Phenolic compounds typically reduce the digestion of carbohydrates and protein [148]. Moreover, interaction with phenols was demonstrated to modify the electrophoretic and chromatographic profile of the protein. The intricate interaction and the unique attraction of phenolic compounds for food matrix components are considered, since the consequence of phenolic food-matrix interaction depends on the individual phenolic chemical used [149]. In summary, the interaction between phenolic chemicals and the food matrix has a substantial influence on the nutritional potential of fortified foods. Thus, the requirement must be regarded throughout the entire design and assessment process. Nevertheless, due to the complexity of these interactions, food science will continue to face these challenges [150].

11.3. Recipes

Most probiotic drinks are naturally made with milk; however, there are plant-based, non-dairy substitutes available. As a result, the beverage samples were produced from germinated WKB seeds using two yoghurt starter cultures [151]. Although the raffinose and stachyose concentrations after fermentation were around 31% and 17% lower, respectively, than before fermentation (2.73 mg kg^{-1} and 0.43 mg kg^{-1}, respectively), in contrast, verbascose was not substantially decreased. Riboflavin, pyridoxine, and niacin levels in each sample were decreased significantly after fermentation (around 88%, 45% and 76%, respectively) [152]. In contrast, the levels of thiamine and cyanocobalamin have not changed much. During the 28-day cold storage period, the water holding capacity shifted, ranging from 49 to 57%. Both within the first 28 days of storage at 60 °C and immediately after fermentation, the initial microbe population reached the minimum therapeutic limit. Drinks with a natural flavor were not well received; however, by adding fruit flavors, the acceptability increased by around 15% [153,154].

11.4. Cooking Methods

11.4.1. Boiling

Among the most popular techniques for minimizing non-nutritional components in KBs, boiling is a necessary treatment [155]. Wang et al. (2023) [156] documented that both the extent and rate of starch digestion increased principally during boiling of chickpea, cowpea, and white kidney beans, by inactivating trypsin and lectin inhibitors in WKB. Both boiling rate and type of bean had a direct impact on the effectiveness of heat treatment in decreasing these non-nutritional components [156]. In comparison to the whole bean, the phytate content in WKB-based spaghetti reduced from 12.90 to 9.25 mg/g post-boiling. It indicates that milling can potentiate boiling to promote the phytic acid breakdown, which in turn improves the absorption of dietary iron. Similarly, soaking and combined hull processing were observed to increase KB seeds' relative water absorption, leading to a significant decrease in the tannin and polyphenol content as well as trypsin inhibition in the boiled products [157,158]. Further discovered that the boiling rate influences the phytic acid retention in KBs. Quick cooking varieties, such as red, cranberry, and yellow KBs, retain higher content of phytic acid than the products that are boiled slowly at a medium heat.

11.4.2. Roasting

The raised temperature during roasting can deactivate non-nutritional constituents more quickly than boiling [159]. Sparvoli et al. (2016) [160] documented that roasting reduced the phytic acid content of red KBs by 16.8%, although this level remained stable after the boiling and soaking procedures. During the roasting process, the amount of KB-derived phytic acid may have decreased due to the formation of insoluble compounds with certain minerals [160].

11.4.3. Microwave

The use of different heat treatments can have combined effects. Microwave treatment, like boiling and baking, has been shown to eliminate non-nutritional components from KBs [129]. Jiasheng Wang et al. (2023) highlighted that combining microwave cooking with hot air drying of pre-paste WKB removes gelatinized surface starch, water, lectins, and the firm seed structure, resulting in improved nutrient accessibility and digestibility [161].

12. Regulatory Considerations

12.1. FDA and Other Regulatory Standards

In the United States, dietary supplements, including those containing white kidney bean extracts, are regulated by the Food and Drug Administration (FDA). FDA regulatory standards mandate accurate labeling and safety for these supplements [162]. Labeling must include essential information on dosage, contraindications, and potential side effects to ensure consumer awareness and transparency. Compliance with these standards is essential for safeguarding public health and maintaining the integrity of these supplements [163].

12.2. Safety and GRAS Status

Most animal and human studies indicate that white kidney bean extract (WKBE) is well tolerated, with no adverse effects at commonly administered doses. Subchronic toxicity tests in rats revealed no observed adverse effect levels (NOAEL) up to 2.5–4.0 g/kg/day, indicating a substantial safety margin for human consumption [164]. Human studies involving doses up to 3000 mg/day for several weeks reported no significant alterations in blood parameters, organ function, or adverse effects [85]. However, inadequate cooking may retain harmful lectins and other non-nutritional components [129].

12.3. Recommended Dosage Ranges

WKBE dosages ranging from 700 to 1000 mg are commonly employed in clinical studies for weight management. These doses are administered three times daily before meals, totaling 2100–3000 mg/day, for weight management [70,163]. Safety trials establish a maximum daily intake of up to 6–10 g for a 70 kg adult, although most supplements recommend lower doses [165,166]. Furthermore, animal studies corroborate these findings, reporting no adverse effects at substantially higher relative doses [164,167–169].

13. Conclusions and Future Directions

White kidney beans (*Phaseolus vulgaris*) represent a promising multifunctional nutraceutical could facilitate the development of innovative obesity management strategies, while serving as a primary nutritional source. WKBE acts synergistically with α-amylase inhibitors, dietary fiber, bioactive phytochemicals, and high-quality protein to enhance satiety, lipid homeostasis, glycemic control, and carbohydrate metabolism. Mechanistic studies elucidate additional pathways involving gut microbiota interactions and nutrient bioavailability, while clinical studies validate their function in controlling caloric uptake and reducing postprandial hyperglycemia. The translational efficacy, dose–response relationships, and long-term safety of WKB across inclusive populations, however, are still largely unknown. The full potential of WKBE as an affordable, accessible, and sustainable food source could be achieved through integration into precision nutrition frameworks and sustainable food systems.

To fully harness the potential of white kidney beans, a comprehensive, integrated research agenda is imperative for future investigations. This agenda should commence with rigorous clinical validation via large-scale, long-term randomized controlled trials to definitively establish the safety, dose–response effects, and efficacy of WKB in weight management. Concurrently, food innovation is needed to develop optimized processing and formulation strategies that enhance the bioavailability of active compounds while minimizing antinutritional factors. To advance mechanistic understanding, multi-omics approaches can elucidate precisely how WKB modulates gut microbiota, lipid regulation, and glucose metabolism at the molecular level. Ultimately, this foundational knowledge will enable precision nutrition by facilitating investigations into individual variability in WKB response, predicated on genetic, metabolic, and microbiome profiles, thereby transforming this legume into a potent, personalized public health intervention.

Funding: This research was funded by the Natural Science Foundation of Jiangsu Province (grant number BK20230313). The APC was also funded by the Natural Science Foundation of Jiangsu Province.

Institutional Review Board Statement: Not applicable.

Informed Consent Statement: Not applicable.

Data Availability Statement: The original contributions presented in the study are included in the article, further inquiries can be directed to the corresponding author.

Acknowledgments: We extend our sincere thanks to our peers and colleagues for their valuable comments and suggestions, which greatly improved the rigor and clarity of this work.

Conflicts of Interest: The authors declare that they have no conflicts of interest.

References

1. Overweight and obese adults. In *Health at a Glance: Latin America and the Caribbean 2023*; OECD Publishing: Paris, France, 2023.
2. Janić, M.; Janež, A.; El-Tanani, M.; Rizzo, M. Obesity: Recent Advances and Future Perspectives. *Biomedicines* **2025**, *13*, 368. [CrossRef] [PubMed]
3. Dudašova-Petrovičova, O.; Đuričić, I.; Ivanović, N.; Dabetić, N.; Dodevska, M.; Ilić, T. Dietary interventions in obesity: A narrative review. *Arh. Za Farm.* **2024**, *74*, 281–297. [CrossRef]
4. Welsh, A.; Hammad, M.; Piña, I.L.; Kulinski, J. Obesity and cardiovascular health. *Eur. J. Prev. Cardiol.* **2024**, *31*, 1026–1035. [CrossRef] [PubMed]
5. Alrayes, M.S.; Altawili, M.A.; Alsuabie, S.M.; Sindi, A.W.; Alharbi, K.M.; Alsalhi, K.M.; Al Alawi, R.M.; Ali, I.D.; Nasser, A.N.; Alabdulrahim, J.M.; et al. Surgical Interventions for the Management of Obesity-Related Joint Pain: A Narrative Review. *Cureus* **2024**, *16*, e59082. [CrossRef]
6. Abdelfattah, D.S.E.; Fouad, M.A.; Elmeshad, A.N.; El-Nabarawi, M.A.; Elhabal, S.F. Anti-Obesity Effect of Combining White Kidney Bean Extract, Propolis Ethanolic Extract and CrPi3 on Sprague-Dawley Rats Fed a High-Fat Diet. *Nutrients* **2024**, *16*, 310. [CrossRef]
7. Misra, S.; Khunti, K.; Goyal, A.; Gable, D.; Armocida, B.; Tandon, N.; Sachdev, P.; Wild, S.H.; Hivert, M.-F.; Beran, D. Managing early-onset type 2 diabetes in the individual and at the population level. *Lancet* **2025**, *405*, 2341–2354. [CrossRef]

8. Schnurr, T.M.; Jakupović, H.; Carrasquilla, G.D.; Ängquist, L.; Grarup, N.; Sørensen, T.I.A.; Tjønneland, A.; Overvad, K.; Pedersen, O.; Hansen, T.; et al. Obesity, unfavourable lifestyle and genetic risk of type 2 diabetes: A case-cohort study. *Diabetologia* **2020**, *63*, 1324–1332. [CrossRef]
9. Seyedjavadi, S.S.; Bagheri, P.; Nasiri, M.J.; Razzaghi-Abyaneh, M.; Goudarzi, M. Fungal Infection in Co-infected Patients With COVID-19: An Overview of Case Reports/Case Series and Systematic Review. *Front. Microbiol.* **2022**, *13*, 888452. [CrossRef]
10. Zhang, Y. Impact of dietary habit changes on college students physical health insights from a comprehensive study. *Sci. Rep.* **2025**, *15*, 9953. [CrossRef]
11. Chandrakanth, M.; Kumar, N.; Sura, C.; Tung, S.; Bilde, T.; Bleu, J. Context-dependent effects of developmental and adult diet on life-history traits in Drosophila melanogaster. *J. Evol. Biol.* **2025**, *38*, 1396–1409. [CrossRef]
12. Popiolek-Kalisz, J. The role of nutrition in cardiovascular protection—Personalized versus universal dietary strategies. *Trends Cardiovasc. Med.* **2025**, *35*, 531–537. [CrossRef] [PubMed]
13. Volek, J.S.; Yancy, W.S.; Gower, B.A.; Phinney, S.D.; Slavin, J.; Koutnik, A.P.; Hurn, M.; Spinner, J.; Cucuzzella, M.; Hecht, F.M. Expert consensus on nutrition and lower-carbohydrate diets: An evidence- and equity-based approach to dietary guidance. *Front. Nutr.* **2024**, *11*, 1376098. [CrossRef] [PubMed]
14. Bhujle, R.R.; Nayak, N.; Gowda, N.A.N.; Pandiselvam, R.; Sunil, C.K. A comprehensive review on influence of millet processing on carbohydrate-digesting enzyme inhibitors and implications for diabetes management. *Crit. Rev. Biotechnol.* **2024**, *45*, 743–765. [CrossRef] [PubMed]
15. Jiang, C.; Li, S.; Su, H.; Zhou, N.; Yao, Y. Kidney Bean Protein Prevents High-Fat and High-Fructose Diet-Induced Obesity, Cognitive Impairment, and Disruption of Gut Microbiota Composition. *Foods* **2024**, *13*, 1718. [CrossRef]
16. Zaffar, A.; Shafi, S.; Fatima, S.; Riyaz, I.; Nabi, A.; Gani, S.; Zargar, S.M.; Mir, R.R.; Nabi, S.; Sofi, P.A. Unveiling Biochemical Dynamics in Common Bean-Pseudocercospora Pathosystem Under Angular Leaf Spot Pressure in Western Himalayan Kashmir. *J. Crop Health* **2024**, *76*, 1343–1356. [CrossRef]
17. Dimopoulou, M.; Vareltzis, P.; Gortzi, O. A Systematic Review of the Twelve Most Popular Bean Varieties, Highlighting Their Potential as Functional Foods Based on the Health Benefits Derived from Their Nutritional Profiles, Focused on Non-Communicable Diseases. *Appl. Sci.* **2024**, *14*, 10215. [CrossRef]
18. Kasar, S.S.; Maheshwari, V.L.; Pawar, P.K. Bioactive α-Amylase Inhibitors: Sources, Mechanism of Action, Biochemical Characterization, and Applications. In *Natural Products as Enzyme Inhibitors: An Industrial Perspective*; Maheshwari, V.L., Patil, R.H., Eds.; Springer: Singapore, 2022; pp. 59–104.
19. Golshany, H.; Helmy, S.A.; Morsy, N.F.S.; Kamal, A.; Yu, Q.; Fan, L. The gut microbiome across the lifespan: How diet modulates our microbial ecosystem from infancy to the elderly. *Int. J. Food Sci. Nutr.* **2024**, *76*, 95–121. [CrossRef]
20. Ntalouka, F.; Tsirivakou, A. Morus alba: Natural and valuable effects in weight loss management. *Front. Clin. Diabetes Healthc.* **2024**, *5*, 1395688. [CrossRef]
21. Sornek, P.; Izdebska, W.; Stanek, J.; Perkowska, K.; Kaźmierczak, A.; Mich, A.; Pawlak, I.; Borkowska, A.; Ciesielski, R.; Kiełb, A. Methods of lowering the glycemic index of food and their underlying mechanisms—A review. *Qual. Sport* **2024**, *20*, 54118. [CrossRef]
22. Paoli, A. The Influence of Physical Exercise, Ketogenic Diet, and Time-Restricted Eating on De Novo Lipogenesis: A Narrative Review. *Nutrients* **2025**, *17*, 663. [CrossRef]
23. Wang, S.; Guo, C.; Xing, Z.; Li, M.; Yang, H.; Zhang, Y.; Ren, F.; Chen, L.; Mi, S. Dietary Intervention With α-Amylase Inhibitor in White Kidney Beans Added Yogurt Modulated Gut Microbiota to Adjust Blood Glucose in Mice. *Front. Nutr.* **2021**, *8*, 664976. [CrossRef] [PubMed]
24. Lee, C.-C.; Liao, Y.-C.; Lee, M.-C.; Cheng, Y.-C.; Chiou, S.-Y.; Lin, J.-S.; Huang, C.-C.; Watanabe, K. Different Impacts of Heat-Killed and Viable Lactiplantibacillus plantarum TWK10 on Exercise Performance, Fatigue, Body Composition, and Gut Microbiota in Humans. *Microorganisms* **2022**, *10*, 2181. [CrossRef]
25. Cabello-Olmo, M.; Oneca, M.; Goñi, S.; Urtasun, R.; Pajares, M.J.; Yavorov-Dayliev, D.; Iturria, I.; Ayo, J.; Encío, I.J.; Barajas, M.; et al. Heat-Inactivated Pediococcus acidilactici pA1c®HI Maintains Glycemic Control and Prevents Body Weight Gain in High-Fat-Diet-Fed Mice. *Int. J. Mol. Sci.* **2025**, *26*, 6408. [CrossRef]
26. Padhiyar, P.N.; Singh, B.P.; Sarkar, P.; Hati, S. Food-Derived Bioactive Peptides With Anti-Diabetic and Antimicrobial Potential: An Updated Review on Current Trends and Challenges. *Sustain. Food Proteins* **2025**, *3*, e70012. [CrossRef]
27. Sannidhi, D.; Abeles, R.; Andrew, W.; Bonnet, J.P.; Vitale, K.; Niranjan, V.; Gulati, M.; Pauly, K.; Moran, R.; Alexander, L.; et al. Lifestyle Medicine for Obesity in the Era of Highly Effective Anti-Obesity Treatment. *Nutrients* **2025**, *17*, 2382. [CrossRef]
28. Jagdish, S.; Rajni, K.; Singh, N.P. Pulse Phytonutrients: Nutritional and Medicinal Importance. *J. Pharm. Nutr. Sci.* **2016**, *6*, 160–171. [CrossRef]
29. Parveen, S.; Jamil, A.; Pasha, I.; Ahmad, F. Pulses: A Potential Source of Valuable Protein for Human Diet. In *Legumes Research*; InTech Open Access Publisher: Rijeka, Croatia, 2022; Volume 2, pp. 1–15.

30. Langyan, S.; Yadava, P.; Khan, F.N.; Bhardwaj, R.; Tripathi, K.; Bhardwaj, V.; Bhardwaj, R.; Gautam, R.K.; Kumar, A. Nutritional and Food Composition Survey of Major Pulses Toward Healthy, Sustainable, and Biofortified Diets. *Front. Sustain. Food Syst.* **2022**, *6*, 878269. [CrossRef]
31. Sinkovič, L.; Pipan, B.; Šibul, F.; Nemeš, I.; Tepić Horecki, A.; Meglič, V. Nutrients, Phytic Acid and Bioactive Compounds in Marketable Pulses. *Plants* **2022**, *12*, 170. [CrossRef]
32. Didinger, C.; Thompson, H.J. The role of pulses in improving human health: A review. *Legume Sci.* **2022**, *4*, e147. [CrossRef]
33. Singh, V.; Yousuf, O.; Prasad, S. Sustainability of Plant Protein. In *The Future of Plant Protein: Innovations, Challenges, and Opportunities*; Kaiser, Y., Yousuf, O., Eds.; Springer: Singapore, 2025; pp. 51–72.
34. Sozer, N.; Holopainen-Mantila, U.; Poutanen, K. Traditional and New Food Uses of Pulses. *Cereal Chem.* **2016**, *94*, 66–73. [CrossRef]
35. Ahmed, W.; Rashid, S. Functional and therapeutic potential of inulin: A comprehensive review. *Crit. Rev. Food Sci. Nutr.* **2017**, *59*, 1–13. [CrossRef] [PubMed]
36. Uebersax, M.A.; Cichy, K.A.; Gomez, F.E.; Porch, T.G.; Heitholt, J.; Osorno, J.M.; Kamfwa, K.; Snapp, S.S.; Bales, S. Dry beans (*Phaseolus vulgaris* L.) as a vital component of sustainable agriculture and food security—A review. *Legume Sci.* **2022**, *5*, e155. [CrossRef]
37. De Pilli, T.; Alessandrino, O. Effects of different cooking technologies on biopolymers modifications of cereal-based foods: Impact on nutritional and quality characteristics review. *Crit. Rev. Food Sci. Nutr.* **2018**, *60*, 556–565. [CrossRef] [PubMed]
38. Dhiman, A.; Thakur, K.; Parmar, V.; Sharma, S.; Sharma, R.; Kaur, G.; Singh, B.; Suhag, R. New insights into tailoring physicochemical and techno-functional properties of plant proteins using conventional and emerging technologies. *J. Food Meas. Charact.* **2023**, *17*, 3845–3873. [CrossRef]
39. Mazumder, K.; Aktar, A.; Biswas, B.; Hossain, M.E. Proteins, Vitamins, Minerals, Cereal, Vegetables, and Beverages as Functional Foods. In *Dietary Supplements and Nutraceuticals*; Mukherjee, B., Ed.; Springer: Singapore, 2025; pp. 1–40.
40. Abdalla, M.A.; Sumon, M.M.; Mühling, K.H. Improvement of cereal- and legume-derived protein quality with selenium and sulfur for plant food production. *J. Sci. Food Agric.* **2024**, *105*, 5611–5623. [CrossRef]
41. Muñoz-Pina, S.; Khvostenko, K.; García-Hernández, J.; Heredia, A.; Andrés, A. Implications of Pleurotus ostreatus solid-state bioprocessing on the nutritional composition, protein structure, and anti-hypertensive and anti-inflammatory potential of local beans (*Phaseolus vulgaris* L.). *Curr. Res. Food Sci.* **2025**, *11*, 101167. [CrossRef]
42. Srenuja, D.; Hema, V.; Anand, M.T.; Mohan, R.J.; Vidyalakshmi, R. Kidney bean: Protein's treasure trove and creates avenues for a healthy lifestyle. *Legume Sci.* **2023**, *5*, e193. [CrossRef]
43. Veber, A.; Zaręba, D.; Ziarno, M. Functional Fermented Beverage Prepared from Germinated White Kidney Beans (*Phaseolus vulgaris* L.). In *Milk Substitutes—Selected Aspects*; IntechOpen: Rijeka, Croatia, 2021; pp. 100–125.
44. Kan, L.; Nie, S.; Hu, J.; Wang, S.; Cui, S.W.; Li, Y.; Xu, S.; Wu, Y.; Wang, J.; Bai, Z.; et al. Nutrients, phytochemicals and antioxidant activities of 26 kidney bean cultivars. *Food Chem. Toxicol.* **2017**, *108*, 467–477. [CrossRef]
45. Marinangeli, C. Methods for Processing Pulses to Optimize Nutritional Functionality and Maximize Amino Acid Availability in Foods and Feeds. *Cereal Foods World* **2020**, *65*, 10-1094. [CrossRef]
46. Wu, X.; Wu, X.; Zhang, X.; Zhang, J.; Yan, X.; Zhang, Q.; Zhang, B. Structural, physicochemical and in vitro digestibility of white kidney bean protein-corn starch complexes under various heat treatments. *Food Res. Int.* **2025**, *200*, 115479. [CrossRef]
47. Jonathan, C.A.; Misha, P.; Sofia Feng, G. Influence of Proprietary White Kidney Bean Protein on Human Postprandial Glucose Response. *Int. J. Res. Med. Clin. Sci.* **2024**, *2*, 83–89. [CrossRef]
48. Meghrabi, S.; Yamani, M.I. Physicochemical and Sensory Characteristics of a New Milk Substitute from Dry White Kidney Bean. *Food Sci. Technol.* **2023**, *11*, 218–225. [CrossRef]
49. Xu, S.; Qin, L.; Mazhar, M.; Zhu, Y. Functional components profile and glycemic index of kidney beans. *Front. Nutr.* **2022**, *9*, 1044427. [CrossRef] [PubMed]
50. Zhang, Z.; Zhang, L.; Yang, Z.; Wang, Y.; Huang, Y.; Zhang, B. Carbohydrate digestion and absorption in human digestive track. In *Carbohydrate Nutrition*; Zhang, B., Dhital, S., Eds.; Academic Press: Cambridge, MA, USA, 2025; pp. 25–46.
51. Cai, M.; Tejpal, S.; Tashkova, M.; Ryden, P.; Perez-Moral, N.; Saha, S.; Garcia-Perez, I.; Serrano Contreras, J.I.; Wist, J.; Holmes, E.; et al. Upper-gastrointestinal tract metabolite profile regulates glycaemic and satiety responses to meals with contrasting structure: A pilot study. *Nat. Metab.* **2025**, *7*, 1459–1475. [CrossRef] [PubMed]
52. Abera, S.; Yohannes, W.; Chandravanshi, B.S.; Roy, D. Effect of Processing Methods on Antinutritional Factors (Oxalate, Phytate, and Tannin) and Their Interaction with Minerals (Calcium, Iron, and Zinc) in Red, White, and Black Kidney Beans. *Int. J. Anal. Chem.* **2023**, *2023*, 1–11. [CrossRef]
53. Guerrero-Martin, C.A.; Ortega-Ramírez, A.T.; Silva-Marrufo, Ó.; Casallas-Martín, B.D.; Cortés-Salazar, N.; Salinas-Silva, R.; Camacho-Galindo, S.; Da Silva Fernandes, F.A.; Guerrero-Martin, L.E.; Paulo de Freitas, P.; et al. Biofortification of Kidney Bean (*Phaseolus vulgaris* L.) Crops Applying Zinc Sulfate and Ferric Sulfate: Pilot Crop in Colombia. *Molecules* **2023**, *28*, 2004. [CrossRef]
54. Cichy, K.; Chiu, C.; Isaacs, K.; Glahn, R. Dry Bean Biofortification with Iron and Zinc. In *Biofortification of Staple Crops*; Kumar, D., Mishra, G.P., Singh, A., Eds.; Springer: Singapore, 2022; pp. 225–270.

55. De la Cruz Lázaro, E.; Félix, J.W.; Sánchez-Chávez, E.; Tosquy-Valle, O.; Preciado-Rangel, P.; Márquez-Quiroz, C. Biofortificación de Frijol (*Phaseolus vulgaris* L.) Variedad Verdín con Quelato y Sulfato de Hierro. *Rev. Terra Latinoam.* **2024**, *42*, e1831. [CrossRef]
56. Hoque, M. A Review on Different Dietary Sources of Important Vitamins and Electrolytes. *Int. J. Res. Publ. Rev.* **2023**, *4*, 731–736. [CrossRef]
57. Vivar-Quintana, A.M.; Absi, Y.; Hernández-Jiménez, M.; Revilla, I. Nutritional Value, Mineral Composition, Fatty Acid Profile and Bioactive Compounds of Commercial Plant-Based Gluten-Free Flours. *Appl. Sci.* **2023**, *13*, 2309. [CrossRef]
58. Lee, Y.H.; Hong, J.-H.; Park, K.H.; Kim, S.-H.; Kim, J.-C.; Kim, D.H.; Park, Y.H.; Lee, K.W.; Kim, J.K.; Kim, K.H. Phytochemical Investigation of Bioactive Compounds from White Kidney Beans (Fruits of *Phaseolus multiflorus* var. Albus): Identification of Denatonium with Osteogenesis-Inducing Effect. *Plants* **2021**, *10*, 2205. [CrossRef]
59. Ebrahim, A.E.; Abd El-Aziz, N.K.; Elariny, E.Y.T.; Shindia, A.; Osman, A.; Hozzein, W.N.; Alkhalifah, D.H.M.; El-Hossary, D. Antibacterial activity of bioactive compounds extracted from red kidney bean (*Phaseolus vulgaris* L.) seeds against multidrug-resistant Enterobacterales. *Front. Microbiol.* **2022**, *13*, 1035586. [CrossRef] [PubMed]
60. Alina, P.; Kiełb Ewa, P.; František, Z. Antimicrobial activity of saponin-containing plants: Review. *J. Dairy Vet. Anim. Res.* **2023**, *12*, 121–127. [CrossRef]
61. Davkova, I.; Zhivikj, Z.; Kukić-Marković, J.; Cvetkovik-Karanfilova, I.; Stefkov, G.; Kulevanova, S.; Karapandzova, M. Natural products in the management of obesity. *Arh. Za Farm.* **2024**, *74*, 298–315. [CrossRef]
62. Fischer, L. Enzymes in Food Production. In *Introduction to Enzyme Technology*; Jaeger, K.-E., Liese, A., Syldatk, C., Eds.; Learning Materials in Biosciences; Springer International Publishing: Cham, Switzerland, 2024; Volume 53, pp. 337–360.
63. Losada-Garcia, N.; Simović, M.; Ćorović, M.; Milivojević, A.; Nikačević, N.; Mateo, C.; Bezbradica, D.; Palomo, J.M. Developing and improving enzyme-driven technologies to synthesise emerging prebiotics. *Green Chem.* **2025**, *27*, 8777–8803. [CrossRef]
64. Rodríguez, L.; Mendez, D.; Montecino, H.; Carrasco, B.; Arevalo, B.; Palomo, I.; Fuentes, E. Role of Phaseolus vulgaris L. in the Prevention of Cardiovascular Diseases—Cardioprotective Potential of Bioactive Compounds. *Plants* **2022**, *11*, 186. [CrossRef]
65. Kashtoh, H.; Baek, K.-H. New Insights into the Latest Advancement in α-Amylase Inhibitors of Plant Origin with Anti-Diabetic Effects. *Plants* **2023**, *12*, 2944. [CrossRef]
66. Liu, T.; Gu, Y.; Waleed, A.-A.; Wang, L.; Li, Y.; Qian, H. Challenges and opportunities in developing low glycemic index foods with white kidney bean α-amylase inhibitor. *Trends Food Sci. Technol.* **2024**, *147*, 104397. [CrossRef]
67. Shi, Z.; Zhu, Y.; Teng, C.; Yao, Y.; Ren, G.; Richel, A. Anti-obesity effects of α-amylase inhibitor enriched-extract from white common beans (*Phaseolus vulgaris* L.) associated with the modulation of gut microbiota composition in high-fat diet-induced obese rats. *Food Funct.* **2020**, *11*, 1624–1634. [CrossRef]
68. Sagor, M.A.T.; Smita, R.M.; Shuvo, A.P.R.; Raihan, S.; Jahan, R.; Simin, F.A.; Rahman, A.; Biswas, S.; Salem, L. The Role of Mineral Deficiencies in Insulin Resistance and Obesity. *Curr. Diabetes Rev.* **2022**, *18*, 19–39. [CrossRef]
69. Zhao, M.-F.; Zhang, X.-G.; Tang, Y.-P.; Zhu, Y.-X.; Nie, H.-Y.; Bu, D.-D.; Fang, L.; Li, C.-J. Ketone bodies promote epididymal white adipose expansion to alleviate liver steatosis in response to a ketogenic diet. *J. Biol. Chem.* **2024**, *300*, 105617. [CrossRef]
70. Jäger, R.; Abou Sawan, S.; Purpura, M.; Grube, B.; Röske, Y.; De Costa, P.; Chong, P.-W. Proprietary alpha-amylase inhibitor formulation from white kidney bean (*Phaseolus vulgaris* L.) promotes weight and fat loss: A 12-week, double-blind, placebo-controlled, randomized trial. *Sci. Rep.* **2024**, *14*, 12685. [CrossRef]
71. Smith, K.; Taylor, G.S.; Walker, M.; Brunsgaard, L.H.; Bowden Davies, K.A.; Stevenson, E.J.; West, D.J. Pre-Meal Whey Protein Alters Postprandial Insulinemia by Enhancing β-Cell Function and Reducing Insulin Clearance in T2D. *J. Clin. Endocrinol. Metab.* **2023**, *108*, e603–e612. [CrossRef] [PubMed]
72. Houghton, D.; Shannon, O.M.; Chater, P.I.; Wilcox, M.D.; Pearson, J.P.; Stanforth, K.; Jordan, C.; Avery, L.; Blain, A.P.; Joel, A.; et al. White kidney bean extract as a nutraceutical: Effects on gut microbiota, alpha-amylase inhibition, and user experiences. *Gut Microbiome* **2023**, *4*, e8. [CrossRef]
73. Nolan, R.; Shannon, O.M.; Robinson, N.; Joel, A.; Houghton, D.; Malcomson, F.C. It's No Has Bean: A Review of the Effects of White Kidney Bean Extract on Body Composition and Metabolic Health. *Nutrients* **2020**, *12*, 1398. [CrossRef]
74. de la Rosa-Millan, J. Lipid-Enriched Cooking Modulates Starch Digestibility and Satiety Hormone Responses in Traditional Nixtamalized Maize Tacos. *Foods* **2025**, *14*, 2576. [CrossRef]
75. Palupi, E.; Nurdin, N.M.; Mufida, G.; Valentine, F.N.; Pangestika, R.; Rimbawan, R.; Sulaeman, A.; Briawan, D.; Filianty, F. High-Fiber Extruded Purple Sweet Potato (*Ipomoea batatas*) and Kidney Bean (*Phaseolus vulgaris*) Extends the Feeling of Fullness. *Pol. J. Food Nutr. Sci.* **2024**, *74*, 82–91. [CrossRef]
76. Li, X.; Bai, Y.; Jin, Z.; Svensson, B. Food-derived non-phenolic α-amylase and α-glucosidase inhibitors for controlling starch digestion rate and guiding diabetes-friendly recipes. *LWT* **2022**, *153*, 112455. [CrossRef]
77. Song, H.; Han, W.; Yan, F.; Xu, D.; Chu, Q.; Zheng, X. Dietary Phaseolus vulgaris extract alleviated diet-induced obesity, insulin resistance and hepatic steatosis and alters gut microbiota composition in mice. *J. Funct. Foods* **2016**, *20*, 236–244. [CrossRef]

78. Razmpoosh, E.; Abdollahi, S.; Sikaroudi, M.K.; Sangsefidi, Z.S.; Zeraattalab-Motlagh, S.; Torabinasab, K.; Hejazi, M.; Meshkini, F.; Motallaei, M.; Soltani, S. The effect of low-fat diets on appetite: A systematic review of randomized clinical trials. *BMC Public Health* **2025**, *25*, 2264. [CrossRef]
79. Thompson, H.; McGinley, J.; Neil, E.; Brick, M. Beneficial Effects of Common Bean on Adiposity and Lipid Metabolism. *Nutrients* **2017**, *9*, 998. [CrossRef] [PubMed]
80. Shi, Z.; Zhang, X.; Zhu, Y.; Yao, Y.; Ren, G. Natural Extracts from White Common Bean (*Phaseolus vulgaris* L.) Inhibit 3T3-L1 Adipocytes Differentiation. *Appl. Sci.* **2020**, *11*, 167. [CrossRef]
81. Feng, Q.; Niu, Z.; Zhang, S.; Wang, L.; Dong, L.; Hou, D.; Zhou, S. Protective Effects of White Kidney Bean (*Phaseolus vulgaris* L.) against Diet-Induced Hepatic Steatosis in Mice Are Linked to Modification of Gut Microbiota and Its Metabolites. *Nutrients* **2023**, *15*, 3033. [CrossRef] [PubMed]
82. Zhang, J.; Jiang, X.; Xu, Q.; Li, W.; Zhang, D. Lipidomic Analysis Reveals the Anti-Obesity and Hepatoprotective Effects of Flavonoid Mimetic Components in Adzuki Beans on High-Fat Diet-Induced Obese Mice. *Foods* **2025**, *14*, 3191. [CrossRef]
83. Barrett, M.L.; Udani, J.K. A proprietary alpha-amylase inhibitor from white bean (Phaseolus vulgaris): A review of clinical studies on weight loss and glycemic control. *Nutr. J.* **2011**, *10*, 34. [CrossRef]
84. Tzeng, H.-P.; Chiu, C.-Y.; Liu, S.-H.; Chiang, M.-T. Improvement of Glycemic Control by a Functional Food Mixture Containing Maltodextrin, White Kidney Bean Extract, Mulberry Leaf Extract, and Niacin-Bound Chromium Complex in Obese Diabetic db/db Mice. *Metabolites* **2022**, *12*, 693. [CrossRef]
85. Wang, S.; Chen, L.; Yang, H.; Gu, J.; Wang, J.; Ren, F. Regular intake of white kidney beans extract (*Phaseolus vulgaris* L.) induces weight loss compared to placebo in obese human subjects. *Food Sci. Nutr.* **2020**, *8*, 1315–1324. [CrossRef]
86. Ciacci, C.; Ciclitira, P.; Hadjivassiliou, M.; Kaukinen, K.; Ludvigsson, J.F.; McGough, N.; Sanders, D.S.; Woodward, J.; Leonard, J.N.; Swift, G.L. The gluten-free diet and its current application in coeliac disease and dermatitis herpetiformis. *United Eur. Gastroenterol. J.* **2015**, *3*, 121–135. [CrossRef]
87. Schaefer, M.; Kühnel, A.; Enge, S. Open-label placebos reduce weight in obesity: A randomized controlled trial. *Sci. Rep.* **2024**, *14*, 21311. [CrossRef]
88. Gheorghiu, O.R.C.; Ciobanu, A.M.; Guțu, C.M.; Dănilă, G.-M.; Nițescu, G.V.; Rohnean, Ș.; Baconi, D.L. Detection of Adulterants in Herbal Weight Loss Supplements. *J. Mind Med. Sci.* **2025**, *12*, 23. [CrossRef]
89. Bhambar, R.S.; Nehete, J.Y.; Bhide, Y.S. Extraction, Characterization and Therapeutic Evaluation of Seeds of Phaseolus vulgaris L. for Weight Management. *Int. J. Pharm. Qual. Assur.* **2022**, *13*, 396–401. [CrossRef]
90. Singhi, H.; Ozturk, O.K. A comprehensive review of kidney bean proteins: Extraction, composition, techno-functional properties, and emerging food applications. *Crit. Rev. Food Sci. Nutr.* **2025**, *65*, 1–31. [CrossRef] [PubMed]
91. Vogelsang-O'Dwyer, M.; Sahin, A.W.; Arendt, E.K.; Zannini, E. Enzymatic Hydrolysis of Pulse Proteins as a Tool to Improve Techno-Functional Properties. *Foods* **2022**, *11*, 1307. [CrossRef] [PubMed]
92. Saad, A.M.; Sitohy, M.Z.; Ahmed, A.I.; Rabie, N.A.; Amin, S.A.; Aboelenin, S.M.; Soliman, M.M.; El-Saadony, M.T. Biochemical and Functional Characterization of Kidney Bean Protein Alcalase-Hydrolysates and Their Preservative Action on Stored Chicken Meat. *Molecules* **2021**, *26*, 4690. [CrossRef]
93. Siegel, J.; Mak, W.S.; German, J.B. Use of Proteolytic Enzymes to Enhance Protein Bioavailability. U.S. Patent Application No. 16/767,535, 26 May 2020.
94. Di Filippo, G.; Melchior, S.; Plazzotta, S.; Calligaris, S.; Innocente, N. Effect of enzymatic hydrolysis with Alcalase or Protamex on technological and antioxidant properties of whey protein hydrolysates. *Food Res. Int.* **2024**, *188*, 114499. [CrossRef]
95. Roy, M.; Sarker, A.; Azad, M.A.K.; Shaheb, M.R.; Hoque, M.M. Evaluation of antioxidant and antimicrobial properties of dark red kidney bean (*Phaseolus vulgaris*) protein hydrolysates. *J. Food Meas. Charact.* **2019**, *14*, 303–313. [CrossRef]
96. Tejasari, T.; Aulia, F.; Agustina, N. Inhibiting Activity of Angiotensin Converting Enzyme-1 by Bean Protein Hydrolysate Genus Phaseolus. *J. Penelit. Pascapanen Pertan.* **2021**, *18*, 75–86. [CrossRef]
97. Yu, S.; Zhang, Y.; Zhang, Y.; Zhang, C.; Liu, X.; Wang, Y.; Meng, F.; Yu, L. Water-Assisted Microwave Processing: Rapid Detoxification and Antioxidant Enhancement in Colored Kidney Beans. *Foods* **2025**, *14*, 3557. [CrossRef]
98. Nickhil, C.; Singh, R.; Deka, S.C.; Srivastava, B. Structural and physicochemical changes in protein and starch during finger millet storage. *Discov. Food* **2025**, *5*, 343. [CrossRef]
99. Mora, L.; Toldrá, F. Advanced enzymatic hydrolysis of food proteins for the production of bioactive peptides. *Curr. Opin. Food Sci.* **2023**, *49*, 100973. [CrossRef]
100. Higa, F.; Nickerson, C.; Nickerson, M.T. Effect of pH shifting and temperature on the functional properties of a commercial pea protein isolate. *Cereal Chem.* **2024**, *102*, 89–101. [CrossRef]
101. Patil, N.D.; Bains, A.; Goksen, G.; Ali, N.; Dhull, S.B.; Khan, M.R.; Chawla, P. Effect of solid-state fermentation on kidney bean flour: Functional properties, mineral bioavailability, and product formulation. *Food Chem. X* **2025**, *27*, 102339. [CrossRef] [PubMed]

102. Dinesh, G.; Krishna, A.S.; Kumar, Y.; Joshi, T.J.; Rao, P.S. Postharvest Processing of Millets: Advancements & Entrepreneurship Development Opportunities. *Future Postharvest Food* **2025**, *2*, 108–123. [CrossRef]
103. Azevedo, P.Z.; Rigolon, T.C.B.; de Souza, B.R.; Stringheta, P.C.; Martins, E.; Campelo, P.H. White bean (*Phaseolus vulgaris* L.) proteins: Extraction methods, physicochemical and techno-functional properties, applications, and challenges for the food industry. *Obs. DE LA Econ. Latinoam.* **2025**, *23*, e10325. [CrossRef]
104. Mekonnen, H.K.; Hassan, S.M.; Ateye, M.D. Nutritional Composition, Utilization, and Processing of Haricot Beans (*Phaseolus vulgaris* L): A Comprehensive Review. *Asian J. Adv. Res. Rep.* **2024**, *18*, 178–187. [CrossRef]
105. Punia, S.; Dhull, S.B.; Sandhu, K.S.; Kaur, M.; Purewal, S.S. Kidney bean (*Phaseolus vulgaris*) starch: A review. *Legume Sci.* **2020**, *2*, e52. [CrossRef]
106. Alfaro-Diaz, A.; Escobedo, A.; Luna-Vital, D.A.; Castillo-Herrera, G.; Mojica, L. Common beans as a source of food ingredients: Techno-functional and biological potential. *Compr. Rev. Food Sci. Food Saf.* **2023**, *22*, 2910–2944. [CrossRef]
107. Finley, J.W.; Holliday, D.; Kim, H.; da Silva Alves, P.L.; Shao, D.-Y.; Bartley, G.; Yokoyama, W. Substitution of cooked kidney beans or ground beef in hypercholesterolemic high fat diets reduces plasma and liver lipids in hamsters. *Food Prod. Process. Nutr.* **2023**, *5*, 20. [CrossRef]
108. Zheng, Y.; Li, Y.; Satija, A.; Pan, A.; Sotos-Prieto, M.; Rimm, E.; Willett, W.C.; Hu, F.B. Association of changes in red meat consumption with total and cause specific mortality among US women and men: Two prospective cohort studies. *BMJ* **2019**, *365*, l2110. [CrossRef] [PubMed]
109. Oz, E.; Aoudeh, E.; Murkovic, M.; Toldra, F.; Gomez-Zavaglia, A.; Brennan, C.; Proestos, C.; Zeng, M.; Oz, F. Heterocyclic aromatic amines in meat: Formation mechanisms, toxicological implications, occurrence, risk evaluation, and analytical methods. *Meat Sci.* **2023**, *205*, 109312. [CrossRef] [PubMed]
110. Seyyedsalehi, M.S.; Mohebbi, E.; Tourang, F.; Sasanfar, B.; Boffetta, P.; Zendehdel, K. Association of Dietary Nitrate, Nitrite, and N-Nitroso Compounds Intake and Gastrointestinal Cancers: A Systematic Review and Meta-Analysis. *Toxics* **2023**, *11*, 190. [CrossRef] [PubMed]
111. Kadyan, S.; Sharma, A.; Arjmandi, B.H.; Singh, P.; Nagpal, R. Prebiotic Potential of Dietary Beans and Pulses and Their Resistant Starch for Aging-Associated Gut and Metabolic Health. *Nutrients* **2022**, *14*, 1726. [CrossRef]
112. Ganesan, K.; Xu, B. Polyphenol-rich dry common beans (*Phaseolus vulgaris* L.) and their health benefits. *Int. J. Mol. Sci.* **2017**, *18*, 2331. [CrossRef]
113. Neji, C.; Semwal, J.; Máthé, E.; Sipos, P. Dough Rheological Properties and Macronutrient Bioavailability of Cereal Products Fortified through Legume Proteins. *Processes* **2023**, *11*, 417. [CrossRef]
114. Lisciani, S.; Marconi, S.; Le Donne, C.; Camilli, E.; Aguzzi, A.; Gabrielli, P.; Gambelli, L.; Kunert, K.; Marais, D.; Vorster, B.J.; et al. Legumes and common beans in sustainable diets: Nutritional quality, environmental benefits, spread and use in food preparations. *Front. Nutr.* **2024**, *11*, 1385232. [CrossRef]
115. Sai-Ut, S.; Ketnawa, S.; Chaiwut, P.; Rawdkuen, S. Biochemical and functional properties of proteins from red kidney, navy and adzuki beans. *Asian J. Food Agro-Ind.* **2009**, *2*, 493–504.
116. Blair, M.W.; Li, H.; Nekkalapudi, L.; Becerra, V.; Paredes, M. Nutritional Traits of Beans (*Phaseolus vulgaris*): Nutraceutical Characterization and Genomics. In *Compendium of Crop Genome Designing for Nutraceuticals*; Kole, C., Ed.; Springer: Singapore, 2023; pp. 611–638.
117. Chattopadhyay, A.; Dey, C.; Mukherjee, S. Unleashing the Positive Impact of Natural Bioactive Compounds in the Therapeutic Strategy for Metabolic Syndrome. In *Bioactive Ingredients for Healthcare Industry Volume 2*; Lahiri, D., Nag, M., Bhattacharya, D., Pati, S., Sarkar, T., Eds.; Springer: Singapore, 2025; pp. 1–53.
118. Ram, S.; Narwal, S.; Gupta, O.P.; Pandey, V.; Singh, G.P. Anti-nutritional factors and bioavailability: Approaches, challenges, and opportunities. In *Wheat and Barley Grain Biofortification*; Gupta, O.P., Pandey, V., Narwal, S., Sharma, P., Ram, S., Singh, G.P., Eds.; Elsevier: Amsterdam, The Netherlands, 2020; pp. 101–128.
119. Bergeron, N.; Chiu, S.; Williams, P.T.; M King, S.; Krauss, R.M. Effects of red meat, white meat, and nonmeat protein sources on atherogenic lipoprotein measures in the context of low compared with high saturated fat intake: A randomized controlled trial. *Am. J. Clin. Nutr.* **2019**, *110*, 24–33. [CrossRef]
120. Berrazaga, I.; Micard, V.; Gueugneau, M.; Walrand, S. The Role of the Anabolic Properties of Plant- versus Animal-Based Protein Sources in Supporting Muscle Mass Maintenance: A Critical Review. *Nutrients* **2019**, *11*, 1825. [CrossRef]
121. Qin, P.; Wang, T.; Luo, Y. A review on plant-based proteins from soybean: Health benefits and soy product development. *J. Agric. Food Res.* **2022**, *7*, 100265. [CrossRef]
122. Velissaridou, A.; Panoutsopoulou, E.; Prokopiou, V.; Tsoupras, A. Cardio-Protective-Promoting Properties of Functional Foods Inducing HDL-Cholesterol Levels and Functionality. *Nutraceuticals* **2024**, *4*, 469–502. [CrossRef]
123. Wu, D.-T.; Li, W.-X.; Wan, J.-J.; Hu, Y.-C.; Gan, R.-Y.; Zou, L. A Comprehensive Review of Pea (*Pisum sativum* L.): Chemical Composition, Processing, Health Benefits, and Food Applications. *Foods* **2023**, *12*, 2527. [CrossRef] [PubMed]

124. Choi, A. Effects of Pea Protein on Satiety, Postprandial Glucose Response and Appetite Hormones: A Literature Review. *Undergrad. Res. Nat. Clin. Sci. Technol. (URNCST) J.* **2022**, *6*, 1–13. [CrossRef]
125. Wang, J.; Kadyan, S.; Ukhanov, V.; Cheng, J.; Nagpal, R.; Cui, L. Recent advances in the health benefits of pea protein (*Pisum sativum*): Bioactive peptides and the interaction with the gut microbiome. *Curr. Opin. Food Sci.* **2022**, *48*, 100944. [CrossRef]
126. Guo, Q.; Luo, J.; Zhang, X.; Zhi, J.; Yin, Z.; Zhang, J.; Zhang, W.; Xu, B.; Chen, L. A Comprehensive Review of the Chemical Constituents and Functional Properties of Adzuki Beans (*Vigna angulariz*). *J. Agric. Food Chem.* **2025**, *73*, 6361–6384. [CrossRef]
127. Dileep, K.C.; Bhatt, K.; Kumar, S.; Sharma, R.; Rana, P.; Thakur, M.; Suthar, P. Kidney Bean Sprouts and Lentil Sprouts. In *Advances in Plant Sprouts*; Kehinde, B.A., Dar, B., Nanda, V., Eds.; Springer International Publishing: Cham, Switzerland, 2023; pp. 201–227.
128. Navin Venketeish, K.S.; Govindarajan, N.; Pandiselvam, R.; Mousavi Khaneghah, A. Unlocking the potential of underutilized legumes: Nutritional benefits, anti-nutritional challenges, and mitigation strategies. *J. Food Meas. Charact.* **2024**, *18*, 9588–9602. [CrossRef]
129. Zhang, Z.; Liu, C.; Wu, S.; Ma, T. The Non-Nutritional Factor Types, Mechanisms of Action and Passivation Methods in Food Processing of Kidney Bean (*Phaseolus vulgaris* L.): A Systematic Review. *Foods* **2023**, *12*, 3697. [CrossRef]
130. Salim, R.; Nehvi, I.B.; Mir, R.A.; Tyagi, A.; Ali, S.; Bhat, O.M. A review on anti-nutritional factors: Unraveling the natural gateways to human health. *Front. Nutr.* **2023**, *10*, 1215873. [CrossRef]
131. Bento, J.A.C.; Ribeiro, P.R.V.; Alexandre, L.M.; Alves Filho, E.G.; Bassinello, P.Z.; de Brito, E.S.; Caliari, M.; Júnior, M.S.S. Chemical profile of colorful bean (*Phaseolus vulgaris* L) flours: Changes influenced by the cooking method. *Food Chem.* **2021**, *356*, 129718. [CrossRef]
132. Li, M.; Wang, B.; Lv, W.; Lin, R.; Zhao, D. Characterization of pre-gelatinized kidney bean (*Phaseolus vulgaris* L.) produced using microwave hot-air flow rolling drying technique. *LWT* **2022**, *154*, 112673. [CrossRef]
133. Choudhary, M.I. (Ed.) *Frontiers in Cardiovascular Drug Discovery*; Bentham Science Publishers: Oak Park, IL, USA, 2022; Volume 6, pp. 1–245.
134. Mehmood, S.; Thirup, S.S.; Ahmed, S.; Bashir, N.; Saeed, A.; Rafiq, M.; Saeed, Q.; Najam-ul-Haq, M.; Khaliq, B.; Ibrahim, M.; et al. Crystal structure of Kunitz-type trypsin inhibitor: Entomotoxic effect of native and encapsulated protein targeting gut trypsin of Tribolium castaneum Herbst. *Comput. Struct. Biotechnol. J.* **2024**, *23*, 3132–3142. [CrossRef]
135. Molino, S.; Pilar Francino, M.; Ángel Rufián Henares, J. Why is it important to understand the nature and chemistry of tannins to exploit their potential as nutraceuticals? *Food Res. Int.* **2023**, *173*, 113329. [CrossRef]
136. Timilsena, Y.P.; Phosanam, A.; Stockmann, R. Perspectives on Saponins: Food Functionality and Applications. *Int. J. Mol. Sci.* **2023**, *24*, 13538. [CrossRef]
137. Ma, Y.; Rang, Y.; Yang, R.; Zhao, W. Effect of white kidney bean extracts on estimated glycemic index of different kinds of porridge. *LWT* **2018**, *96*, 576–582. [CrossRef]
138. Uawongwattana, A.; Posridee, K.; Promyo, K.; Thaeomor, A.; Oonsivilai, R. Evaluation of Phaseolus vulgaris Extract in a Rat Model of Cafeteria-Diet-Induced Obesity: Metabolic and Biochemical Effects. *Foods* **2025**, *14*, 2038. [CrossRef]
139. Zare, L.; Karimi, J.; Sanaei-Nasab, S.; Tahmouzi, S.; Feizollahi, E.; Mollakhalili-Meybodi, N. Effective strategies for producing low glycemic index wheat bread and their impact on technological properties: A review. *J. Food Meas. Charact.* **2025**, *19*, 5222–5241. [CrossRef]
140. Guerrero-Becerra, L.; Morimoto, S.; Arrellano-Ordoñez, E.; Morales-Miranda, A.; Guevara-Gonzalez, R.G.; Feregrino-Pérez, A.A.; Lomas-Soria, C. Polyphenolic Compounds in Fabaceous Plants with Antidiabetic Potential. *Pharmaceuticals* **2025**, *18*, 69. [CrossRef]
141. Lange, E.; Kęszycka, P.K.; Pałkowska-Goździk, E.; Billing-Marczak, K. Comparison of Glycemic Response to Carbohydrate Meals without or with a Plant-Based Formula of Kidney Bean Extract, White Mulberry Leaf Extract, and Green Coffee Extract in Individuals with Abdominal Obesity. *Int. J. Environ. Res. Public Health* **2022**, *19*, 12117. [CrossRef] [PubMed]
142. Rani, M.; Yannam, S.K.; LakshmiA, J. Utilization of Plant Protein Hydrolysates in an Instant Porridge Mix with Improved Protein Digestibility and Palatability. *ACS Food Sci. Technol.* **2025**, *5*, 2743–2753. [CrossRef]
143. Ucar, E.M.; Kizil, M. A Review on Healthier Bread Making: Focusing on the Underlying Mechanisms for Decreasing Glycemic Index. *Food Rev. Int.* **2025**, *41*, 1469–1508. [CrossRef]
144. Sawant, S.S.; Park, H.-Y.; Sim, E.-Y.; Kim, H.-S.; Choi, H.-S. Microbial Fermentation in Food: Impact on Functional Properties and Nutritional Enhancement—A Review of Recent Developments. *Fermentation* **2025**, *11*, 15. [CrossRef]
145. Amin, W.S.; Amin, W.S.; Dewidar, O. A new approach in the production of a novel gluten-free probiotic beverage using broken white beans, broken rice, and whey. *Egypt. J. Chem.* **2024**, *67*, 2119–2131. [CrossRef]
146. Shahidi, F.; Pan, Y. Influence of food matrix and food processing on the chemical interaction and bioaccessibility of dietary phytochemicals: A review. *Crit. Rev. Food Sci. Nutr.* **2021**, *62*, 6421–6445. [CrossRef]
147. Wojtunik-Kulesza, K.; Oniszczuk, A.; Oniszczuk, T.; Combrzyński, M.; Nowakowska, D.; Matwijczuk, A. Influence of In Vitro Digestion on Composition, Bioaccessibility and Antioxidant Activity of Food Polyphenols—A Non-Systematic Review. *Nutrients* **2020**, *12*, 1401. [CrossRef]

148. Pinarli, B.; Simge Karliga, E.; Ozkan, G.; Capanoglu, E. Interaction of phenolics with food matrix: In vitro and in vivo approaches. *Mediterr. J. Nutr. Metab.* **2020**, *13*, 63–74. [CrossRef]
149. Sęczyk, Ł.; Gawlik-Dziki, U.; Świeca, M. Influence of Phenolic-Food Matrix Interactions on In Vitro Bioaccessibility of Selected Phenolic Compounds and Nutrients Digestibility in Fortified White Bean Paste. *Antioxidants* **2021**, *10*, 1825. [CrossRef]
150. Sik, B.; Székelyhidi, R.; Lakatos, E.; Kapcsándi, V.; Ajtony, Z. Analytical procedures for determination of phenolics active herbal ingredients in fortified functional foods: An overview. *Eur. Food Res. Technol.* **2021**, *248*, 329–344. [CrossRef]
151. Cichońska, P.; Ziarno, M. Legumes and Legume-Based Beverages Fermented with Lactic Acid Bacteria as a Potential Carrier of Probiotics and Prebiotics. *Microorganisms* **2021**, *10*, 91. [CrossRef]
152. Tian, C.; Yang, J.; Zeng, Y.; Zhang, T.; Zhou, Y.; Men, Y.; You, C.; Zhu, Y.; Sun, Y.; Zhou, N.-Y. Biosynthesis of Raffinose and Stachyose from Sucrose via an In Vitro Multienzyme System. *Appl. Environ. Microbiol.* **2019**, *85*, e02306-18. [CrossRef] [PubMed]
153. Ziarno, M.; Zaręba, D.; Ścibisz, I.; Kozłowska, M. Texture and water holding capacity of oat drinks fermented with lactic acid bacteria, bifidobacteria and Propionibacterium. *Int. J. Food Prop.* **2023**, *27*, 106–122. [CrossRef]
154. Voelker, A.L.; Taylor, L.S.; Mauer, L.J. Effect of pH and concentration on the chemical stability and reaction kinetics of thiamine mononitrate and thiamine chloride hydrochloride in solution. *BMC Chem.* **2021**, *15*, 47. [CrossRef] [PubMed]
155. Khrisanapant, P.; Leong, S.Y.; Kebede, B.; Oey, I. Effects of hydrothermal processing duration on the texture, starch and protein in vitro digestibility of cowpeas, chickpeas and kidney beans. *Foods* **2021**, *10*, 1415. [CrossRef]
156. Wang, Z.; Fan, M.; Hannachi, K.; Li, Y.; Qian, H.; Wang, L. Impact of red kidney bean protein on starch digestion and exploring its underlying mechanism. *Int. J. Biol. Macromol.* **2023**, *253*, 127023. [CrossRef]
157. Une, S.; Nonaka, K.; Akiyama, J. Effects of hull scratching, soaking, and boiling on Antinutrients in Japanese red sword bean (*Canavalia gladiata*). *J. Food Sci.* **2016**, *81*, C2398–C2404. [CrossRef]
158. Wiesinger, J.A.; Cichy, K.A.; Glahn, R.P.; Grusak, M.A.; Brick, M.A.; Thompson, H.J.; Tako, E. Demonstrating a nutritional advantage to the fast-cooking dry bean (*Phaseolus vulgaris* L.). *J. Agric. Food Chem.* **2016**, *64*, 8592–8603. [CrossRef]
159. Godrich, J.; Rose, P.; Muleya, M.; Gould, J. The effect of popping, soaking, boiling and roasting processes on antinutritional factors in chickpeas and red kidney beans. *Int. J. Food Sci. Technol.* **2023**, *58*, 279–289. [CrossRef]
160. Sparvoli, F.; Laureati, M.; Pilu, R.; Pagliarini, E.; Toschi, I.; Giuberti, G.; Fortunati, P.; Daminati, M.G.; Cominelli, E.; Bollini, R. Exploitation of common bean flours with low antinutrient content for making nutritionally enhanced biscuits. *Front. Plant Sci.* **2016**, *7*, 928. [CrossRef]
161. Wang, J.; Liu, C.; Zheng, X.; Hong, J.; Sun, B.; Liu, M. The structural integrity of endosperm/cotyledon cells and cell modification affect starch digestion properties. *Food Funct.* **2023**, *14*, 6784–6801. [CrossRef] [PubMed]
162. Bailey, R.L. Current regulatory guidelines and resources to support research of dietary supplements in the United States. *Crit. Rev. Food Sci. Nutr.* **2018**, *60*, 298–309. [CrossRef] [PubMed]
163. Gupta, M.K.; Balodi, V.; Tripathi, A.D.; Rizwana; Agarwal, A.; Nirmal, N.P. The importance of labeling and certification: Ensuring quality and transparency. In *Unleashing the Power of Functional Foods and Novel Bioactives*; Grumezescu, A.M., Holban, A.M., Eds.; Academic Press: Cambridge, MA, USA, 2025; pp. 449–465.
164. Ta, A.; Pn, A.; An, O.; No, B.; Po, U.; On, N.; Oc, P.; A, I. Effects of Raw and Cooked Aqueous and Methanol Extracts of *Phaseolus vulgaris* (Kidney Beans) on Renal Function in Albino Wistar Rats. *Univers. J. Pharm. Res.* **2020**, *5*, 6–11. [CrossRef]
165. Qin, G.; Wang, F.; Liang, H.; Tang, S.; Shekh, K.; Wang, Y.; Li, B.; Dong, B.; Wen, P. Subchronic Study of a White Kidney Bean (*Phaseolus vulgaris*) Extract withα-Amylase Inhibitory Activity. *BioMed Res. Int.* **2019**, *2019*, 9272345. [CrossRef]
166. Ferreira, K.C.; Bento, J.A.C.; Caliari, M.; Bassinello, P.Z.; Berrios, J.D.J. Dry bean proteins: Extraction methods, functionality, and application in products for human consumption. *Cereal Chem.* **2021**, *99*, 67–77. [CrossRef]
167. Islam, S.S.; Adhikary, S.; Mostafa, M.; Hossain, M. Vegetable Beans: Comprehensive Insights into Diversity, Production, Nutritional Benefits, Sustainable Cultivation and Future Prospects. *OnLine J. Biol. Sci.* **2024**, *24*, 477–494. [CrossRef]
168. Singh, R.P. Kidney bean starch: Composition, structure, properties, and modifications. In *Non-Conventional Starch Sources*; Singh, J., Kaur, L., Eds.; Academic Press: Cambridge, MA, USA, 2024; pp. 439–465.
169. Kelly, J.D.; Varner, G.V.; Chilvers, M.I.; Cichy, K.A.; Wright, E.M. Registration of 'Coho' light red kidney bean. *J. Plant Regist.* **2020**, *14*, 134–138. [CrossRef]

MDPI AG
Grosspeteranlage 5
4052 Basel
Switzerland
Tel.: +41 61 683 77 34

Foods Editorial Office
E-mail: foods@mdpi.com
www.mdpi.com/journal/foods

www.ingramcontent.com/pod-product-compliance
Lightning Source LLC
Chambersburg PA
CBHW061504210926
51850CB00008B/1822
* 9 7 8 3 7 2 5 8 7 7 8 7 4 *